Edge Artificial Intelligence

This book provides an in-depth exploration of the integration of cloud and edge computing in artificial intelligence, addressing critical topics such as hybrid architectures, real-time data processing, security, and practical artificial intelligence applications.

- Offers a detailed exploration of foundational concepts and practical applications of edge artificial intelligence, making it a comprehensive resource for understanding this emerging technology.
- Features real-world case studies from various sectors such as smart cities, industrial applications, healthcare, and agriculture, providing practical insights into how edge artificial intelligence is being applied.
- Discusses the latest trends and future directions in edge artificial intelligence, including advancements in edge hardware and software, federated learning, and the potential impact of blockchain technology.
- Addresses critical issues like security threats, privacy-preserving techniques, and the importance of explainable artificial intelligence.
- Highlights case studies and real-world examples to illustrate the tangible benefits of edge artificial intelligence.

This book is primarily written for senior undergraduate students, graduate students, and academic researchers in the fields of electrical engineering, electronics and communications engineering, computer science and engineering, and information technology.

Edge AI in Future Computing

Series Editors:
Arun Kumar Sangaiah, SCOPE, VIT University, Tamil Nadu
Mamta Mittal, G. B. Pant Government Engineering College, Okhla, New Delhi

Internet of Things
Frameworks for Enabling and Emerging Technologies
Bharat Bhushan, Sudhir Kumar Sharma, Bhuvan Unhelkar, Muhammad Fazal Ijaz, and Lamia Karim

Soft Computing
Recent Advances and Applications in Engineering and Mathematical Sciences
Pradip Debnath, Oscar Castillo, and Poom Kumam

Computational Statistical Methodologies and Modeling for Artificial Intelligence
Priyanka Harjule, Azizur Rahman, Basant Agarwal, and Vinita Tiwari

Industry 5.0 for Smart Healthcare Technologies
Utilizing Artificial Intelligence, Internet of Medical Things and Blockchain
Edited by Sherin Zafar, S. N. Kumar, A. Ahilan, and Gulsun Kurubacak Cakir

Computational Intelligence Algorithms for the Diagnosis of Neurological Disorders
Edited by S.N. Kumar, Sherin Zafar, and Sameena Naaz

Soft Computing
Engineering Applications
Pradip Debnath and Binod Chandra Tripathy

Edge Artificial Intelligence
Transforming Computing from Cloud to Edge
Edited by Manju Khari, Priya Gupta, and Chandan Pal Singh

For more information about this series, please visit: https://www.routledge.com/Edge-AI-in-Future-Computing/book-series/EAIFC

Edge Artificial Intelligence

Transforming Computing from Cloud to Edge

Edited by
Manju Khari, Priya Gupta,
and Chandan Pal Singh

CRC Press
Taylor & Francis Group
Boca Raton London New York

CRC Press is an imprint of the
Taylor & Francis Group, an **informa** business

First edition published 2027
by CRC Press
2385 NW Executive Center Drive, Suite 320, Boca Raton FL 33431

and by CRC Press
4 Park Square, Milton Park, Abingdon, Oxon, OX14 4RN

CRC Press is an imprint of Taylor & Francis Group, LLC

ISBN: 978-1-041-06378-0 (hbk)
ISBN: 978-1-041-06379-7 (pbk)
ISBN: 978-1-003-63514-7 (ebk)

DOI: 10.1201/9781003635147

Typeset in Sabon
By Apex CoVantage, LLC

Contents

About the Editors

Manju Khari serves as a professor in the School of Computer and Systems Sciences at Jawaharlal Nehru University in New Delhi (JNU), India. Her research contributions span information security, software security testing, computer networks, the Internet of Things, cyber forensics, and deep learning. Before joining JNU, she worked with Netaji Subhas University of Technology, East Campus, under the auspices of the Government of National Capital Territory (NCT) Delhi. Dr. Khari is a Ph.D. holder in computer science and engineering from the prestigious National Institute of Technology Patna. She also holds a master's in information security (M. Tech) from Guru Gobind Singh Indraprastha University in Delhi, India. Her prolific research portfolio includes over 100 published papers in premier refereed national and international journals and conferences, including highly regarded publishers such as IEEE, ACM, Springer, Inderscience, and Elsevier. Dr. Khari has also co-authored two books published by NCERT and co-edited 10 edited books, further solidifying her status as a leading scholar in her field. Dr. Khari is serving as a reviewer/technical program committee member for various international conferences. Dr. Khari's reputation as a rigorous and meticulous scholar has earned her positions as associate editor/guest editor for Springer, Wiley, and Elsevier books and a reviewer for several leading international journals. Dr. Khari's expertise is not limited to academia, as she is also a highly valued member and research consultant of several noteworthy projects in collaboration with prominent institutions such as All India Institute of Medical Sciences (AIIMS), the Department of Biotechnology (DBT), Department of Science and Technology (DST), and the Council of Scientific and Industrial Research–National Physical Laboratory (CSIR–NPL) through the Ministry of Consumer and Legal Affairs, India. These projects are invaluable to advancing research and development in her field, further underscoring her reputation as a trailblazer in computer science and engineering.

Priya Gupta is presently working as an associate professor at the Atal Bihari Vajpayee School of Management and Entrepreneurship at Jawaharlal Nehru University (JNU), New Delhi, India. Specializing in business analytics, data visualization, and various programming languages, she has made over 20 years of significant contributions through her teaching and research. Dr. Gupta holds a Ph.D. from Birla Institute of Technology (BIT) Mesra, India, focusing on customer relationship management systems in banking and has a strong educational background, including an M.Phil. in computer science, a Master of Computer Applications (MCA), and a Post Graduate Diploma in Information Technology Management (PGDITM, Gold Medalist). Her career includes roles in both academia and industry, starting as Assistant Software Engineer at the Center for Railway Information System (CRIS), and later teaching at institutions like Maharaja Agrasen College, Vivekanand College, and Shaheed Sukhdev College of Business Studies before joining JNU. Dr. Gupta's research covers a wide range of topics, including business analytics, compiler design, and AI applications in finance and healthcare. She has published extensively in reputable journals and authored several books. Her projects include empirical research on the National Education Policy 2020 and the Skill India Mission, and she has been involved in innovation projects funded by the University of Delhi. Her patents include a smart grid network for electricity supply in smart cities and an IoT-based electric vehicle charging system. Dr. Gupta has also chaired sessions and organized committees for numerous international conferences, reflecting her leadership in the academic community.

Chandan Pal Singh is a distinguished educator, accomplished book editor, and internationally certified Global Citizenship Education (GCED) and Digital Citizenship trainer with over 18 years of international teaching experience in computer science and information technology across Kazakhstan, India, Indonesia, China, and Singapore. Renowned for his expertise in integrating educational technology, AI, and machine learning into teaching practices, Chandan has co-authored and edited numerous influential books and published over 50 research papers in leading international journals. As an active member of various educational societies and a certified trainer under Asia-Pacific Centre of Education for International Understanding (APCEIU), UNESCO, Chandan is committed to advancing innovative instructional methods, fostering global citizenship, and promoting academic excellence.

Contributors

Poonam Arora, CSIR–National Physical Laboratory, New Delhi, India.

Daphne Grace Backiam Nathaniel, Department of Computer Science & Engineering, Alliance University, Bengaluru, India.

Arunima Banerjee, Department of Computer Science & Engineering, Alliance University, Bengaluru, India.

Trilok Bhardwaj, CSIR–National Physical Laboratory, New Delhi, India.

Nidhika Chauhan, Chandigarh University, Mohali, Punjab, India.

Renu Dalal, University School of Automation & Robotics, GGSIPU, East Delhi Campus, India.

Apurva Dalela, School of Computer and Systems Sciences, Jawaharlal Nehru University, New Delhi, India.

Ishanvi Goswami, Shyama Prasad Mukherji College (For Women), University of Delhi, India.

Mayank Gupta, University School of Automation & Robotics, GGSIPU, East Delhi Campus, India.

Priya Gupta, Atal Bihari Vajpayee School of Management and Entrepreneurship, Jawaharlal Nehru University, New Delhi, India.

Archi Jain, Shyama Prasad Mukherji College (For Women), University of Delhi, India.

Nitin Jain, School of Computer Science Engineering & Technology, Bennett University, Greater Noida, Uttar Pradesh, India.

Jyoti Dagar, University School of Automation & Robotics, Guru Gobind Singh Indraprastha University, East Delhi Campus, Delhi, India.

Shakir Khan, College of Computer and Information Sciences, Imam Mohammad Ibn Saud Islamic University (IMSIU), Riyadh, Saudi Arabia.

Manju Khari, School of Computer and Systems Sciences, Jawaharlal Nehru University, New Delhi, India.

Deepak Kumar, Tata Consultancy Services, Boston, MA, USA.

Kapil Kumar, School of Computer and System Sciences, Jawaharlal Nehru University, New Delhi, India.

Anmol Kumar Pandit, University School of Automation & Robotics, GGSIPU, East Delhi Campus, India.

Asha Kurian, Department of Computer Science & Engineering, Alliance University, Bengaluru, India.

S. Uma Maheswari, Department of Information Technology, C. Abdul Hakeem College of Engineering & Technology, Vellore, India.

Sakshi Mahipal, Swami Shraddhanand College, University of Delhi, India.

Sandhya Malikar Patel, CSIR–National Physical Laboratory, New Delhi, India.

Udit Misra, Department of Computer Science, North Carolina State University, USA.

M. Mohammed Mustafa, Department of AIDS, Sri Eshwar College of Engineering, Coimbatore, India.

Kunal Nayak, University School of Automation & Robotics, Guru Gobind Singh Indraprastha University, East Delhi Campus, Delhi, India.

Arvind Panwar, School of Computing Science & Engineering, Galgotias University, Greater Noida, India.

Komal Pathak, University School of Automation & Robotics, Guru Gobind Singh Indraprastha University, East Delhi Campus, Delhi, India.

Kumari Seema Rani, Shyama Prasad Mukherji College (For Women), University of Delhi, India.

S. Saranya, Department of Computer Science Engineering, C. Abdul Hakeem College of Engineering & Technology, Vellore, India.

Meera Sharma, Swami Shraddhanand College, University of Delhi, India.

Neha Sharma, Bharati Vidyapeeth College of Engineering, Paschim Vihar, New Delhi, India.

Ishita Singhal, Maharaja Surajmal Institute of Technology, I.P. University, New Delhi, India.

Ritu Singhal, Indraprastha College for Women, Delhi, India.

V. B. Singh, School of Computer and Systems Sciences, Jawaharlal Nehru University, India.

Ananya Singhal, Maharaja Surajmal Institute of Technology, I.P. University, New Delhi, India.

Urvashi Sugandh, Department of Artificial Intelligence & Machine Learning, Raj Kumar Goel Institute of Technology, Ghaziabad, India.

Swetha C. B., Department of Computer Science & Engineering, Alliance University, Bengaluru, India.

Arpit Tanwar, University School of Automation & Robotics, GGSIPU, East Delhi Campus, India.

Ankit Tayal, University School of Automation & Robotics, GGSIPU, East Delhi Campus, India.

Kulshrest Tiwari, University School of Automation & Robotics, GGSIPU, East Delhi Campus, India.

Srijan Tripathi, Chandigarh University, Mohali, Punjab, India.

Sunita Yadav, School of Computing Science & Engineering, Galgotias University, Greater Noida, India.

Preface

The rapid convergence of artificial intelligence (AI), cloud computing, and edge computing has redefined the technological landscape, transforming how data is generated, processed, and utilized in real time. Together, these technologies are enabling a paradigm shift from centralized architectures to distributed, intelligent, and autonomous systems. Their impact is now visible across a wide spectrum of domains—including smart cities, healthcare, industry, finance, and cybersecurity demonstrating their potential to drive innovation while advancing sustainability, resilience, and societal well-being.

This book brings together cutting-edge research, conceptual frameworks, and real-world applications from distinguished scholars and practitioners. It seeks to provide a holistic understanding of how AI, cloud, and edge technologies are collectively shaping the present and steering the future. The chapters are thoughtfully structured to guide readers from foundational concepts to practical implementations and emerging research directions.

The initial chapters set the technological foundation. Chapter 1 introduces edge AI as a transformative bridge between cloud and edge infrastructures. Chapter 2 outlines the fundamentals of edge computing, focusing on reduced latency, responsiveness, and improved security. Together, these chapters establish the transition from traditional centralized models to decentralized intelligence. The discussion then advances toward smart urban systems.

Chapters 3 to 5 explore the application of edge AI in smart cities, presenting conceptual frameworks and global case studies ranging from waste management in Barcelona to scalable architectures that ensure privacy and efficiency. These chapters highlight the evolution of smart cities into AI-driven, citizen-centric ecosystems.

The middle section dives into industrial and healthcare innovations. Chapter 6 addresses Industry 4.0 and the integration of cyber-physical systems with AI, while acknowledging critical cybersecurity risks. Chapter 7 discusses cloud–edge computing innovations in healthcare. Chapter 8 presents predictive modelling of COVID-19 in India, emphasizing data-driven policymaking and preparedness.

The book further explores security, finance, and analytical methodologies leveraged through AI: Chapter 9 introduces ML-based fraud detection models for credit card transactions. Chapter 10 compares botnet detection algorithms in IoT environments. Chapter 11 presents an innovative integration of Data Envelopment Analysis (DEA) with edge computing for evaluating healthcare efficiency. Chapter 12 examines big data sentiment analysis using Hadoop and Python, demonstrating how large-scale analytics can shape informed decision-making.

The final section turns to the critical dimensions of trust, transparency, and security in edge AI: Chapter 13 explores deep transfer learning at the edge for resource optimization.

Chapter 14 introduces the Trustworthy and Explainable edge AI (TE2AI) framework, emphasizing transparency and reliability. Chapters 15 and 16 provide insights into security and privacy challenges, supported by real-world case studies and mitigation strategies.

This volume presents a multidisciplinary perspective, uniting theory, application, and practical relevance. It aims to serve as a valuable resource for researchers, industry experts, practitioners, and policymakers seeking to harness the potential of AI, cloud, and edge technologies in building secure, intelligent, and sustainable ecosystems. We hope that this book not only contributes to advancing scientific and industrial knowledge but also inspires future innovations in the rapidly growing landscape of intelligent computing.

Editors
Professor Manju Khari
Dr. Priya Gupta
Dr. Chandan Pal Singh

Chapter 1

Edge AI

Transforming Computing from Cloud to Edge

M. Mohammed Mustafa, S. Uma Maheswari, and S. Saranya

1.1 OVERVIEW OF AI, CLOUD COMPUTING, AND EDGE COMPUTING AS TRANSFORMATIVE TECHNOLOGIES

AI technology, cloud-based computing, and on-device or edge-level processing have turned into key parts of the tech world today. AI lets machines handle jobs that used to need human smarts, like making decisions, working with natural language, and doing predictive analytics. Cloud computing changes how we store data and run apps by offering scalable, cheap options over the internet [1]. Edge computing speeds up real-time responses and cuts down on latency by handling data right near where it comes from. All together, these things are changing whole industries, allowing for smarter apps and better data handling.

1.2 INTRODUCTION TO TRANSFORMATIVE TECHNOLOGIES

AI technology, cloud-based computing, and on-device or edge-level processing are big forces that change how businesses run, how data gets processed, and how people deal with digital stuff. They work on their own and together too, which opens up smarter apps, real-time decisions, and good use of data in all kinds of areas. AI technology focuses on getting machines to do tasks that require human intelligence. That includes decision-making, understanding natural language, and predictive analysis.

Edge computing handles main issues like cutting response times, saving battery, lowering bandwidth costs, and boosting data privacy and security [2]. Cloud computing shifts old school computing by putting apps and services on the internet. Key perks are scalability when we need it, cost savings, and access from anywhere, so it's a must have for organizations and people now. Edge computing builds on that cloud stuff, processing data close to where it starts, like in IoT devices and sensors. It cuts down on needing central systems, so things process faster and respond better.

DOI: 10.1201/9781003635147-1

1.3 EDGE COMPUTING AND NEEDS

This closeness cuts latency, improves response times, and makes better use of bandwidth. With more IoT devices and real-time apps popping up, on-device or edge-level processing is crucial for dealing with huge amounts of data and making sure it's processed on time.

1.3.1 Edge intelligence or edge AI

Edge intelligence or edge AI mixes artificial intelligence into on-device or edge-level processing systems, so devices can do tough tasks like image recognition, natural language processing, and predictive analytics right there locally. Combining AI with edge processing lets organizations make decisions quicker, keep data more private, and not depend so much on cloud setups.

1.3.2 Moving deep learning applications to the edge

Deep learning models take a lot of computing power, but they are getting adapted for edge spots more and more. Methods like model compression, quantization, and pruning help neural networks run well on edge devices that don't have many resources [3]. These steps make it possible for apps like facial recognition, anomaly detection, and autonomous navigation. Running them at the edge means less need for central servers, so operations get faster and more efficient. Industry folks and researchers are putting a lot more into on-device or edge-level processing lately, where computing and storage are right at the internet's edge near mobile devices or sensors.

1.4 ON-DEVICE AI AND INFERENCE AT THE EDGE

On-device AI means putting pre-trained models straight on edge devices for local inference jobs. This cuts data sending, boosts privacy, and keeps real-time performance going. For example, smartphones with on-device AI can process voice commands, identify objects in photos, and give augmented reality experiences. This local handling reduces the push for cloud connections, which improves user experience and data security.

1.5 EDGE INTELLIGENCE ENABLES AI DEMOCRATIZATION

By dropping the computing and connection hurdles for AI, edge intelligence opens up advanced tech to more people [4]. It lets small businesses, startups,

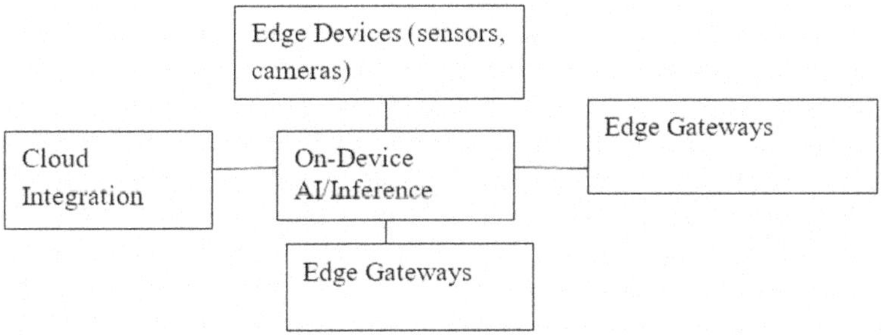

Figure 1.1 Diagram Edge AI Workflow.

and far off communities use AI without big spends on cloud infrastructure. Figure 1.1 shows the flow of the edge AI process. This kind of access sparks innovation, closes tech divides, and pushes for fair chances in adopting technology.

1.6 OVERVIEW OF CLOUD AND EDGE COMPUTING IN AI

1.6.1 Introduction to cloud and edge computing

Cloud computing gives us that virtual setup for utility stuff. It pulls together monitoring gear, storage units, analytics tools, visualization setups, and client delivery options. The cost-effective side of cloud computing helps with full service delivery. Businesses and users can grab applications on demand from anywhere. Cloud and on-device or edge-level processing go hand in hand in AI tech. They complement each other pretty well. On-device or edge-level processing lets us handle data right there locally. It does real-time stuff close to where the data comes from. That cuts down latency a lot.

1.6.2 The growing importance of AI

AI systems depend more and more on huge datasets and tricky algorithms. Bringing in cloud and edge tech creates this balanced way to work. It plays to the strengths of both to boost performance and scalability in AI apps [5]. On-device or edge-level processing pushes tasks from the network core out to the edge. People see it as a solid solution these days. Edge AI, or edge intelligence, is this new field mixing disciplines. It is getting a ton of attention fast. Cloud computing brings application services over the internet in a fresh way. It boosts the power of mobile systems too. That means more advanced and quick responses.

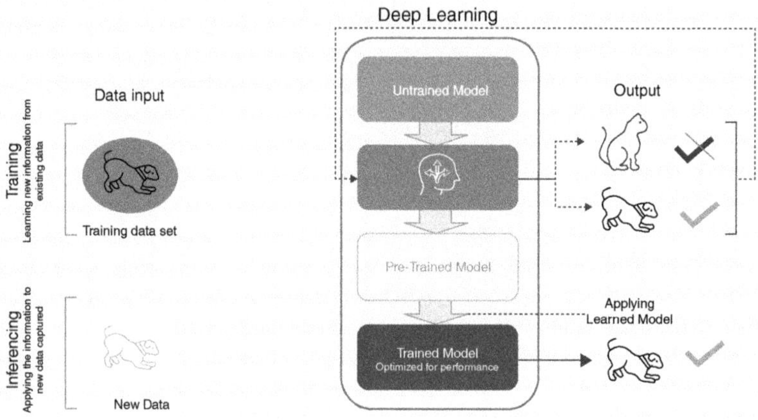

Figure 1.2 Cloud and Edge Integration.

1.6.3 Benefits of integration

Mixing cloud and on-device or edge-level processing helps organizations get better efficiency in computing. It allows seamless real-time decisions. Plus, it adapts well to different use cases. Figure 1.2 showing cloud and edge integration. It has centralized AI model training in the cloud.

Then localized inference happens at the edge. Next-gen cloud systems are getting more seamless and widespread. They are everywhere now. This comes from trends in distributed, diverse, ad hoc cloud setups and their architectures.

1.7 HYBRID ARCHITECTURES

1.7.1 Defining hybrid architectures

Hybrid setups in AI mix cloud and edge tech to build a dynamic computing space. They spread tasks between cloud and edge nodes. It depends on what resources are available and latency needs for communications.

1.7.2 Components of hybrid architectures

The main parts include cloud servers for storing data and training models. Edge devices handle data pre-processing and inference. Communication networks keep data flowing smoothly between the layers.

1.7.3 Real-world implementations

Hybrid architectures are in things like autonomous vehicles. Edge devices take care of immediate processing there. The cloud holds historical data for long-term looks. First-person view shots in videos are not smooth on purpose [6]. That ramps up immersion for viewers. They usually get captured by someone walking with regular camera gear.

1.8 REAL-TIME DATA PROCESSING

1.8.1 The need for real-time AI

Real-time processing matters a lot in areas like healthcare, autonomous driving, and smart cities. Decisions have to happen right away. Edge computing steps in to process data near the source.

1.8.2 Challenges in real-time processing

Things like latency, bandwidth limits, and power issues get in the way. To beat them, we need good data splitting and optimized protocols between cloud and edge.

1.8.3 Collaborative data processing

Hybrid systems let us collaborate on processing. Time-sensitive stuff stays at the edge. Complex analysis goes to the cloud later. It balances speed and accuracy. Current platforms for on-device or edge-level processing show how emerging tech will hit future IoT apps.

1.9 SECURITY CHALLENGES

1.9.1 Understanding security risks

Mixing cloud and on-device or edge-level processing brings specific security issues. Think data breaches, unauthorized access, and weak spots in networks.

1.9.2 Protecting data integrity

We need encryption, authentication, and strong access controls to keep data safe as it moves between cloud and edge.

1.9.3 Strategies for secure integration

We should use secure development practices, keep things updated, and follow industry standards to cut risks. We also should harden edge devices against physical and network attacks especially.

1.10 PRACTICAL AI APPLICATIONS AND FUTURE PERSPECTIVES

1.10.1 Current applications

Cloud-edge AI systems are changing industries. Examples include predictive maintenance in manufacturing, personalized diagnostics in healthcare, and low-latency immersive gaming.

1.10.2 Emerging trends

Better AI algorithms, 5G connections, and edge hardware open doors for smarter autonomous systems. Quantum computing with hybrids could tackle tough problems.

1.10.3 Future prospects

Tech will make hybrid cloud-edge systems more advanced. That leads to AI everywhere. These systems will learn in real time, adapt to changes, and collaborate globally. Mobile on-device or edge-level processing extends cloud services to network edges via base stations. It works in mobile, wireless, and wired setups with software and hardware near users.

1.11 CORE CHARACTERISTICS OF AI, CLOUD, AND EDGE COMPUTING

- **AI capabilities**—Machine learning models look at patterns and predict things. Natural language processing handles human–computer talks via speech and text. Computer vision runs stuff like facial recognition and self-navigation.
- **Cloud computing features**—Pay-as-we-go cuts initial costs. Elastic scaling fits changing workloads. Virtualization makes hardware use efficient.
- **Edge computing benefits**—Local processing drops latency. It boosts reliability by cutting internet dependence.

1.12 INTERACTION OF TECHNOLOGIES

Integrating AI, cloud computing, on-device, or edge-level processing builds a strong base for new ideas. Organizations use their strengths for scalable, efficient, responsive solutions to various needs. AI in the cloud gets power and storage from infrastructure for training models well, like AI analytics platforms using cloud datasets. Edge-enabled AI puts algorithms on devices for real-time work. Think facial recognition on phones and predictive maintenance in factories. Edge handles sensitive timing and smooths workflow.

1.12.1 Real-world applications of transformative technologies

In healthcare, AI diagnostics check cloud-stored medical images. Edge devices watch patient vitals in real time. Wearables give quick feedback and cut emergency response times. In retail, AI looks at consumer behavior with cloud systems for personalized marketing. Edge computing adds in-store real-time analytics on foot traffic. In smart cities, edge IoT sensors grab traffic and pollution data in real time [7]. Cloud technology aggregates and analyzes data for better urban planning. Benefits of AI, cloud, and edge integration include enhanced efficiency.

Real-time processing reduces delays and boosts productivity. Cost savings from scalable solutions skip big infrastructure spends. Scalability and adaptability handle varying loads for steady performance. Data security needs strong cloud encryption and controls. Latency and bandwidth issues grow with data volumes, so efficient processing matters. Regulatory compliance with data laws is key for these tech users. Future trends point to deeper AI, cloud, and edge ties [8]. Things like AI-driven edge devices, quantum in cloud, and decentralized AI open new industry doors. By using strengths and synergy, these techs make smarter, faster, and more efficient systems. Innovation will push boundaries in personal and industrial uses.

Rapid advancement in communications technology and the mobile device boom have brought on-device or edge-level processing as a hot new paradigm. AI apps thrive with deep learning breakthroughs and hardware tweaks.

1.12.1.1 Evolution from traditional servers to cloud and edge

Traditional servers were costly, centralized, and limited in scaling. Cloud technology brought elastic storage, on-demand scale, and lower ops costs. Edge processes locally to cut latency and bandwidth. This shift enables apps like self-driving cars and smart cities more accessible responsive computing.

1.12.1.2 The edge cuts latency by processing near source

Cloud latency comes from sending data over the internet to centers. Autonomous vehicles use edge for local sensor data and instant decisions for safety. It is ideal for gaming and health monitoring in real time.

1.12.1.3 Daily life integration and business transformation

Daily life sees AI and cloud computing blend in for convenience and personalization. Virtual assistants like Siri and Alexa use cloud AI for command interpretation and quick responses.

1.13 BUSINESS TRANSFORMATION WITH CLOUD AND EDGE COMPUTING

1.13.1 The shift to cloud-based solutions

Cloud computing changes business ops with flexibility, scalability, and cost-efficiency. Software as a Service like Zoom, Salesforce, and Microsoft Teams helps collaboration and customer management. Platform as a Service like Google Cloud and AWS lets we build apps without server hassle. Infrastructure as a Service gives virtual resources to scale infrastructure as needed. Retail analytics does real-time foot traffic for inventory tweaks. Manufacturing IoT at edge gives quick machinery feedback and stops downtime. Healthcare uses real-time monitors to speed medical responses. Business agility and resilience come from cloud and edge combo. Scalable solutions adjust resources on demand. Disaster recovery with cloud backups and edge redundancy keeps things going in outages. Customer engagement gets personal with AI-cloud solutions.

1.13.2 Examples of business transformation

E-commerce uses cloud AI rec engines for personal shopping. Edge streamlines checkouts. Finance has edge fraud detection for local transaction checks and better security. Smart cities use IoT sensors for data collection and cloud analysis for urban plans.

1.13.2.1 The road ahead

Businesses adopting cloud and edge more can use AI insights to cut costs and boost efficiency. Tighter integration is coming and driving industry innovation.

1.13.3 Benefits of AI-cloud integration for businesses

Integrating AI with cloud gives cost-effectiveness and cuts infrastructure costs. Enhanced productivity automates routine tasks. Reliability keeps downtime low. AI Customer Relationship Management Systems (CRMs) such as Salesforce analyze behavior for personal experiences and optimize ops.

1.14 ADVANCED APPLICATIONS AND INNOVATION

1.14.1 Dynamic AI-cloud integration

AI handles vast datasets well with cloud's scalable setup, leading to big advancements. Benefits include streamlined ops, better decisions, and top customer experiences. Large storage plus AI analytics keeps businesses agile, competitive [9]. Evolution has progressed from basic automation to neural nets in distributed spots, as AI cloud keeps growing and pushing industry innovation.

1.14.2 Core components of AI-cloud integration

AI algorithms in cloud are sophisticated ones on platforms process data, spot patterns, and give insights. Machine learning (ML) models update and scale for needs. Cloud infrastructure runs AI tasks efficiently with VMs, serverless, and keeps ops smooth under load. Data management with cloud tools for storage, streaming, and encryption makes secure AI space.

1.15 THE SYNERGY OF AI, CLOUD COMPUTING, AND EDGE COMPUTING

1.15.1 Transforming industries and daily life

- **The convergence of AI and cloud computing**—AI processes big datasets for insights, fits cloud's scalable infrastructure perfectly. This combo brings a new computational power, accessibility, and efficiency era.
- **Benefits of integration**—AI with cloud platforms streamlines ops, boosts decisions, and improves customer experiences. Cloud storage with AI analysis helps organizations stay competitive in changing markets.
- **Evolution and growth**—From simple automation to advanced neural nets in distributed setups, AI cloud has exploded. It shows potential for innovation, solving tough industry challenges.

- **Key applications across industries**—Healthcare-AI cloud changes diagnostics, treatment, and monitoring. Cloud predictive analytics spots diseases early. Edge supports real-time wearables for monitoring.
- **Retail**—Retailers use AI cloud for behavior analysis, inventory, and personal shopping. Cloud AI processes data for targeted marketing and loyalty.
- **Manufacturing**—Predictive maintenance, supply chain, automation benefits. Edge real-time monitoring cuts downtime and costs.
- **Real-time analytics**—AI cloud gives instant business insights for improved decisions and quick strategy adaptations.

1.15.2 Challenges in AI-cloud integration

- **Computational complexity**—AI models need power, which strains cloud. Optimize algorithms, use GPUs to fix. Data privacy: Cloud storage risks access. Strong privacy, audits vital. Integration hurdles. Mixing AI with old infrastructure is tricky. Plan well and check compatibility for smooth shifts.
- **Financial services**—Enhances fraud detection, reduces risk. Cloud AI spots anomalies and predicts trends.
- **Education**—AI-cloud personalizes learning. Adaptive platforms analyze performance and give tailored resources and feedback for better outcomes.
- **Collaborative ecosystems**—Tech firms, academia, and government partnerships drive AI-cloud innovation. Shared resources and expertise speed advances.
- **Open-source initiatives**—Democratize tech, as small organizations get cutting-edge tools. Community inputs improve ongoing community development. Dev communities ensure enhancements to platforms. Training and workshops skill up pros for these techs.
- **Enhancing interoperability**—Future apps need it for seamless platform integration.

1.16 EDGE AI TRANSFORMING COMPUTING FROM CLOUD TO EDGE

1.16.1 Introduction to edge AI

This decentralized way speeds data processing and real-time decisions near the source. Devices like smartphones, IoT, and industrial robots handle local data. It cuts latency and bandwidth. It is key for high-response apps in healthcare, transport, and manufacturing. Real-time uses like self-driving cars and smart health devices get immediate replies. Local processing boosts privacy and compliance. Powerful edge devices and optimized AI algos speed this shift the evolution of edge computing [10]. Started as basic local processing, now adds AI for smart decisions. First to cut latency, boost network efficiency.

Shows shift to decentralized, intelligent systems for modern challenges.

1.16.1.1 Transforming computing from cloud to edge

The journey from central cloud to edge marks a paradigm shift in data handling. Cloud AI gives scale and cost-efficiency. Edge AI fixes latency and bandwidth issues by computing near sources. The rest of this chapter explores transition, benefits, challenges, and the future.

1.16.2 Key features of edge AI

Real-time processing—Edge AI does immediate analysis, crucial for self-driving cars and industrial automation. Cars analyze surroundings in real time for safe navigation. Robots monitor lines and prevent issues. Feedback ensures efficiency and safety in dynamic spots.

1.16.3 Components of edge AI systems

Key components include

- **Edge devices**—Sensors, cameras, and wearables collect and pre-process data. Embedded AI does basic analytics on them.
- **Edge gateways**—Aggregate from multiple devices, with extra processing and secure transmit to cloud or centers.
- **AI models**—Pre-trained, optimized for edge handle recognition, and predictive. Quantization techniques for efficiency.
- **Hardware accelerators**—GPUs, AI chips boost performance for heavy tasks.
- **Connectivity modules**—5G, Wi-Fi for seamless edge-cloud comms when needed.
- **Security mechanisms**—Encryption, auth protocols, secure boot protect systems.
- **Management software**—Platforms monitor, update, maintain edge, give health, performance insights.

1.16.4 Applications in healthcare

- **Remote patient monitoring**—Edge AI wearables watch real-time, analyze vitals, alert providers to issues. Sensors spot heart rhythms, notify docs for timely help.
- **Diagnostic assistance**—Portable devices with edge AI analyze images instantly, faster diagnosis in remote areas. Cuts treatment delays, improves outcomes, saves resources.
- **Personal health devices**—Smartwatches, wearables track heart rate, activity, give health insights.
- **Local processing**—Enhances privacy, real-time feedback.

1.16.5 Industrial applications of edge AI

- **Smart manufacturing**—Edge AI enables predictive maintenance and quality control. Real-time analysis spots defects, optimizes processes, and cuts downtime.
- **Energy management**—Smart grids with edge AI optimize consumption and regulate machines and sustainability. Energy management helps meet regulations and reduce costs.
- **Supply chain optimization**—Edge AI boosts visibility and decisions. Sensors track inventory and conditions, boost efficient logistics, and protect perishables.

1.17 CLOUD AND EDGE COMPUTING BRIDGING CENTRALIZED AND DECENTRALIZED PARADIGMS

Evolution of computing from mainframe central power to cloud flexibility, now edge real-time perks, plays a major role. These shape the tech future by handling data processing, storage, and analysis needs. The rest of this chapter looks at cloud and edge basics, characteristics, apps, and synergy. Virtualized infrastructure offers scale, cost-effectiveness, and global access.

1.17.1 Key features of cloud computing

Scalability platforms let organizations adjust resources dynamically for demands. E-commerce sites can scale up in peak seasons. Global accessibility enables the availability of data and apps from anywhere, thereby supporting remote collaboration and business continuity.

1.17.2 Applications of cloud computing

Cloud computing supports big data analytics, ML, enterprise resource planning (ERP), and content delivery. Finance, healthcare, and media rely heavily on cloud for their operations.

1.17.3 Limitations of cloud computing

Challenges include latency from central processing, internet dependence, and privacy concerns.

1.17.4 Key features of edge computing

Reduced latency: Local processing gives fast responses for real-time use like self-driving. Bandwidth optimization: Local transfers cut big transmissions to servers. Enhanced privacy: Sensitive data at edge reduces exposure, which is good for health and finance industries.

1.17.5 Applications of edge computing

Edge computing excels in immediate insights fields. IoT devices like thermostats and wearables process locally for efficiency. Healthcare real-time monitors analyze patient data at edge for alerts. Manufacturing benefits from predictive maintenance and optimization with edge sensors.

1.17.6 Challenges in edge computing

Adoption is slowed by infra costs, interoperability, and security needs.

- Cloud vs edge computing—Both enhance capabilities but differ in approach and use cases. Key differences: Centralization vs decentralization. Cloud uses central centers, edge local. Latency: Edge gives real-time service by skipping cloud delays. Scalability: Cloud is unlimited and edge is constrained by decentralization.
- Complementary roles—Not competing, they complement each other. Edge handles real-time sensitive operations, which cloud is suited for big analytics, storage, and ML training.
- Use case example—In self-driving, edge processes sensors locally for navigation. Cloud stores and analyzes traffic patterns for route improvements over time.

1.18 BENEFITS OF HYBRID ARCHITECTURES

- Flexibility—Organizations balance local and centralized computing based on application needs.
- Resilience—Hybrids give backups and continuity in outages.
- Cost optimization—Distributed workloads cut bandwidth and storage costs.

1.18.1 Example applications

- Smart cities—Edge in traffic systems processes locally for decisions. Cloud aggregates trends for planning.
- Retail—Smart shelves, point of scale use edge for immediate customer insights. Cloud analyzes purchases for inventory.

1.19 FUTURE TRENDS AND SYNERGY

- Convergence of cloud and edge drives innovation and trends shape roles. AI at the edge: Enhanced devices with AI enable smart real-time services like maintenance and fraud prevention. 5G integration: Rollout amplifies edge with low latency, high speed for AR/VR and self-driving.

- Sustainability: Less transmission to servers cuts energy use and is eco-friendly. Edge-to-cloud continuum: Seamless integration leverages both strengths for holistic solutions.

1.20 APPLICATION ORIENTATION—FROM CLOUD TO EDGE

Transitioning from cloud to edge shifts the tech landscape to decentralized, processing near the source. It is driven by real-time needs, privacy, performance. Real-time insights and performance enhancement: Handling data near the origin like sensors, cameras, and IoT cuts latency considerably. It is critical for immediate action industries, self-driving, health monitoring, and automation. Self-driving cars use edge for road analysis, obstacle detection, and quick decisions without a server wait time. In health, wearables monitor vitals and alert pros to issues in real time.

1.21 ENHANCED PRIVACY AND SECURITY

Another major benefit of on-device or edge-level computing is that it can enhance security and privacy. Sensitive information may be locally processed without being sent over to centralized servers, minimizing chances of data interception while being sent. This localized nature supports strict data-protection laws, like GDPR, and addresses increasingly consumer-driven demands for privacy-friendly technology. For instance, home automation devices can process user data locally so that sensitive information stays under the control of the user.

1.21.1 The synergy of edge and cloud computing

Although on-device or edge-level processing has many advantages, it is not meant to make cloud-based computing a relic of the past. Rather, the two operate in tandem, leveraging the strengths of each to bring about a hybrid model. The cloud offers scalable storage and compute power, which is best suited for long-term data processing and machine learning model training. On the other hand, on-device or edge-level processing addresses time-critical tasks and takes pressure off network resources. This combination makes it possible for businesses to strike a balance between responsiveness, scalability, and efficiency.

1.21.2 Spurring innovation across industries

The combined strengths of edge and cloud computing are powering innovation in a wide range of industries. In manufacturing, edge systems streamline production lines through real-time monitoring and predictive maintenance. In retail, they improve customer experiences by enabling personalized recommendations and streamlined inventory management. Even in remote locations, on-device or edge-level processing enables access to high-tech technologies, including telemedicine and e-learning, by reducing reliance on high-speed internet.

1.21.3 The future of digital transformation

With ongoing advances in technology, the synergy between edge and cloud-based computing will determine the next phase of digital transformation. The harmony between centralized and decentralized approaches will enable companies to innovate, streamline operations, and enrich user experiences. For people, these developments hold out the promise of smarter, more responsive systems that blend into everyday life. From smart cities to cutting-edge AI-based applications, the future will revolve around leveraging the complementary capabilities of both edge and cloud-based computation to usher in new opportunities.

In conclusion, the transition from cloud to on-device or edge-level processing represents a significant step toward decentralized and efficient systems. Edge AI's real-time capabilities, combined with the cloud's scalability, offer unparalleled opportunities for innovation and optimization. Together, they form the backbone of a smarter, more connected world, driving progress across industries and improving lives globally.

1.22 CONCLUSION

Cloud-based computing, on-device or edge-level processing, and artificial intelligence are transforming data processing, storage, and usage. From enabling smarter applications to facilitating real-time decision-making, these technologies are crucial in the current digital era. As edge intelligence evolves, it opens the doors to a world where everyone has access to AI, spurring innovation and inclusivity in industries and communities. AI, cloud computing, and on-device or edge-level processing are the cornerstones of contemporary digital transformation. Through their complementary strengths and synergy potential, these technologies make systems smarter, faster, and more efficient. With advancing innovation, their convergence will continue to redefine potential in industries and daily life.

- **Synergy of transformational technologies**—The converging of AI, cloud computing, and on-device or edge-level processing is revolutionizing industries and daily life. Their convergence fuels innovation, real-time decision-making, and operating efficiency.
- **Future directions**—For solving cyber security issues, driving regulatory compliance, and ensuring top-down ethical AI deployment, it will be imperative as these technologies continue to mature. Companies need to keep on riding AI-powered strategies in order to innovate and stay competitive.
- **Key takeaways**—The strength of convergence of AI, cloud, and on-device or edge-level processing is scalability, personalization, and

real-time responsiveness. These technologies are set to define the future of industries, society, and world economies, opening new doors for innovation and growth.

Cloud-to-on-device or edge-level processing is a major progression toward decentralized and efficient systems. Edge AI enables real-time insights for industries, improving performance and privacy. The collaboration of edge and cloud will propel the future wave of digital transformation, unlocking new uses and more intelligent systems. On-device or edge-level processing and the cloud constitute revolutionary changes in how data gets processed and leveraged. Their interplay provides unprecedented ability to streamline operations, improve user experiences, and create innovation across industries. With improving technology, the equilibrium between decentralized and centralized patterns will characterize the next wave of digital transformation, enabling businesses as well as individuals.

REFERENCES

[1] Shi, W., Cao, J., Zhang, Q., Li, Y., & Xu, L. (2016). Edge computing: Vision and challenges. *IEEE Internet of Things Journal,* 3(5), 637–646. https://doi.org/10.1109/JIOT.2016.2579198

[2] Satyanarayanan, M. (2017). The emergence of on-device or edge-level processing. *Computer,* 50(1), 30–39. https://doi.org/10.1109/MC.2017.9

[3] Gubbi, J., Buyya, R., Marusic, S., & Palaniswami, M. (2013). Internet of Things (IoT): A vision, architectural elements, and future directions. *Future Generation Computer Systems,* 29(7), 1645–1660. https://doi.org/10.1016/j.future.2013.01.010

[4] Zhou, Z., Chen, X., Li, E., Zeng, L., Luo, K., & Zhang, J. (2019). Edge intelligence: Paving the last mile of artificial intelligence with on-device or edge-level processing. *Proceedings of the IEEE,* 107(8), 1738–1762. https://doi.org/10.1109/JPROC.2019.2918951

[5] Li, Y., & Wang, L. (2020). A survey of on-device or edge-level processing: Principles, applications, and challenges. *ACM Computing Surveys,* 53(2), 1–34. https://doi.org/10.1145/3378673

[6] Kumar, K., & Lu, Y.-H. (2010). Cloud computing for mobile users: Can offloading computation save energy? *Computer,* 43(4), 51–56. https://doi.org/10.1109/MC.2010.98

[7] Varghese, B., & Buyya, R. (2018). Next generation cloud-based computing: New trends and research directions. *Future Generation Computer Systems,* 79, 849–861. https://doi.org/10.1016/j.future.2017.09.020

[8] Premsankar, G., Di Francesco, M., & Taleb, T. (2018). Edge computing for the Internet of Things: A case study. *IEEE Internet of Things Journal,* 5(2), 1275–1284. https://doi.org/10.1109/JIOT.2018.2805263

[9] Abbas, N., Zhang, Y., Taherkordi, A., & Skeie, T. (2018). Mobile on-device or edge-level processing: A survey. *IEEE Internet of Things Journal,* 5(1), 450–465. https://doi.org/10.1109/JIOT.2017.2750180

[10] Deng, S., Zhao, H., Fang, B., Yin, J., Dustdar, S., & Zomaya, A. Y. (2020). Edge intelligence: The confluence of on-device or edge-level processing and artificial intelligence. *IEEE Internet of Things Journal,* 7(8), 7457–7469. https://doi.org/10.1109/JIOT.2020.2984887

Chapter 2

Edge Computing Fundamentals

Trilok Bhardwaj, Sandhya Malikar Patel, Poonam Arora, and Manju Khari

2.1 CLOUD TO EDGE COMPUTING: WAVES OF INNOVATION

2.1.1 Introduction to cloud computing

Cloud computing is one of the most groundbreaking innovations among the computing services. The concept and model of cloud computing have transformed the network access for storing, processing, and accessing data and applications across the internet rather than relying on local servers or personal devices [1, 2]. The cloud has become the backbone of modern digital transformation, handling everything from large data management to application hosting. It has provided a centralized platform which enables users and organizations to use computing resources—such as servers, storage, databases, networking, software, and analytics—on demand, with high scalability and flexibility. In essence, the word "cloud" metaphorically refers to the internet, and "cloud computing" basically entails providing computer services online [3]. In fact, the development of cloud computing is not typical, but it is a blend of the expansion of computer technology and communication technology to a certain stage. Cloud computing technology is the product of a collection of various technologies.

In 1961, John McCarthy proposed the concept of utility computing, which finally gave the origin of cloud computing [4]. His remark, '*Computation may someday be organised as a public utility*,' gradually became the basis of a new and important industry. Salesforce led the way in offering remote, customizable services to businesses in the late 1990s [5] and demonstrated to the world that software could be purchased as a service. The computer world quickly accepted the concept of service. Later in 2002, the Amazon Web Services (AWS) platform was launched, which offered businesses services including remotely configurable storage, computing resources, and business operations [6]. For example, Amazon Web Services EC2 (Elastic Compute Cloud) enabled its customers to create and destroy custom-sized web-based virtual machines on demand, meeting the "elasticity" criteria of the cloud [7, 8]. It was in the year 2006 that the CEO of Google, Eric

DOI: 10.1201/9781003635147-2

Schmidt, came up with an evolved idea and proposed the concept of "cloud computing" [9]. Also, NIST-USA launched a document in 2011 [10] which characterizes the important aspects of cloud computing. This document is intended to serve as a means for broad comparisons of cloud services and deployment strategies and to provide a baseline for discussion from what cloud computing is to how to best use cloud computing.

One of the most significant aspects of cloud computing is its on-demand availability. Users can get the resources they require whenever they need them, without owning or managing any physical infrastructure. This has greatly reduced the cost and complexity of IT management for enterprises. In place of investing in costly technologies, including software, businesses can pay for what they need via a pay-as-you-go approach or subscription-based services.

Based on service models, cloud computing is generally divided into three main categories: i. Infrastructure as a Service (IaaS), ii. Platform as a Service (PaaS), and iii. Software as a Service (SaaS) [10, 11].

- IaaS provides virtualized computing resources over the internet. Companies like AWS, Microsoft Azure, DigitalOcean, and Google Cloud offer services where users can rent virtual servers, storage, and networking capabilities on a pay-as-you-go basis.
- PaaS offers a platform allowing customers to develop, run, and manage applications without the complexity of building and maintaining infrastructure. Here, the developers can focus on coding and innovation using the tools and resources provided by the PaaS providers like Heroku, AWS Elastic Beanstalk, and MS Azure.
- SaaS delivers and distributes the software applications over the internet, typically on a subscription basis. Popular examples include Google Workspace, Microsoft 365, and Salesforce. It is cloud-based delivery, which eliminates the need for local installations. Users generally subscribe to the applications on a recurring basis depending on the requirement, which is cost-effective and allows flexibility and accessibility as well.

Cloud computing offers various deployment models, which are exclusive and suitable for different needs [2, 11, 12]. In principle, a cloud deployment model describes the infrastructure for the deployment and determines the ownership and control over that infrastructure. Different types of cloud computing deployment models are described next.

2.1.1.1 Public cloud

A public cloud is one where cloud infrastructure services are made available to the general public or significant industry groups via the internet, e.g., web hosting and email services. As the public cloud is accessible to anybody, it can be less secure and cannot be customized as per the need of an individual. However, this model does not require dedicated infrastructure management, as the service provider does all the maintenance tasks, and hence it requires minimal investment.

2.1.1.2 *Private cloud*

Due to its exclusive focus on a single organization, the private cloud model offers more control and security than the public deployment method. Since no hardware is shared in this deployment technique, complete control and customization are possible. It is also known as the interior cloud at times. The customer is responsible for managing the infrastructure, which frequently necessitates a secure environment with strong firewalls that must be handled by qualified individuals, thus raising the costs.

2.1.1.3 *Hybrid cloud*

The hybrid model integrates both public and private cloud deployment models, thus allowing organizations to leverage the benefits of both. This model gives a single cloud infrastructure that is aimed at increasing flexibility and deployment options for the business. Hybrid cloud enables sharing data and applications between private and public cloud environments. Companies can keep sensitive data in their private cloud while leveraging the capabilities of the public cloud.

2.1.1.4 *Community cloud*

This model is a lesser-known and lesser-adopted model which allows sharing of infrastructure at a lower cost. A community cloud is a cost-effective, mutually shared deployment architecture among enterprises within a specific community, such as government agencies, banks, or other commercial entities. Security, privacy, and performance are common concerns among members of the community. Third-party providers internally host and maintain community clouds.

Cloud computing offers various advantages. One benefit of using cloud computing is its scalability feature, which allows companies and businesses to adapt their resources based on varying demands. This flexible trait of cloud computing provides an economical and reasonable approach without investing in large upfront capital expenditures. Additionally, cloud computing offers secure and reliable services, with providers offering features such as data backups, disaster recovery, and compliance with industry regulations. These aspects aid in enhancing confidence in the business operations with secure data, minimizing the risk of data breaches or disruptions. But there are several drawbacks and challenges associated with cloud computing. It possesses several concerns, like data privacy, downtime risks, and vendor lock-in issues, which need to be addressed in effective ways. Companies can resolve these issues not only by carefully verifying and evaluating service providers but also by scrutinizing their terms of service. Regulatory obligations need to be thoroughly verified before the deployment of any model for the effective compliance and data protection measures. The vendor lock-in problem in cloud computing arises due to challenges in transferring data or services between cloud providers. These challenges are faced due to various incompatibilities, migration fees which are sometimes high and proprietary technologies in a single cloud provider mainly. These issues demand a multi-cloud strategy aimed at removing these risks by distributing dependencies and continuing services. By adopting the multi-cloud framework, the risk of downtime can be reduced significantly, thus providing protection from financial losses and any kind of operational disruptions.

Despite its many advantages, edge computing comes with challenges. Managing a distributed network of edge devices can be complex and requires a reliable and robust infrastructure, upkeep, and monitoring. In addition, there are concerns over the standardization, interoperability, and scalability of devices, as well as the requirement for more advanced cybersecurity measures to safeguard the wider attack surface that is created by decentralized devices.

Edge computing complements cloud computing and works hand-in-hand with it rather than replacing it [13, 14]. Although the cloud is utilized for more complex processing, storage, and analysis, the edge is responsible for handling real-time, vital activities in many different types of systems. The utilization of the hybrid approach can provide businesses with the opportunity to maximize not only their performance but also their operating expenses.

In conclusion, edge computing represents a paradigm shift in our conceptualization of data processing. By bringing computation closer to where data is generated, it offers faster response times, reduced costs, and enhanced security. Edge computing will play a vital role in enabling smarter, quicker, and more responsive systems across various industries as technologies such as the Internet of Things (IoT), fifth-generation (5G), and artificial intelligence continue to advance.

2.1.2 Transformation from cloud computing to edge computing

The rapid growth of the wireless industry and implementation of new technologies over the past few decades has seen a rapid migration from on-premise data centers to cloud servers. However, with the increasing number of Industrial Internet of Things (IIoT) applications and devices, performing computation at either data centers or cloud servers is neither feasible nor efficient. Edge computing marks a significant change in how data is processed, stored, and managed in the modern computing environments.

The transformation from cloud computing to edge computing marks a significant shift in how data is processed, stored, and managed in modern computing environments. While cloud computing revolutionized the IT landscape by enabling scalable, on-demand access to computing resources via centralized data centers, it also introduced limitations in latency, bandwidth, and data sovereignty. Edge computing emerged as a complementary approach to overcome these limitations by bringing computation closer to where data is generated—at the "edge" of the network.

Cloud computing has served as the backbone of digital transformation for over a decade, allowing businesses to scale infrastructure, reduce IT costs, and deploy applications globally. However, as the number of connected devices and sensors has exploded with the rise of the IoT [15], sending all that data to centralized cloud servers has become increasingly inefficient. Real-time applications such as autonomous vehicles, augmented reality, industrial automation, and remote healthcare require ultra-low latency, reliable processing, and immediate decision-making challenges that cloud computing alone cannot always meet effectively.

This need for real-time responsiveness has driven the transition toward edge computing [16], where data is processed near its source, such as on local edge servers, gateways, or even directly on smart devices. This proximity significantly reduces latency, conserves bandwidth, and ensures faster decision-making. In contrast to cloud computing's centralized model, edge computing embraces a decentralized approach that allows for more responsive and context-aware systems.

The introduction of 5G networks has provided faster speeds and lower latency compared to earlier generations, which is a key factor driving the transformation to edge computing. The applications of 5G requiring quick responses, like robotic surgery, smart traffic systems, and immersive virtual experiences, can be achieved flawlessly in conjunction with edge computing. A hybrid cloud-edge architecture is emerging as the new norm by using edge devices for crucial projects and shifting large-scale or complex processing to the cloud.

Another important factor supporting this major transformation is data privacy and being in compliance with regulatory authorities. The concerns like safekeeping of data, storage, their location, and complying with regulations like General Data Protection Regulation (GDPR) [17, 18] are forcing the organizations to look for potential ways to process and store vital data locally. All these concerns are addressed by edge computing by reducing the amount of sensitive data to be transmitted to the cloud, which not only enhances control but also ensures compliance.

Despite these benefits of edge computing, it does not replace cloud computing; rather, it complements cloud computing by enhancing its potential. The aim of this transformation is to attain an ideal balance between centralized and decentralized processing rather than to pick one over the other. While edge computing is an excellent option required for local and time-sensitive tasks, cloud computing is still the best choice for those projects requiring enormous processing power, long-term storage, and worldwide access.

The transformative shift to edge from cloud computing signifies and aids the growing needs of the highly interconnected world. Edge computing has become a vital factor for the emerging digital infrastructures in the developed technology era, anticipating an increasing demand of localized and secured data with real-time processing. The cloud and edge computing together create a highly potent ecosystem, enabling organizations a faster, smarter, and adaptive framework to work with.

2.1.3 Limitations of cloud computing

Cloud computing has revolutionized the methods by which individuals and organizations store, access, and process data, offering advantages like scalability, cost-effectiveness, adaptability, and minimal infrastructure management.

However, cloud computing comes with inherent limitations, besides its extensive use and several benefits which can affect performance, security, compliance, and overall operational efficiency, particularly in certain use cases or industries.

The most significant constraint with cloud computing is its latency. Relying on centralized data centers, which may be located at a distant place from the user or any data-generating device, often delays the communication in cloud computing. For faster and real-time processing applications, like remote surgeries, industrial automation, and smart vehicular systems, this communication delay of even a few milliseconds might result in operational complications and safety hazards. This constraint has triggered the evolution and thereby the utility of edge computing, which lowers latency by allowing the computation task to be done nearer to the data source.

Another constraint which affects the efficiency of cloud services is its dependence on internet connectivity. The concept of cloud computing demands a stable, secure, and high-speed internet connection. Cloud-based applications and services suffer a major operational lag in areas with unstable internet connections or during network failures. This interrupted service not only results in financial loss as productivity is hampered but may also possess unsettling conditions like data compromise if it remains unaddressed.

A critical threat faced in the cloud computing framework is data security and privacy concerns. The third-party service providers process the data stored in the cloud, which introduces data breaches and lack of control due to unauthorized access. Though the cloud providers invest significantly to ensure the security of data, cyberattacks are unavoidable and affect any operational system. Also, the organizations must inspect and scrutinize the policies adopted by the cloud providers to comply with the agreed-upon security guidelines. This measure significantly reduces the risk of data privacy for companies dealing with confidential and sensitive information.

Another aspect of the data security issue associated with cloud computing is the data sovereignty and compliance. Various policies pertaining to data storage, transfer, and privacy have been formulated and adopted by different countries across the globe to mitigate this issue. The General Data Protection Regulation (GDPR) in Europe [17, 18], the Health Insurance Portability and Accountability Act (HIPAA) and California Consumer Privacy Act (CCPA) in the United States [19, 20], the Digital Personal Data Protection Act of 2023 (DPDP) in India [21], and ISO/IEC 27001 [22] are a few examples of the regulations that organizations using global cloud services must verify that they are in compliance with. Heavy fines and strict legal issues may be faced by the organizations for violating these guidelines. Sometimes organizations and businesses are prohibited from storing data in certain cloud regions thus restraining the flexible trait of cloud operation.

A major concern for the organizations utilizing cloud computing is the risk of vendor lock-in. It is a challenging situation where the providers often use proprietary technologies, APIs, and management tools that bind the clients to their platforms. This dependency restricts the migration to another platform once the workloads, applications, and data are deeply integrated into a specific provider's framework. This results in limited service operations and can leave organizations vulnerable to price increases. Such circumstances restrict organizations from adapting to varying market conditions and hinder innovation.

Lastly, cost management still remains a persistent challenge in cloud computing. Although cloud services are often more economical and profitable than the conventional work frames, the overheads can increase rapidly without careful oversight. This issue is significant for the small businesses and startups where limited financial resources are available, which need strict cost control. The unpredicted usage spikes, idle or underutilized resources, and reliance on premium service tiers can contribute to significant budget overruns.

2.1.4 Conclusion

In conclusion, cloud computing inherits various limitations and drawbacks even though it provides strong and adaptable measures for a variety of digital requisites. Adopting cloud-based solutions entails careful consideration of significant issues like latency, data security, secured internet dependence, compliance, vendor lock-in, and cost unpredictability. Organizations must evaluate their particular needs, seek out the potential risks involved, and develop proactive approaches to mitigate them in order to fully utilize the benefits of cloud services while minimizing their drawbacks.

2.2 INTRODUCTION TO EDGE COMPUTING ARCHITECTURES

2.2.1 Edge computing architecture

Edge computing is a distributed computing model that moves data processing and storage closer to where it is needed—typically near the data source or end users. This approach has significantly minimized latency, reduced bandwidth usage, improved response times, and enhanced data privacy and reliability. With the prediction that the number of IoT devices and real-time applications will reach 77.5 billion [23], numerous organizations have proposed various IoT architectures, considering a range of perspectives [24]. In this rapidly evolving landscape, the architecture of edge computing is becoming a critical component for efficient and high-performance systems.

This section explores the architecture of edge computing, including its components, layers, communication models, advantages, and use cases.

2.2.2 Understanding edge computing architecture

Edge computing is an umbrella term that incorporates a variety of architectural techniques that aim to bring processing and storage resources in close proximity to end users and devices [25, 26]. Edge computing is a notion that encompasses these many architectural approaches. In the context of edge computing, IoT devices transmit data to edge devices, which then receive all the data from IoT devices and transmit the most important information to cloud computing [15].

A well-structured edge computing architecture is a distributed network that typically consists of numerous layers [27]. This type of architecture extends the capabilities of the cloud to the edge network by integrating edge nodes that link end devices to the cloud. The edge architecture is composed of multiple layers, each of which serves a distinct purpose in the data lifecycle. Data collection, processing, analytics, storage, and transmission are all functions that are performed by these layers cooperating with one another.

2.2.3 Key layers of edge computing architecture

Edge computing brings computation and storage resources to the nearer location to increase response times and save bandwidth. Edge computing architecture is a service-based and layered solution that is often identified as an intermediary layer composed of edge nodes (edge tier) and defines specific capabilities for other layers, including the upper layer known as the cloud layer and the lower layer also known as the edge devices layer [15, 28].

2.2.3.1 Edge devices layer

This is the **outermost layer**, where data is generated and/or consumed. It includes sensors, actuators, smartphones, wearable devices, industrial machines, surveillance cameras, and other IoT devices. These devices collect raw data and may perform simple processing tasks like filtering or compression before forwarding the data to an edge node.

Some modern edge devices have built-in computing capabilities (known as "smart edge devices") and can perform real-time processing using lightweight machine learning models or rule-based decision engines.

2.2.3.2 Edge node layer

Edge nodes are intermediary computing devices that sit between the edge devices and central cloud or data centers. These nodes include local servers, gateways, routers, or even micro data centers deployed on-site (e.g., in a factory or retail store).

Edge nodes perform more advanced processing, such as data aggregation, analytics, anomaly detection, and localized decision-making. In many applications, edge nodes act autonomously, reducing the need to send data to the cloud, thereby saving bandwidth and reducing latency.

2.2.3.3 Fog layer

Fog computing is sometimes considered an extension or a layer between the edge and the cloud. It provides a regional layer of computing power and storage. Fog nodes can manage multiple edge nodes and perform tasks like orchestration, load balancing, security enforcement, and temporary data storage. Fog computing is particularly useful in large-scale distributed environments like smart cities, where centralized processing would cause delays or overloads.

2.2.3.4 Cloud layer

The cloud continues to play an important part in the overall architecture, despite the fact that edge computing is primarily concerned with decentralization. The cloud layer is equipped with substantial capabilities for the processing and storage of data.

The current trend in cloud layer development is based on the core of cloud computing technology and the ability of edge computing. This creates a complete elastic cloud platform at the edge that includes computing, network, storage, security, and other capabilities. People utilize the cloud for jobs that don't need to be done right away, for long-term storage, for analyzing complex data, for centralized control, for software upgrades, and for training models. The results of cloud-based processing may be returned to the edge for implementation.

2.2.4 Core components of edge computing architecture

- **Edge Gateways:** Edge gateways are crucial components in edge computing, as these act as the bridge between edge devices (e.g., sensors and IOT devices) and edge/cloud networks generally termed as higher-level systems [25]. They handle protocol translation, encryption, and data preprocessing. Edge gateways can manage network connectivity, ensuring reliable communication between edge devices and the cloud.
- **Local Compute Infrastructure:** The physical hardware and software located at or near the edge of a network, known as local compute infrastructure in edge computing, facilitate data processing in close proximity to the source of data generation or user interaction. This infrastructure comprises sensors, gateways, and servers that collaborate to manage data processing, storage, and analysis on a local level, thereby decreasing dependence on centralized cloud resources [27]. The **Local Compute Infrastructure** includes mini-servers, GPU/TPU chips, and edge clusters capable of running analytics, AI/ML models, and real-time applications.
- **Storage Modules:** This module consists of the temporary storage at the edge and is essential for buffering, caching, or holding time-series data. Edge storage is the practice of storing data at the network's edge, rather than relying exclusively on centralized cloud storage, in order to be closer to the source [27, 29]. By distributing storage resources, this method reduces latency, improves processing speed, and enhances data security. This encompasses storage that is directly integrated into devices such as IoT sensors, cameras, or even laptops and mobile devices that are used by field workers. These are dedicated devices that are located at the network's edge, where they offer storage and processing capabilities to adjacent devices.
- **Security Modules:** Edge security is a type of decentralized security that takes place on or close to end-user devices, or at the "edge" of the network. Although complex, edge security can refer to either the security systems on a network's edge or the systems used to protect an edge network [25]. Security modules for edge computing are specialized pieces of hardware or software made to safeguard resources and data at the "edge" of a network, which is closer to where data is created or used, as opposed to a centralized data center. In decentralized systems, security modules are key components that safeguard networks, applications, and edge devices against various threats. Decentralized systems require distributed protection environments that ensure

reliability and confidence, unlike centralized architectures, where security can often be managed from a single point. To guarantee secure operation, these modules integrate intrusion detection systems (IDS), firewalls, authentication frameworks, and encryption services, which are deployed locally at various nodes. Such a localized approach to security enhances resilience against attacks and reduces the risks associated with the single points of failure, thereby ensuring that the system remains robust and operational even under adverse conditions.

- **Orchestration and Management Platforms:** The effective deployment and management of scalable applications and services across distributed edge computing environments, adopting the edge computing orchestration and management platforms, are crucial. These transformative and robust platforms enable organizations to take full control of their edge infrastructure by automating key processes, such as provisioning, configuration, monitoring, and security management, across potentially thousands of edge devices. By orchestrating various intricate tasks which include resource scaling, application deployment, and workflow management across diverse edge locations, these platforms ensure that services and applications operate with unwavering reliability and maximum efficiency in the distributed edge environment [30].

  Moreover, they enhance operational efficiency by streamlining the critical functions, like software upgrades, configuration management, security oversight, and comprehensive monitoring of infrastructure and edge devices. By adopting these advanced measures, user-organizations can represent themselves as not only adapting to the demands of emerging digital requirements but also taking the lead in innovation and efficiency.

2.2.5 Communication and data flow in edge architecture

By using a distributed architecture to enable communication and data flow, edge computing moves data processing closer to the point of origin. This entails a number of elements, including servers, gateways, and edge devices, cooperating to reduce latency and bandwidth consumption. For more in-depth analysis and storage, data mainly moves from edge devices to local processing units (edge nodes/gateways), and then optionally to edge servers or cloud platforms [25, 27]. Communication in edge computing architecture typically follows a **bottom-up** flow from the edge devices to the cloud and a **top-down** flow from cloud to edge:

- **Data Generation:** Begins at the edge devices where the raw data is originated before transmission to the edge nodes or the cloud for processing. This directly affects how systems are designed for speed, storage, and intelligence.

- **Local Processing:** Simple or real-time tasks are processed at the edge.
- **Edge Node Aggregation:** Aggregated data is analyzed for insights or decision-making.
- **Data Filtering:** Only necessary or abnormal data is forwarded to higher layers (fog or cloud).
- **Cloud Processing:** Used for model training, historical analysis, and storage.

Protocols such as MQTT, CoAP, HTTP/HTTPS, and AMQP are used for communication between layers, depending on device capabilities and network constraints.

2.2.6 Technical benefits of edge computing architecture

A considerable technical advantage has emerged from distributed data processing and analysis occurring at the network edge, where the data originates, instead of between cloud servers. A significant technical advantage is the dependable, ultra-low latency, high bandwidth, and real-time data accessibility and processing capabilities of the computing paradigm as briefly mentioned as follows:

- **Low Latency:** Reduces delays by processing data locally, which is crucial for applications like autonomous vehicles or remote surgery.
- **Reduced Bandwidth Usage:** Limits the need to send massive volumes of data to the cloud, conserving network resources.
- **Increased Reliability:** Systems can operate independently of cloud availability.
- **Enhanced Security and Privacy:** Sensitive data can be processed locally, reducing exposure to external threats.
- **Real-Time Decision-Making:** Ideal for dynamic environments requiring fast reactions.

Because of the aforementioned benefits in the edge computing architecture (EA), organizational systems work faster and respond quickly. Also, the fact that the users are close to the place where the information is processed takes some of the load off the cloud server, which improves the quality of service (QoS). From an organizational point of view, edge computing has the technical ability to make big data analytics more efficient by collecting data from mobile devices and processing it at smart gateways or indirectly through sink nodes in large architectures [25, 27]. By separating host and location identifications, the method makes it easier for mobile devices to connect directly to the network. The technology is also aware of its location, which lets mobile users or devices on the network find out how close they are to each other. This lets devices find their nearest edge server (Figure 2.1).

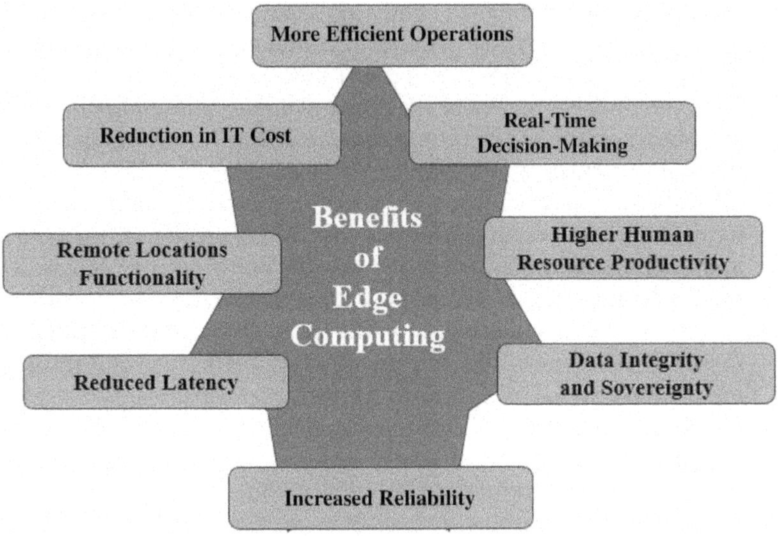

Figure 2.1 Edge computing in distributed and real-time computing environments.

2.2.7 Challenges in edge computing architecture

Despite its benefits, edge computing architecture faces several challenges:

- **Scalability:** Effectively managing thousands or even millions of distributed nodes is undeniably complex. It is essential to create edge architecture that not only scales seamlessly with rising data volumes but also accommodates the expanding array of devices. By designing a robust and scalable edge infrastructure that can effortlessly handle intense and rising data loads and by adopting modular and flexible systems that facilitate the seamless expansion with the rising and evolving needs, this challenge can be effectively addressed. This strategic approach is crucial for overcoming the complexities of modern data management and ensuring future readiness.
- **Security:** Decentralization in the edge computing architecture significantly increases the attack surface; hence, it is critical to protect the distributed data from various cyber threats. This major concern demands the implementation of a robust and strong encryption to secure data both at rest and in transit. Also, deploying advanced authentication and access control mechanisms is essential to secure edge devices and networks. This effectively enhances the security feature in a decentralized environment and hence ensures the integrity of the data.
- **Interoperability:** It is very challenging to incorporate and integrate different hardware and communication protocols. This issue can be overcome by carefully adopting and implementing standardized

communication protocols and interoperability frameworks. This approach enables all edge devices to communicate seamlessly with one another and also connect to central systems in a more efficient way, which finally leads to a more effective and unified technological platform.

- **Maintenance:** In the edge computing architecture, managing the maintenance of numerous edge devices may be complex and expensive. To mitigate this issue, reliable and robust automated systems are required to monitor and control the activity. Also, predictive maintenance approaches can be adopted to identify the unforeseen problems before they turn into serious issues. This proactive strategy ensures that the systems function smoothly and reliably in the rapidly changing technological environment by improving overall efficiency and safeguarding the integrity of operations.
- **Resource Constraints:** Effectively managing large volumes of data locally is a complex and costly task, which requires careful planning and strategic execution. To address this challenge successfully, it is imperative to establish robust data management frameworks that guarantee consistent data handling. By employing a sophisticated data handling technique, storage space can be optimized and the processing speed can be improved. Adopting these strategies will not only streamline data management but also empower the organizations to make more informed decisions based on accurate, accessible data.

2.2.8 Use cases of edge computing architecture

Edge computing is transforming numerous industries by strategically placing computation and data storage near the data source. This approach significantly improves processing speed and enables real-time responses, leading to greater efficiency and better decision-making. Key use cases include autonomous vehicles, smart grids, predictive maintenance, in-hospital patient monitoring, and various applications in manufacturing, healthcare, and smart cities [27, 30]. A detailed study on the use cases of edge computing is provided in Section 2.5, with the following summary.

- **Smart Cities:** Smart cities use digital technologies to improve the quality and performance of urban services such as energy, transportation, healthcare, and public safety [31]. **Edge computing** plays a crucial role in enabling such cities to operate efficiently and in real time using distributed sensors and edge gateways.
- **Industrial IoT:** By integrating smart sensors, edge computing, data analytics, and communication technologies into industrial processes, IIoT aims to improve efficiency, productivity, safety, and decision-making across sectors like manufacturing, energy, logistics, and agriculture including predictive maintenance, quality control, and robotics in manufacturing environments [32].

- **Healthcare:** Remote patient monitoring, wearable diagnostics, and telemedicine with real-time feedback [33].
- **Retail:** Personalized promotions, inventory management, and cashier-less checkouts using edge cameras and AI.
- **Autonomous Vehicles:** Processing sensor and vision data locally to make split-second navigation decisions.
- **Agriculture:** Real-time monitoring of soil conditions, crop health, and weather data using field-deployed edge devices.

2.2.9 Future trends in edge computing architecture

The emerging trends in edge computing architecture are significantly influenced by the convergence of 5G technology, artificial intelligence and machine learning (AI/ML), and the rapid growth of IoT devices. This convergence is set to revolutionize computing by fostering a decentralized model where data processing occurs closer to its source, dramatically enhancing response times and minimizing latency. This approach enhances response times and reduces latency. As organizations increasingly adopt these technologies, we can anticipate a major shift toward more distributed, intelligent, and secure edge deployments. These systems often utilize hybrid edge-cloud architectures, which facilitate seamless integration between local data processing at the edge and centralized resources in the cloud. This flexibility enables organizations to optimize their workloads based on specific performance and compliance requirements.

2.2.10 Conclusion

Edge computing architecture is a transformative shift from centralized cloud models to distributed, localized processing. By pushing computation closer to the data source, it delivers faster response times, enhanced privacy, and more efficient use of network resources. The architecture consists of multiple interconnected layers—from edge devices and nodes to the cloud—and involves a range of technologies that work together to create intelligent, responsive, and scalable systems.

As edge computing matures, it will play a foundational role in next-generation applications across industries, enabling the full potential of IoT, AI, and real-time systems. Designing robust and flexible edge computing architectures will be key to unlocking this future.

2.3 HOW EDGE COMPUTING WORKS: A COMPREHENSIVE

Edge computing is an innovative distributed computing model that processes data closer to the source where it is generated—at the "edge" of the network—rather than relying entirely on centralized cloud data centers. This approach reduces latency, enhances real-time responsiveness, improves bandwidth efficiency, and addresses privacy and data sovereignty concerns. Understanding how edge computing works requires an exploration of its architecture, data flow, components, and operational mechanisms across different applications.

2.3.1 The fundamental idea behind edge computing

At its core, edge computing is about decentralizing data processing. Traditionally, data generated by devices (like sensors, smartphones, and IoT systems) is sent to a centralized data center or cloud server for analysis and storage. This model works well for many applications, but it becomes problematic in scenarios requiring low latency, large-scale data processing, or limited network availability.

Edge computing addresses these challenges by moving computation closer to the data source. Instead of transmitting all raw data to the cloud, devices or nearby edge nodes perform processing tasks locally and send only essential information upstream if necessary.

2.3.2 Key components in edge computing

Several components work together in an edge computing environment:

2.3.2.1 Edge devices

These are the data-generating units that operate at the very edge of the network. Examples include smart sensors, cameras, industrial machines, medical devices, autonomous vehicles, and mobile phones. Many edge devices today have some degree of computing power, allowing them to pre-process or analyze data on the spot.

2.3.2.2 Edge nodes or gateways

These are intermediary devices positioned between edge devices and cloud infrastructure. Edge nodes typically consist of local servers, routers, or specialized appliances with greater processing capability than edge devices. They aggregate data, execute analytics, perform filtering, and may even host lightweight machine learning models.

2.3.2.3 Network connectivity

Reliable and efficient communication channels—using Wi-Fi, Ethernet, 4G, 5G, or other protocols—connect edge devices to edge nodes and, when necessary, to cloud platforms. Network design affects speed, latency, and overall performance of the edge computing system.

2.3.2.4 *Cloud and data centers*

Though edge computing minimizes reliance on the cloud, centralized cloud servers still play a vital role in long-term storage, complex processing, and model training. The cloud acts as a control and coordination center, managing software updates, backups, and global data analysis.

2.3.3 How edge computing works: step-by-step process

Edge computing revolutionized data processing by strategically placing it closer to its sources, usually near IoT devices, sensors, or end-user applications. This advanced method removes the limitations and drawbacks of the traditional centralized cloud servers and ensures the real-time data processing. The benefits of edge computing are numerous: it considerably reduces latency, enhances operational efficiency, and significantly minimizes the necessity of transmitting vast amounts of data to the cloud. The implementation of edge computing not only signifies a technological advancement, but it also empowers the companies to remain competitive by fulfilling emerging demands and opportunities in an increasingly data-centric world.

Now, let's explore the intricate step-by-step workings involved in edge computing networks.

2.3.3.1 *Step 1: Data generation*

An edge computing architecture comprises various components, including IoT devices, sensors, cameras, actuators, and other linked edge devices. These edge devices generate and collect large amounts of raw data. For instance, thousands of sensors are required in a small company to continuously monitor the critical parameters like temperature, vibration, and pressure of various machines. This real-time data collection enhances the operational efficiency and also enables the user to implement timely repairs and maintenance. By proactively identifying issues, organizations and businesses can significantly improve performance and effectively prevent costly equipment failures.

2.3.3.2 *Step 2: Local processing at the edge device*

Local processing at the edge means processing and analyzing data on devices that are close to the data source instead of sending all the data to a central server for processing. Edge devices make it possible to process data locally, which cuts down on the time it takes to get results. This is very important for real-time applications like self-driving cars and industrial automation. These edge devices may filter and process data before sending it to the cloud, which means that only useful information is sent. This strategy efficiently reduces the amount of data that needs to be sent, which helps save bandwidth. Local processing also helps lower the danger of data breaches because it means that sensitive information doesn't have to be sent to the cloud. This is especially important in industries where tight data sovereignty rules apply.

2.3.3.3 *Step 3: Processing at the edge node*

Data that requires more advanced processing or aggregation is sent to an edge node. At this stage:

- Data from multiple devices may be combined.
- Real-time analytics are performed to detect patterns or anomalies.
- Local dashboards may be updated with current system metrics.
- Commands may be sent back to edge devices to take action.

For example, if multiple sensors detect unusual heat levels in a manufacturing line, the edge node might temporarily shut down a machine and alert maintenance personnel.

2.3.3.4 *Step 4: Data filtering and transmission*

Not all data needs to be transmitted to the cloud; in fact, certain information is stored securely on local systems to enhance privacy and optimize performance. By selectively choosing what to upload, we can safeguard sensitive data while ensuring that our cloud resources are utilized efficiently. Edge nodes can determine what is important by:

- Only abnormal readings or summarized statistics are transmitted.
- Data that's not time-sensitive can be sent in batches during off-peak hours.
- Sensitive or regulatory-protected data may be stored locally to meet compliance requirements.

2.3.3.5 *Step 5: Cloud interaction*

When broader or historical analysis is needed, or when a machine learning model must be retrained with new data, edge nodes transmit selected data to the cloud. The cloud performs tasks like:

- Advanced analytics across multiple sites.
- Centralized management and monitoring.
- Model training and global optimization.

The cloud may send back new models, updated configurations, or system updates to edge nodes and devices.

2.3.4 Edge AI: machine learning at the edge

Edge computing is a significant paradigm shift that is reshaping the internet and applications landscapes by bringing data processing closer to the source of the data. This strategic evolution has the potential to enhance efficiency, responsiveness, and the respect of privacy. Starting from mostly cloud-based solutions, more and more applications are now pushed along the computing continuum, closer and closer to edge devices. The emergence of artificial intelligence (AI) and its significant demand for training data has made the use of edge devices for training and inference a clear subsequent development. While exploring the synergy of AI and edge computing, it is crucial to address the unique challenges the integration of edge computing and intelligence presents [34]. A major development in edge computing is the deployment of AI models directly on edge devices or nodes. This is known as edge AI. The following explain how edge AI works:

- A model is trained using cloud resources with massive datasets.
- The trained model is compressed and optimized using tools like TensorFlow Lite or ONNX.
- The model is deployed on edge hardware for inference (decision-making).
- The edge device processes input data in real time and produces immediate results without cloud involvement.

Examples of edge AI include:

- Facial recognition on smartphones.
- Defect detection in manufacturing.
- Predictive maintenance on turbines.

Edge AI stands out due to its ability to deliver low latency and robust privacy protection, making it a compelling alternative to cloud-based AI. This technology processes data closer to the source, which not only speeds up response times but also safeguards sensitive information. As a result, ensuring power efficiency and computational effectiveness becomes crucial for the successful implementation of edge AI solutions.

2.3.5 Security in edge computing

Security is a critical part of edge computing operations. Since data is processed across multiple decentralized devices, the attack surface expands.

How security is managed:

- **Encryption** of data both in transit and at rest.
- **Authentication and authorization** for devices and users.

- **Secure boot and firmware updates** to ensure devices aren't compromised.
- **Anomaly detection** at the edge to identify suspicious behavior quickly.

Edge computing also helps mitigate risk by limiting data exposure—since not all data is sent to the cloud, opportunities for interception are reduced.

2.3.6 Benefits of edge computing in practice

- **Reduced Latency:** Edge computing architecture intends to minimize the round-trip time for data transmission between devices and centralized cloud servers [35] which is a prime requirement for critical applications like robotics or AR/VR.
- **Efficient Bandwidth Use:** when a large amount of data is generated, higher bandwidth can result in transmission delay. This issue is addressed by edge computing by reducing the need to send large data volumes to centralized cloud servers, which optimizes the use of network bandwidth. Less data sent across networks means lower costs and congestion.
- **Improved Reliability:** This architecture ensures that critical functions can continue even if the cloud is slow, unreachable, or offline, hence systems can operate independently of cloud availability.
- **Data Sovereignty:** In edge computing, data is often processed locally—on-site, near the source. This robust architecture ensures compliance with data protection laws and privacy regulations.
- **Scalable Design:** Edge computing delivers a robust and scalable architecture by effectively distributing computational power across edge devices, gateways, and local servers. This strategic distribution ensures efficient resource allocation and enables seamless horizontal scaling.

2.3.7 Conclusion

Edge computing works by decentralizing data processing, allowing data to be analyzed and acted upon near its source. Through a combination of edge devices, edge nodes, local networks, and cloud platforms, edge computing enables real-time responsiveness, efficient bandwidth usage, and greater control over data privacy. Whether enabling autonomous vehicles, enhancing industrial systems, or supporting smart cities, edge computing is redefining how modern digital infrastructure operates.

As technology continues to evolve—particularly with the growth of 5G and edge AI—edge computing will become even more powerful and integral to the future of computing. Understanding how edge computing works is essential for designing resilient, intelligent, and efficient systems in an increasingly connected world.

2.4 EDGE COMPUTING TO SUPPORT USER APPLICATIONS

Edge computing is rapidly reshaping how data is processed and used in the modern digital landscape. By shifting data processing from centralized cloud data centers to the edge of the network—closer to where data is generated—edge computing enables faster response times, reduced bandwidth consumption, and greater autonomy for devices. This decentralized approach is particularly useful in real-time, data-intensive, or latency-sensitive scenarios.

As technology evolves and the number of connected devices grows, edge computing is becoming an essential part of digital transformation strategies across numerous sectors. This section explores the most impactful applications of edge computing in various industries, illustrating how it is driving innovation, efficiency, and smarter decision-making.

2.4.1 Smart cities

Edge computing plays a foundational role in building smarter, more efficient urban environments. In smart cities, large volumes of data are generated by IoT devices such as traffic sensors, surveillance cameras, environmental monitors, and public utilities.

Key Applications:

- **Traffic Management:** Edge devices installed at intersections process traffic data locally to adjust signal timings in real time, easing congestion and improving flow.
- **Public Safety:** Surveillance cameras equipped with edge-based AI can detect unusual activities (e.g., crowd formation, violence, or unattended bags) and alert authorities instantly.
- **Waste Management:** Smart bins with edge sensors monitor fill levels and optimize waste collection routes to reduce operational costs and carbon emissions.

By processing data locally, smart cities minimize the latency and bandwidth required to transmit information to centralized systems, allowing for faster, autonomous action and better urban resource management.

2.4.2 Industrial IoT and manufacturing

Manufacturing and industrial facilities generate enormous data from equipment, sensors, and production lines. Edge computing enables real-time monitoring, predictive maintenance, and automation, significantly enhancing operational efficiency.

Key Applications:

- **Predictive Maintenance:** Edge nodes analyze vibration, temperature, and usage data from machines to detect early signs of wear or failure, reducing downtime and maintenance costs.
- **Quality Control:** Cameras and sensors at the edge perform on-the-fly defect detection using AI algorithms, ensuring that only quality products move forward in the production process.
- **Worker Safety:** Wearables and sensors monitor environmental conditions and worker vitals, instantly notifying supervisors of any risks, such as gas leaks or heat stress.

Edge computing supports high-speed industrial automation by eliminating the need for cloud-based decision-making, making operations safer and more reliable.

2.4.3 Healthcare and remote patient monitoring

In healthcare, speed and accuracy are crucial. Edge computing empowers medical devices and health monitoring systems to perform real-time analysis and decision-making close to the patient.

Key Applications:

- **Wearable Health Monitors:** Devices like smartwatches and modern fitness trackers utilize the power of edge processing to analyze vital health metrics, including heart rate, oxygen saturation, and physical activity levels. These devices are designed to promptly alert the users or healthcare providers if any anomalies are detected, thus promoting proactive healthcare management and ensuring timely medical intervention.
- **Remote Diagnostics:** Sophisticated medical equipment like portable ultrasound and imaging devices that are integrated with edge AI enables real-time diagnostic feedback. This remarkable technology is a boon for individuals in rural or underdeveloped areas, where access to constant internet connectivity may be limited. By providing swift and accurate medical assessments in such regions, these devices not only enhance healthcare accessibility but also support the delivery of critical medical services where they are most needed.
- **Hospital Operations:** Edge computing enables hospitals to manage all the connected devices and equipment and seamless monitoring of inventory levels, thus enhancing the optimization of patient flows in real time. This approach improves the healthcare facilities by swiftly responding to dynamic conditions, ensuring that critical resources are efficiently utilized and patient care is streamlined for better outcomes.

By reducing reliance on cloud networks and ensuring data privacy, edge computing enhances the accessibility and responsiveness of healthcare systems.

2.4.4 Retail and customer experience

Retailers are using edge computing to transform in-store experiences and improve inventory and operations management.

Key Applications:

- **Smart Shelves and Inventory Tracking:** Advanced edge sensors are designed to precisely track inventory in real time. They intelligently monitor product activity, promptly detecting when items are picked up or when stock levels go below a certain limit. This proactive approach effortlessly minimizes out-of-stock scenarios and also streamlines inventory management, ensuring that shelves are always well-stocked and ready for customers, enhancing the shopping experience.
- **Customer Behavior Analysis:** Advanced cameras utilizing sophisticated edge AI algorithms can be helpful to have deep insights into customer behavior within the store. These cameras can track their movement and can monitor the amount of time spent by them in the particular zone of the store. With such strategic data, retailers can optimize product placements, catering to the needs of customers, thus improving customer satisfaction.
- **Self-Checkout Systems:** The new and advanced edge devices make the transactions very quick and efficient. The customers can experience a seamless, contactless checkout process with significantly reduced wait times.

These innovations allow retailers to deliver seamless, personalized shopping experiences while enhancing operational efficiency.

2.4.5 Autonomous vehicles and transportation

Autonomous vehicles rely heavily on real-time data processing. Edge computing is critical in enabling these systems to make split-second decisions.

Key Applications:

- **Navigation and Control:** Self-driving cars use onboard edge processors to analyze data from cameras, radar, lidar, and GPS for obstacle detection, lane keeping, and path planning.
- **Fleet Management:** Edge-enabled telematics systems provide real-time location, fuel usage, and engine diagnostics for logistics companies, allowing efficient routing and predictive maintenance.

- **Traffic Coordination:** Smart traffic signals and road infrastructure with edge capabilities can communicate with connected vehicles to reduce congestion and improve safety.

Without edge computing, the latency involved in sending data to the cloud for processing would make autonomous driving impractical and unsafe.

2.4.6 Energy and utilities

The energy sector is adopting edge computing to improve grid management, optimize energy usage, and integrate renewable sources.

Key Applications:

- **Smart Grids:** Edge devices monitor consumption and generation in real time, balancing supply and demand more efficiently and reducing outages.
- **Renewable Energy:** Wind turbines and solar panels use edge computing to predict energy production, adjust settings based on weather conditions, and identify maintenance needs.
- **Oil and Gas Monitoring:** Remote edge systems analyze pressure, temperature, and flow rates in pipelines or wells, enabling faster response to anomalies or leaks.

By enabling localized decision-making, edge computing makes energy infrastructure more intelligent and resilient.

2.4.7 Agriculture and precision farming

Agricultural operations benefit greatly from edge computing, especially in remote areas where connectivity is limited.

Key Applications:

- **Soil and Crop Monitoring:** Sensors in fields collect data on moisture, pH, and temperature, processed locally to guide irrigation and fertilization.
- **Livestock Tracking:** Wearable devices on animals monitor health and behavior, alerting farmers to illness or abnormal activity.
- **Drone Surveillance:** Drones equipped with edge processors analyze aerial imagery for crop health, pest detection, and yield estimation in real time.

Edge computing supports sustainable farming practices by ensuring timely decisions and efficient resource use, even in connectivity-constrained environments.

2.4.8 Media and entertainment

In content delivery and live event management, edge computing improves user experience by reducing delays and buffering.

Key Applications:

- **Live Video Streaming:** Edge servers cache and process video closer to users, minimizing latency and network load.
- **AR/VR Applications:** Immersive experiences require real-time responsiveness; edge computing ensures smooth rendering and interaction.
- **Interactive Gaming:** Gaming platforms use edge nodes to process data locally for faster gameplay and reduced lag, essential for multiplayer and cloud gaming.

With user demand for high-quality, low-latency content growing, edge computing enables scalable and responsive delivery.

2.4.9 Finance and banking

Edge computing enhances the speed and security of financial transactions and customer interactions.

Key Applications:

- **ATM and Kiosk Security:** Edge devices detect tampering or unusual activity in real time.
- **Fraud Detection:** Local processing of transactions allows edge systems to flag suspicious behavior instantly, reducing risk.
- **Branch Automation:** In-branch devices handle customer verification, document scanning, and service customization with minimal cloud interaction.

These applications not only improve security but also reduce operational costs and enhance customer experience.

2.4.10 Conclusion

Edge computing is transforming how industries operate by bringing intelligence to the source of data. From smart cities and healthcare to agriculture and finance, its applications are both diverse and impactful. By enabling local processing, edge computing reduces latency, ensures data privacy, and empowers real-time decision-making.

As 5G networks, AI, and IoT technologies continue to advance, edge computing will become even more central to digital innovation. Organizations that leverage its capabilities stand to gain improved efficiency, lower operational costs, and the agility needed to succeed in an increasingly connected world.

2.5 APPLICATIONS OF EDGE COMPUTING USING 5G AND IOT

Edge computing and 5G are two transformative technologies that, when combined, unlock new possibilities across various industries. Edge computing and the IoT form a powerful combination that is transforming industries by enabling real-time data processing and smarter device interactions. While edge computing focuses on processing data closer to its source (the "edge" of the network), 5G provides the ultra-fast, low-latency, and high-bandwidth connectivity required for real-time communication between devices and edge nodes. IoT refers to the network of interconnected devices—such as sensors, cameras, vehicles, and machines—that generate vast amounts of data. Traditionally, this data would be sent to the cloud for processing. However, with edge computing, much of this processing is done locally—on or near the IoT devices—allowing for faster response times, reduced bandwidth use, and improved reliability (Figure 2.2).

Together, these technologies enable rapid data analysis, seamless user experiences, and support for advanced applications such as autonomous

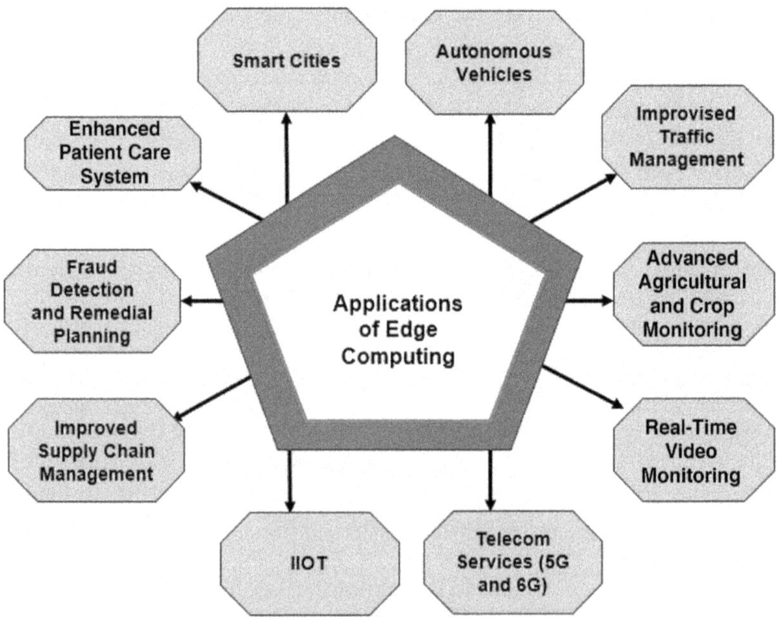

Figure 2.2 Applications of Edge Computing across healthcare, smart cities, transportation, agriculture, telecommunications, and industrial IoT.

systems (transportation), smart cities, healthcare and remote monitoring, industrial automation and smart manufacturing, agriculture, and immersive technologies.

2.5.1 Autonomous vehicles and transportation

One of the most impactful applications of edge computing with 5G is in autonomous vehicles. These vehicles rely on real-time data from sensors, cameras, and radar to navigate safely. With 5G's ultra-low latency and edge computing's local processing, cars can instantly detect and respond to road conditions, traffic signals, and potential hazards without relying solely on cloud communication. Furthermore, edge nodes installed along roads or in traffic systems can share environmental data between vehicles (V2X communication), improving safety and traffic flow.

2.5.2 Smart cities

Edge computing and 5G both are foundational technologies for smart cities. By placing edge servers close to IoT devices like traffic lights, surveillance cameras, and pollution sensors, cities can analyze and respond to events in real time. For example, edge-enabled traffic management systems can adjust signal timings based on real-time traffic data, while public safety cameras with AI processing at the edge can detect accidents or suspicious activity instantly. 5G ensures fast and reliable connectivity between these devices and edge infrastructure. IoT-enabled infrastructure like traffic lights, street lamps, and waste bins generate continuous data that helps manage urban systems more efficiently. With edge computing, this data is processed locally at edge nodes placed throughout the city. Applications include:

- **Traffic Optimization:** Real-time traffic data from cameras and sensors is processed at the edge to dynamically adjust signal timings.
- **Public Safety:** Edge-enabled surveillance cameras use AI to detect unusual behavior or emergencies and alert authorities instantly.
- **Environmental Monitoring:** Sensors measure air quality or noise levels and trigger actions such as adjusting traffic flow or notifying residents of health risks.

These applications enable real-time responses and reduce the need to send massive data volumes to a central server.

2.5.3 Smart homes and buildings

IoT devices in homes and buildings—such as thermostats, lights, security cameras, and appliances—generate data used to automate and optimize energy use, comfort, and security. Edge computing enables these systems to function even without a constant cloud connection. For example:

- Advanced security systems utilize the power of edge-based architecture to detect and respond to intrusions in real time, ensuring unparalleled protection and immediate action against potential threats. With the ability to analyze data at the source, these systems provide robust and proactive defenses, ensuring peace of mind in an increasingly complex security landscape.
- Energy management systems can adjust heating, lighting, or ventilation based on occupancy or environmental conditions. This localized processing enhances privacy, reliability, and responsiveness.

2.5.4 Healthcare and remote monitoring

With 5G and edge computing, healthcare can move beyond hospitals and into homes and rural areas. Wearable health monitors and portable diagnostic devices can process patient data locally, using 5G to quickly send critical updates to healthcare providers when needed. For example, in emergency care, ambulance-based edge nodes can analyze patient vitals in transit and transmit real-time data to hospitals, allowing medical teams to prepare in advance. This enables faster response and potentially life-saving interventions.

IoT medical devices such as wearable heart monitors, glucose sensors, and connected inhalers continuously collect health data. Edge computing allows this data to be analyzed immediately on the device or nearby gateway. This enables timely alerts for abnormal readings without waiting for cloud processing. In hospitals, edge-based IoT systems track equipment, monitor patient vitals, and streamline operations, helping healthcare providers respond faster and improve patient outcomes.

2.5.5 Augmented reality (AR) and virtual reality (VR)

AR and VR applications demand ultra-low latency and high throughput to deliver smooth, immersive experiences. Edge computing processes data close to the user, reducing lag, while 5G ensures high-speed data transmission. This is particularly useful in gaming, remote collaboration, and training simulations. For instance, field technicians can use AR headsets powered by edge computing and 5G to receive live guidance or overlay instructions while repairing complex machinery.

2.5.6 Industrial automation and smart manufacturing

Edge computing and 5G make it possible to monitor and manage machines in real time in industrial settings. Sensors in production lines can look at performance data at the edge to find problems or inefficiencies right away. 5G's dependable, low-latency connection lets people and machines work together to control machines and robots from a distance, which makes them more efficient and cuts down on downtime. In factories, IoT sensors are built inside machinery to keep an eye on things like performance, temperature, vibrations, and other important factors. Edge computing processes this data on-site in real time, which makes predictive maintenance and fast anomaly identification possible. For instance, if a machine starts to overheat, edge devices can either turn it off automatically or let maintenance workers know before it breaks down. This cuts down on downtime, makes things safer, and boosts production in factories.

2.5.7 Agriculture

In precision agriculture, IoT sensors monitor soil moisture, temperature, humidity, and crop health. With edge computing, farmers receive immediate recommendations for irrigation, fertilization, or pest control. Drones with edge AI can scan fields and detect issues like disease or weeds without needing cloud access, enabling faster intervention and higher crop yields.

2.5.8 Conclusion

Edge computing enhances the power of IoT by enabling faster decision-making, reducing latency, saving bandwidth, and improving reliability. Across industries from manufacturing and healthcare to agriculture and smart cities this combination is revolutionizing how systems operate and respond to the world around them. The combination of edge computing and 5G opens up powerful new applications that require real-time processing, minimal latency, and reliable connectivity. Together, they are revolutionizing industries by enabling smarter, faster, and more efficient operations, laying the groundwork for a truly connected and intelligent future.

2.6 CHALLENGES OF EDGE COMPUTING

Edge computing is a paradigm that brings computation and data storage closer to the location where it is needed, to improve response times and save bandwidth. As the number of IoT devices grows exponentially and applications increasingly demand real-time performance, edge computing offers a powerful alternative to traditional cloud models. However, despite its promise, edge computing is not without significant challenges. These challenges span multiple dimensions—technical, operational, and strategic—and they must be addressed for edge computing to reach its full potential.

This section explores the primary challenges of edge computing, categorizing them under infrastructure, security, scalability, management, interoperability, cost, and regulatory issues.

2.6.1 Infrastructure challenges

2.6.1.1 *Limited computational resources*

Unlike centralized cloud data centers with vast computational capabilities, edge devices are often resource-constrained. They may lack the processing power, memory, and storage to handle complex analytics, AI models, or big data processing. This can limit the types of applications that can run effectively at the edge.

2.6.1.2 *Power and cooling constraints*

Edge computing nodes are often deployed in remote or outdoor environments—on cell towers, factory floors, vehicles, or street lights. In such cases, maintaining consistent power and adequate cooling can be a challenge. Battery life and energy efficiency become critical, especially for mobile and rugged edge devices.

2.6.1.3 *Network reliability*

Although edge computing reduces the need for constant connectivity with the central cloud, many applications still require communication across networks. In remote areas or mobile settings, network reliability and bandwidth can fluctuate, causing interruptions in data flow and service delivery.

2.6.2 Security and privacy challenges

2.6.2.1 *Distributed attack surface*

One of the biggest challenges in edge computing is security. Unlike the centralized cloud, where defenses are consolidated, edge computing spreads data and computation across a wide range of devices and nodes. This creates a larger attack surface and more potential points of vulnerability [36].

2.6.2.2 *Data privacy and sovereignty*

Edge computing often involves processing sensitive data—such as health records, financial transactions, or personal identifiers—at or near the point of generation. Ensuring privacy and compliance with data protection regulations (like GDPR or HIPAA) becomes complex when data is processed in diverse, decentralized locations.

2.6.2.3 *Lack of unified security standards*

There is currently no universal standard for securing edge devices and networks. Vendors often implement proprietary solutions, leading to inconsistencies in authentication, encryption, firmware updates, and intrusion detection.

2.6.3 Scalability and manageability

2.6.3.1 *Device management complexity*

Managing a few cloud servers is far simpler than managing thousands—or even millions—of edge devices spread across different geographies. Each device requires monitoring, patching, software updates, and sometimes physical maintenance, all of which present significant logistical challenges.

2.6.3.2 *Orchestration and deployment*

Deploying applications across edge environments requires new orchestration strategies [30]. While tools like Kubernetes help manage cloud-native workloads, orchestrating containers or microservices across low-power, intermittent, or heterogeneous edge devices adds a layer of complexity.

2.6.3.3 *Firmware and software updates*

Updating software in cloud systems is often automated and centralized. In edge computing, ensuring timely updates across all nodes—especially in disconnected or remote environments—is challenging and can lead to inconsistencies or security vulnerabilities.

2.6.4 Interoperability and standardization

2.6.4.1 *Diverse hardware and software ecosystem*

Edge devices vary greatly in terms of hardware specifications, operating systems, and communication protocols. This heterogeneity makes it difficult to create standardized software stacks that can run across all edge environments. Application developers often need to build multiple versions of the same application, increasing complexity and cost.

2.6.4.2 *Vendor lock-in*

Many edge solutions are often proprietary, which binds the users to specific vendors' hardware and software platforms. This reduces operational flexibility and stifles innovation by limiting choices and capabilities. Over time, these vendor dependencies can become challenging and costly when trying to switch to a different vendor or adopt more versatile solutions.

2.6.4.3 *Lack of universal standards*

Despite the rapid growth of edge computing, the industry still faces the need for universally accepted standards for edge architectures, APIs, data exchange formats, and security protocols. This lack of standardized frameworks leads to a fragmented ecosystem, which hampers seamless integration and curtails innovation. Addressing this critical gap by establishing comprehensive and widely adopted standards is vital for encouraging collaboration, enhancing interoperability, and ultimately driving the transformative impact of edge computing.

2.6.5 Data management challenges

2.6.5.1 Data lifecycle and storage

Edge computing involves generating large volumes of data at the edge. Deciding which data to process locally, which to store temporarily, and which to send to the cloud is a complex issue. Improper data management can lead to increased latency, bandwidth waste, or even data loss.

2.6.5.2 Data consistency and synchronization

Ensuring that data remains consistent across distributed edge devices, especially when some are offline or intermittently connected, is a major challenge. Syncing data without duplication or corruption, and reconciling local with central databases, requires robust algorithms and systems.

2.6.5.3 Data quality

Edge data can be noisy, incomplete, or inconsistent, especially when generated by sensors in uncontrolled environments. Processing low-quality data at the edge can lead to inaccurate results and poor decision-making.

2.6.6 Cost and ROI considerations

In order to promote the commercialization of edge computing, it is important to explore effective ways to reduce the cost of edge computing networks. The architectural framework of edge computing complements the cloud computing model by deploying computation, storage, and network resources to the edge locations of wireless access networks, empowering end devices to run resource-intensive applications. Mathematical models for cost reduction thus supporting optimization objectives and optimization methods have been developed by researchers and the effectiveness have also been evaluated [37–41]. However, it is important to weigh associated costs and return on investment (ROI) to determine its true value for an organization which is briefly described as follows:

2.6.6.1 Capital investment

Deploying edge infrastructure often involves a significant upfront cost, including the purchase of edge devices, servers, networking equipment, and specialized sensors. For smaller businesses, this capital investment may be a barrier to adoption.

2.6.6.2 *Operational costs*

Operating and maintaining distributed edge networks can be cost-intensive. Organizations must consider maintenance, power consumption, software licensing, network fees, and human resources required to support edge deployments.

2.6.6.3 *Uncertain ROI*

In many cases, especially for early adopters, the ROI in edge computing is unclear. Measuring ROI is difficult because the benefits—such as lower latency, improved user experience, or data privacy—are not always directly quantifiable.

2.6.7 Application development and deployment

2.6.7.1 *Tooling and skills gap*

Edge computing requires new tools and specialized skill sets. Traditional developers may lack the critical expertise needed to effectively optimize applications for resource-constrained environments, deal with intermittent connectivity, or operate real-time systems. Therefore, specialized training in edge-native development practices is essential to equip them with the essential knowledge and skills to harness the full potential of edge computing, ensuring they remain competitive and innovative in the field.

2.6.7.2 *Debugging and monitoring*

Debugging software on a cloud server is relatively straightforward. Debugging software on an edge device that's in a remote location, intermittently connected, or physically inaccessible presents unique challenges. Real-time monitoring tools are still evolving for edge ecosystems.

2.6.7.3 *Real-time requirements*

Many edge computing applications—like autonomous vehicles, smart manufacturing, or telemedicine—have stringent real-time requirements. Meeting these demands requires extremely efficient, low-latency computing systems that can operate reliably even under pressure.

2.6.8 Reliability and fault tolerance

2.6.8.1 *Device failures*

Edge devices are often deployed in harsh environments or remote locations and are prone to wear, environmental damage, or battery drain. Building fault-tolerant systems that can continue operating despite individual device failures is critical but difficult.

2.6.8.2 Redundancy and backup

Ensuring data and service availability in case of device or network failures is more complex in decentralized environments. Redundancy strategies at the edge must be carefully designed to avoid inefficiency and unnecessary costs.

2.6.8.3 Monitoring system health

Continuously monitoring system health and diagnosing issues across a widespread edge deployment requires sophisticated telemetry and diagnostics systems that can operate with minimal bandwidth and high automation.

2.6.9 Regulatory and compliance issues

2.6.9.1 Compliance with data laws

Edge computing often involves processing and storing sensitive data locally. Ensuring that these practices comply with jurisdiction-specific regulations (e.g., GDPR in Europe, CCPA in California) can be complex, especially for global organizations.

2.6.9.2 Auditability and transparency

Providing audit trails, logging, and transparency in distributed edge environments is more complicated than in centralized systems. This is critical for regulated industries like healthcare, finance, and government.

2.6.9.3 Physical security of edge devices

Edge devices in the field may not be physically secure. They can be stolen, tampered with, or damaged. Ensuring that data stored on such devices is encrypted and inaccessible to unauthorized users is essential.

2.6.10 Environmental and sustainability concerns

2.6.10.1 Energy consumption

Although individual edge devices may consume less power than data centers, the overall energy footprint of maintaining thousands or millions of such devices globally can be significant. Sustainable deployment strategies are needed to minimize environmental impact.

2.6.10.2 *Electronic waste*

As edge devices are replaced, upgraded, or fail, they contribute to the growing problem of electronic waste. Ensuring responsible recycling and lifecycle management is a growing concern for sustainability-minded organizations.

2.7 CONCLUSION

Edge computing offers revolutionary capabilities with unparalleled speed, responsiveness, and localized intelligence. It caters to the needs of the evolving use cases across various sectors—from autonomous vehicles and smart cities to healthcare and manufacturing—thus empowering groundbreaking applications. However, the shift to edge architectures also presents substantial challenges that must be carefully mitigated.

Hardware limitations, security vulnerabilities, scalability issues, and the need for regulatory compliance are a few challenges that are faced by the users, which demand a fundamental rethinking of the design, deployment, and management of IT infrastructure. To unlock the full potential of edge computing, it is essential to tackle this obstacle through enhanced tools, standardized practices, comprehensive workforce development, and strategic design.

As technology advances, innovative and sophisticated solutions, such as AI-optimized edge chips, advanced orchestration platforms, 5G integration, and robust cybersecurity frameworks, will play a crucial role in overcoming many of these hurdles. For organizations adopting edge computing, a strategic and well-informed approach needs to be implemented, thus balancing the innovation it enables against the complexity it introduces.

REFERENCES

[1] Armbrust, M., Fox, A., Griffith, R., Joseph, A. D., Katz, R., Konwinski, A., Lee, G., Patterson, D., Rabkin, A., Stoica, I., & Zaharia, M. (2010). A view of cloud computing. *Communications of the ACM, 53*(4), 50–58.
[2] Buyya, R., Broberg, J., & Goscinski, A. M. (2013). *Cloud computing: Principles and paradigms*. Wiley.
[3] NHTC Ltd. (2022). *Cloud computing technology*. Springer.
[4] Garfinkel, S. (1999). *Architects of the information society: Thirty-five years of the laboratory for computer science at MIT* (H. Abelson, Ed.). MIT Press.
[5] Salesforce. (*n.d.*). *Cloud computing: Definition, importance, benefits.*
[6] Ramirez, D. (2008). *Amazon web services: History & overview presentation* [SlideShare slides].
[7] Balduzzi, M., Zaddach, J., Balzarotti, D., Kirda, E., & Loureiro, S. (2012, March 26–30). A security analysis of Amazon's elastic compute cloud service. In *Proceedings of the 27th annual ACM symposium on applied computing (SAC '12), security track* (pp. 1427–1434). ACM.

[8] Aymerich, F. M., Fenu, G., & Surcis, S. (2008). An approach to a cloud computing network. *In Proceedings of the applications of digital information and web technologies (ICADIWT). IEEE; Mell, P., & Grance, T. (2011). The NIST definition of cloud computing (NIST special publication 800-145).* National Institute of Standards and Technology.

[9] Rountree, D. (2014). Chapter 4: Cloud service models. In D. Rountree (ed.), *The basics of cloud computing* (pp. 49–70). Elsevier.

[10] Ometov, A., Molua, O. L., Komarov, M., & Nurmi, J. (2022). A survey of security in cloud, edge, and fog computing. *Sensors, 22(3), Article 927.*

[11] Satyanarayanan, M., Bahl, P., Cáceres, R., & Davies, N. (2009, October). The case for VM-based cloudlets in mobile computing. *IEEE Pervasive Computing,* 8(4), 14–23.

[12] Zwolenski, M., & Weatherill, L. (2014). The digital universe: Rich data and the increasing value of the Internet of Things. *Australian Journal of Telecommunications and the Digital Economy, 2(3), 47.*

[13] Voigt, P., & Von dem Bussche, A. (2017). *The EU general data protection regulation (GDPR): A practical guide.* Springer.

[14] European Parliament, & Council of the European Union. (2016). Regulation (EU) 2016/679 of the European Parliament and of the Council of 27 April 2016 on the protection of natural persons with regard to the processing of personal data and on the free movement of such data, and repealing Directive 95/46/EC (general data protection regulation). *Official Journal of the European Union*, L 119, 1–88.

[15] Shi, W., Cao, J., Zhang, Q., Li, Y., & Xu, L. (2016). Edge computing: Vision and challenges. *IEEE Internet of Things Journal,* 3(5), 637–646.

[16] Shi, W., Pallis, G., & Xu, Z. (2019). Edge computing [scanning the issue]. *Proceedings of the IEEE*, 107(8), 1474–1481.

[17] U.S. Department of Health and Human Services. (2013, January 25). Modifications to the HIPAA privacy, security, enforcement, and breach notification rules under the health information technology for economic and clinical health act and the genetic information nondiscrimination act; Other modifications to the HIPAA rules. *Federal Register,* 78(17), 5566–570.

[18] Harding, E. L., Vanto, J. J., Clark, R., Hannah Ji, L., & Ainsworth, S. C. (2019). Understanding the scope and impact of the california consumer privacy act of 2018. *Journal of Data Protection & Privacy,* 2(3), 234–253.

[19] Government of India. (2023). *The digital personal data protection act, 2023 (act no. 22 of 2023).* Ministry of Law and Justice (Legislative Department).

[20] International Organization for Standardization & International Electrotechnical Commission. (2022, October 25). *ISO/IEC 27001:2022—information security, cybersecurity and privacy protection: Information security management systems—requirements* (3rd ed.). ISO.

[21] Statista. (2018). *Internet of Things (IoT) connected devices installed base worldwide from 2015 to 2025.* Statista.

[22] Gubbi, J., Buyya, R., Marusic, S., & Palaniswami, M. (2013). Internet of Things (IoT): A vision, architectural elements, and future directions. *Future Generation Computer Systems,* 29(7), 1645–1660.

[23] Lone, A. N., Mustajab, S., & Alam, M. (2023). A comprehensive study on cybersecurity challenges and opportunities in the IoT world. *Security and Privacy, 6(6), Article e318.*

[24] Cao, K., Liu, Y., Meng, G., & Sun, Q. (2020). An overview on edge computing research. *IEEE Access, 8*, 85714–85728.
[25] Bandi, A., & Hurtado, J. A. (2022). Edge computing as an architectural solution: An Umbrella review. In R. Patgiri, S. Bandyopadhyay, M. D. Borah, & V. Emilia Balas (eds.), *Edge analytics. Lecture notes in electrical engineering* (Vol. 869). Springer, Singapore.
[26] Iorga, M., Feldman, L., Barton, R., Martin, M. J., Goren, N., & Mahmoudi, C. (2018). *Fog computing conceptual model (NIST special publication 500-325).* National Institute of Standards and Technology.
[27] Hazra, A., Kalita, A., & Gurusamy, M. (2024). Meeting the requirements of Internet of Things: The promise of edge computing. *IEEE Internet of Things Journal,* 11(5), 7474–7498
[28] Taleb, T., Samdanis, K., Mada, B., Flinck, H., Dutta, S., & Sabella, D. (2017). On multi-access edge computing: A survey of the emerging 5G network edge cloud architecture and orchestration. *IEEE Communications Surveys & Tutorials,* 19(3), 1657–1681.
[29] Bonomi, F., Milito, R., Zhu, J., & Addepalli, S. (2012, August). Fog computing and its role in the Internet of Things. In *Proceedings of the first edition of the MCC workshop on mobile cloud computing* (pp. 13–16). Association for Computing Machinery.
[30] Sahoo, S., Sahoo, K. S., Sahoo, B., & Gandomi, A. H. (2023). A learning automata based edge resource allocation approach for IoT-enabled smart cities. *Digital Communications and Networks,* 5, 1258–1266.
[31] Amin, S. U., & Hossain, M. S. (2021). Edge intelligence and Internet of Things in healthcare: A survey. *IEEE Access,* 9, 45–59.
[32] Zhou, Z., Chen, X., Li, E., Zeng, L., Luo, K., & Zhang, J. (2019). Edge intelligence: Paving the last mile of artificial intelligence with edge computing. *Proceedings of the IEEE,* 107(8), 1738–1762.
[33] Mogul, J. C., & Kompella, R. R. (2015, May). Inferring the network latency requirements of cloud tenants. In *Proceedings of the 15th workshop on hot topics in operating systems (HotOS XV)* (pp. 1–6). Association for Computing Machinery.
[34] Hu, Y. C., Patel, M., Sabella, D., Sprecher, N., & Young, V. (2023). Denial-of-service attacks in edge computing layers: Taxonomy, vulnerabilities, threats and solutions. *Ad Hoc Networks,* 152, Article 103322.
[35] Nikolaou, P., Sazeides, Y., Lampropulos, A., Guilhot, D., Bartoli, A., Papadimitriou, G., Chatzidimitriou, A., Gizopoulos, D., Tovletoglou, K., Mukhanov, L., Karakonstantis, G., Kleanthous, M., & Prat, A. (2022). Total cost of ownership perspective of cloud vs. edge deployments of IoT applications. In G. Karakonstantis & C. J. Gillan (eds.), *Computing at the EDGE: New challenges for service provision* (pp. 141–161). *Springer, Cham.*
[36] Cao, G., Wu, F., Ling, Y., Zhou, Y., & Xie, Q. (2024). *Cost optimization in edge computing: A survey. Artificial Intelligence Review, 57, Article 312.*
[37] Nigam, S., Sugandh, U., & Khari, M. (2022). The integration of blockchain and IoT edge devices for smart agriculture: Challenges and use cases. In *Advances in computers* (Vol. 127, pp. 507–537). Elsevier.
[38] Vimal, S., Khari, M., Dey, N., Crespo, R. G., & Robinson, Y. H. (2020). Enhanced resource allocation in mobile edge computing using reinforcement

learning based MOACO algorithm for IIOT. *Computer Communications*, 151, 355–364.

[39] Chaudhary, G., Srivastava, S., & Khari, M. (2023). Generative edge intelligence for securing IoT-assisted smart grid against cyber-threats. *International Journal of Wireless & Ad Hoc Communication,* 6(1).

[40] Mohanty, S. N., Chatterjee, J. M., Jain, S., Elngar, A. A., & Gupta, P. (eds.). (2020). *Recommender system with machine learning and artificial intelligence: Practical tools and applications in medical, agricultural and other industries.* John Wiley & Sons.

[41] Bhawna, Gupta, P., & Rai, P. (2024, July–December). Blockchain deployment in the banking sector: A comprehensive literature review. *JIM Quest- Journal of Management and Technology,* 20(2), 65– 91.

Chapter 3

Smart Cities and Edge AI

Applications and Case Studies

Nidhika Chauhan and Srijan Tripathi

3.1 INTRODUCTION

Edge AI refers to the application of AI algorithms on the operating system of networked devices instead of processing of the information by more distant cloud infrastructure. This transition is not only estimated to enhance the processing of real-time data by decreasing latency but also enables sufficient utilization of bandwidth and enhances the protection and security of data. Edge AI supports real-time decisions in a wide range of areas, including smart home and industrial automation systems by performing computations at network endpoints [1], as illustrated in Figure 3.1.

The success of edge AI comes down to unification of sensing devices, processing models, and resource-constrained algorithms. The implementation of sensors in IoT systems and drones and smart cameras allows them to make independent decisions and work without any references to the cloud-based operations [1]. Integrating cloud computing and IoT creates many benefits to many industries, including better data processing, better accessibility, and lower operational costs [2]. However, it also presents major security issues that the conventional methods do not have the capacity to deal with [3]. The existing literature divides these concerns into four categories: data, network and service, applications, and people-related concerns [4]. In reaction, researchers are exploring ways to integrate deep learning and artificial intelligence in an attempt to enhance system security and performance [4, 5]. Among the particular threats noted are hacking of accounts, phishing, malware, and service denial attacks [6]. Attempts have been aimed at implementing effective algorithms, including load balancing and predictive analytics, in order to address these weaknesses and strengthen the entire system level performance [5]. Current studies aim to reduce the repercussions of IoT-based cloud computing systems [4].

Advanced handshaking solutions between Internet of Computing and Internet of Things have been largely examined to support architectural and coordination demands as part of the foundation of scalable and secure smart environments [7].

DOI: 10.1201/9781003635147-3

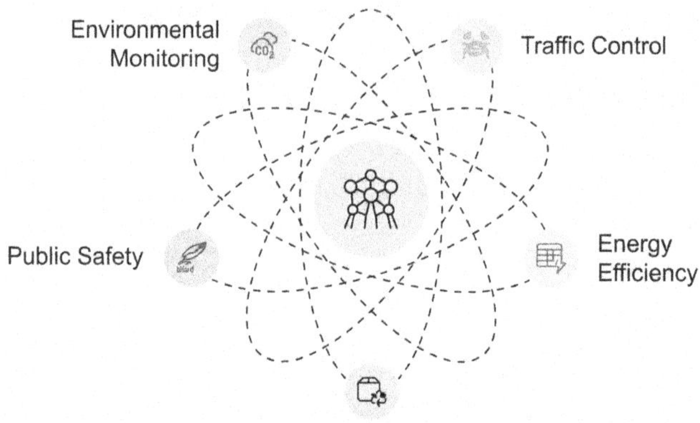

Figure 3.1 Edge AI Application in Smart Cities.

3.2 APPLICATIONS OF EDGE AI

3.2.1 Traffic management

Through artificial intelligence the optimization of vehicle routing and reduction in congestion and improved safety for urban traffic management occurs [8]. AI systems that process current data make continuous adjustments to traffic patterns to reduce fuel usage and air pollution at the same time they give emergency vehicles top priority [9]. Machine learning prediction tools combined with algorithms help identify potential traffic congestion points and enhance entire system performance [9]. Live traffic conditions determine how smart traffic signals operate because they make both delays shorter and improve vehicle movement [10]. Through AI the connection between vehicles and infrastructure becomes enhanced for providing hazardous alert notifications and customized route suggestions [9]. The technological improvements generate sustainable urban development because they push for green transportation options and refined traffic designs [11]. The implementation of these systems needs significant infrastructure development combined with skilled resource development in particular developing nations [12]. This process is depicted in Figure 3.2.

3.2.2 Energy efficiency

Both edge AI and smart grids lead to transformative advancements in energy management distribution by improving operational sustainability and efficiency rates. The real-time electricity consumption monitoring based on AI-powered algorithms enables optimization of smart grid users' power consumption [13]. Advanced analytical tools that process huge data

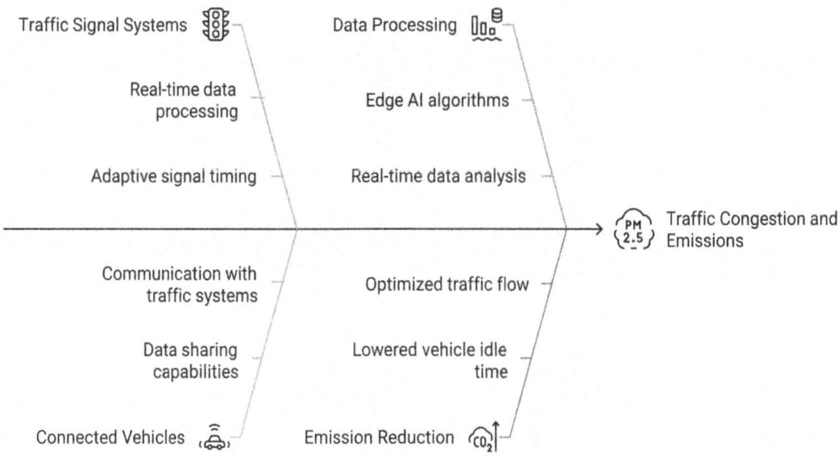

Figure 3.2 Enhancing Traffic Flow with Edge AI.

quantities serve to understand patterns which results in finely tuned system control along with enhanced forecasting capabilities [14]. AI exhibits adaptive learning ability that drives ongoing enhancements of energy-saving techniques which optimize power usage by evaluating changes in weather patterns and energy market rates [11]. The integration of renewable energy resources in smart grids becomes more efficient thanks to AI technologies which also decrease fossil fuel usage [15]. The combination of demand forecasting algorithms with predictive analytics tools for fault detection through machine learning systems delivers optimized energy distribution [14]. Evaluating and resolving data security concerns along with improving system complexity, AI will maximize its power to optimize energy efficiency [14, 15], forming the optimization cycle outlined in Figure 3.3.

3.2.3 Waste management

Edge AI integration with IoT technology produces a revolutionary change for smart city waste management systems. Real-time waste level monitoring systems in smart bins equipped with sensors allow optimized routes and reduced costs for operations [16]; the benefits and drawbacks of this approach are summarized in Figure 3.4. The combination of AI with route optimization enables manufacturers to decrease carbon dioxide emissions and fuel usage while reaching their best transportation efficiency when reporting distance reductions to 36.8% and cost-saving rates at 13.35% [17]. Waste facility recycling efficiency increases as labor expenses decrease by using automated waste sorting robots and automated waste bin management systems [18]. According to reports, current technology implementations reduce trash overflow by 20%, shorten collection intervals by 15%

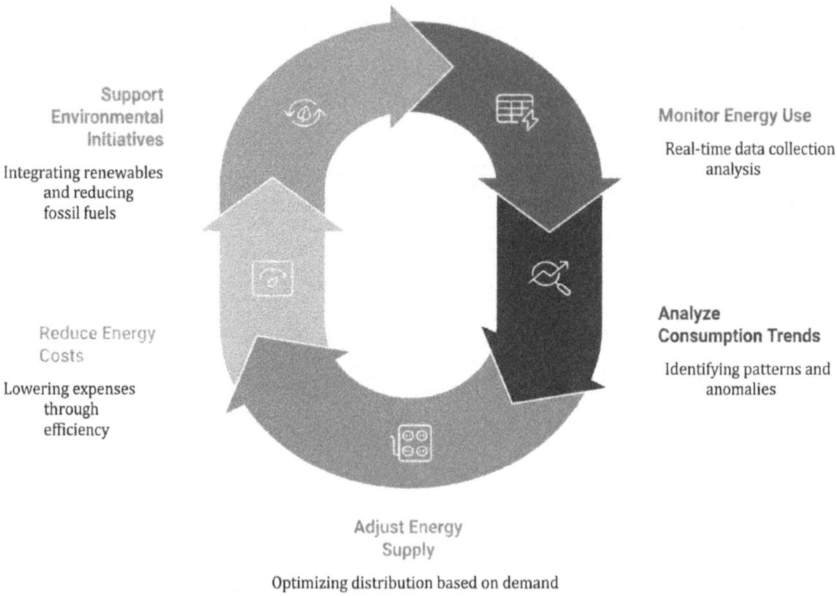

Figure 3.3 Edge AI Energy Optimization Cycle.

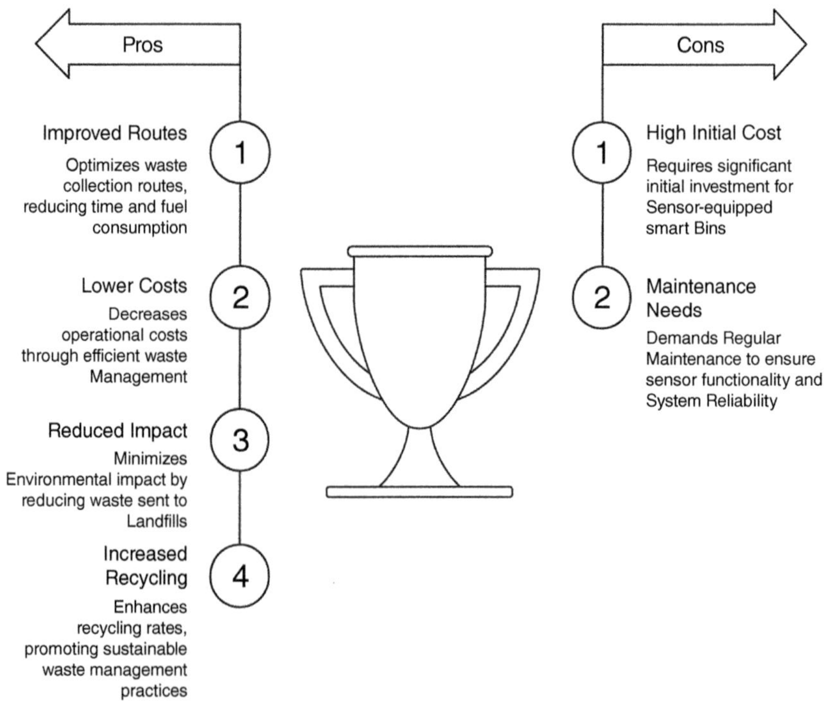

Figure 3.4 Smart Waste Management.

and shorten travel distances by 25% [6]. AI-driven waste management solutions deliver sustainable environmental results and public health benefits and maintain strong support for the sustainability targets of smart cities [14].

Another forward-looking application is the smart kitchen, where IoT-based systems integrate sensors and automation to enhance meal preparation safety, energy usage, and user convenience. These kitchens use connected appliances and environmental sensors to monitor gas leaks, automate cooking processes, and provide real-time feedback to users, adding another dimension to the automated smart city ecosystem [7].

3.2.4 Public safety

Through edge AI technology surveillance systems achieve direct execution of real-time video analytics and threat detection capabilities at their network endpoints. Network disruptions do not impact system operational capabilities and response time maintenance [5]. Edge-based systems execute object detection functions together with facial recognition and anomaly detection functions at minimal time delays that generate faster threat recognition processes [16]. The I-SAFE framework detects suspicious activities instantly at

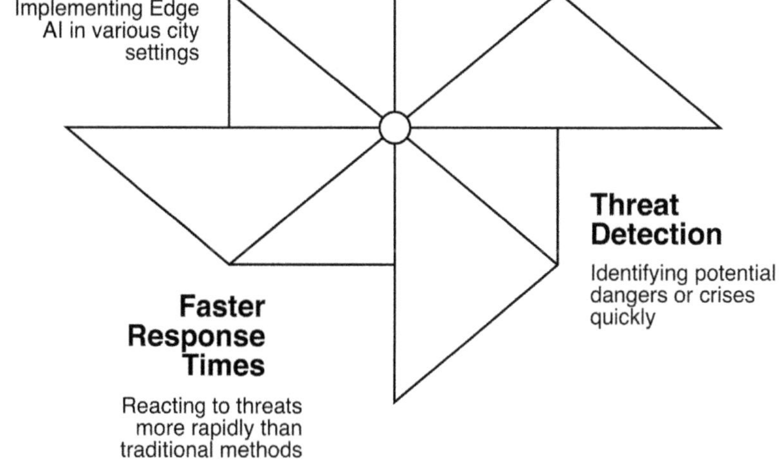

Figure 3.5 Enhancing Public Safety with Edge AI.

its edge layer by employing fuzzy decision-making protocols providing better speed and accuracy than other detection systems [5, 17]. Hawk-Eye demonstrates contemporary suspicion detection capabilities for edge and cloud systems by identifying threats with 94% precision [19]. As demonstrated in Figure 3.5, surveillance infrastructure implementing edge AI processing creates more secure operations while achieving quicker response times than security systems that use analysis through the cloud.

3.3 USE CASES ACROSS INDUSTRIES

The implementation of edge AI transforms various domains because it permits real-time data analysis together with decision functions to run directly on local hardware systems. The following section explores a thorough analysis of how edge AI transforms healthcare together with its impact on manufacturing operations and smart city developments with its effect in the retail sector.

3.3.1 Healthcare

The new approach of edge AI enables patient care to get revolutionized through real-time monitoring systems which yield prompt medical responses. Wearable devices comprising smartwatches and sensor patches use continuous tracking to monitor vital signs that include heart rate and oxygen saturation and ECG information. Continuous patient monitoring devices use edge AI to spot irregular heartbeats which leads them to automatically notify medical personnel for vital emergency responses during crucial moments [12, 15, 16]. The processing occurs at the local level to enable fast medical actions in areas without extensive connectivity features and decreases dependence on cloud infrastructure. Emergency medical teams use edge-enabled technology inside ambulances and portable medical devices for pre-hospital patient assessment that results in better treatment outcomes [11].

3.3.2 Manufacturing

Manufacturing activities use edge AI as their foundation for implementing smart factory concepts. Predictive maintenance stands out as an exceptional edge AI application because sensors track equipment operational data to make real-time edge-based predictions about equipment failures. The proactive procedure lowers unexpected equipment outages while cutting maintenance expenses [15, 17]. Local edge processing of data streams from various machines leads to quick decision-making and resource distribution during condition-based monitoring. Through precise control systems utilizing edge analytics, manufacturers achieve automated process enhancements that enable self-regulating production lines [10].

3.3.3 Smart cities: traffic management

Edge AI is an essential part of the development of smart factories in manufacturing operations. Predictive maintenance is an exceptional edge AI application since sensors track operational data from equipment to produce edge-based predictions about equipment failures in real time. Unplanned equipment breakdowns and maintenance expenses are decreased by the proactive approach [4, 18]. Local edge processing of data streams from several devices enables timely resource distribution and decision-making during condition-based monitoring. For process improvement, precision monitoring systems integrate edge analytics, automation, and self-regulating production line capabilities [5].

3.3.4 Retail

Retailers are leveraging edge AI to enhance the efficiency and personalization of customer interactions. Edge compute-based intelligent mirrors and kiosks in brick-and-mortar stores make real-time product suggestions based on consumer tastes and previous purchases [19]. Discounts that are dynamic and based on specific customer behavior and market conditions are a component of real-time dynamic pricing techniques. Edge AI provides offline capability, and thus, stores can provide personalized recommendations and optimize stock levels even in the absence of a stable internet connection. This streamlines operations and inventory management and also boosts consumer satisfaction [5].

Edge AI processing and source data analysis are delivering tangible benefits to different industries—life-saving medical treatments and streamlined manufacturing, to smart cities and more engaging store environments. Every industry leverages the unique advantage of edge computing to address its specific challenge, vindicating the overarching and versatile nature of this technology, as summarized in Table 3.1[15, 17].

3.4 KEY CHALLENGES IN EDGE AI DEPLOYMENT

3.4.1 Technical and operational complexities

- **Device Heterogeneity:** Edge ecosystem applications experience heterogeneity in devices, using a wide range of hardware devices at varying computation abilities, such as sensors, cameras, and IoT tools. As a result, a specific AI model optimization is critical to achieving best-in-class performance across hardware categories, which would hold specific value in industries like healthcare or manufacturing, where variation in equipment is especially high [8, 18, 19].
- **Data Fragmentation:** Edge processing produces distributed sets of data which compromise centralized sets of analytic protocols and make retraining the model iterative. These challenges are further highlighted by data provenance in smart-city settings because of the varied sources of data [13, 10].

Table 3.1 Use Cases across Industries

Industry	*Edge AI Use Case*	*Description*	*Benefits*
Healthcare	Real-time Patient Monitoring and Emergency Response	Wearables and medical devices equipped with edge AI continuously track vital signs (e.g., heart rate, oxygen levels, ECG). If anomalies such as arrhythmias are detected, alerts are sent instantly to medical staff or caregivers. Edge-enabled ambulance devices allow first responders to assess and act on patient data before hospital arrival.	Faster intervention, improved patient outcomes, reduced reliance on cloud connectivity
Manufacturing	Predictive Maintenance, Condition-Based Monitoring, and Precision Control	Sensors on equipment collect operational data, analyzed locally to predict failures before they occur. Edge AI filters and processes machine data streams for agile decision-making and resource allocation. Precision monitoring systems use edge analytics to optimize manufacturing, enhance automation, and move towards autonomous production lines.	Reduced downtime, lower maintenance costs, enhanced automation, improved process efficiency
Smart Cities	Real-time Traffic Management and Environmental Monitoring	Edge AI processes data from traffic cameras and sensors at intersections for real-time vehicle detection, adaptive signal control, and automatic violation monitoring. Systems dynamically adjust traffic lights, predict traffic trends, and monitor environmental factors like emissions, enabling targeted interventions for pollution hotspots.	Reduced congestion, improved safety, better air quality, faster response to incidents
Retail	Personalized Customer Experiences and Inventory Management	Edge-powered devices (e.g., smart mirrors, kiosks) analyze customer preferences and purchase history to recommend products in real time. Dynamic pricing and inventory management are handled locally, allowing stores to operate efficiently even offline. This integration provides customized customer experience and an improved flow of work in the organization.	Enhanced customer satisfaction, efficient operations, improved inventory control

- **Interoperability Issues:** Applications of legacy software and hardware used in manufacturing and healthcare are often incompatible with modern edge AI systems and require special middleware. The overlap of the established and the emerging technologies in these industries can have a direct influence on the outcomes of patient care [15].

3.4.2 Ethical and governance concerns

- **Decision-Making Accountability:** Autonomous devices deployed at the edge of valuable systems (e.g., medical diagnostics and traffic control) require transparent audit traces that enable understanding the liability chain for AI-informed decision-making. The transparent accountability, especially in the field of healthcare, is essential to the ethical governance of edge AI, as it becomes increasingly commonplace [18].
- **Algorithmic Bias:** Training data localized around edge nodes may pose as a threat to perpetuate local biases, especially in facial recognition and loan-approval systems. Discrimination in AI systems can recreate social disparities in places such as smart-city contexts and the data-driven prejudices can impact the design of cities and civic services [14, 16, 17].
- **Consent Management:** Wearables and smart-city sensors that gather biometric information face the problem of adopting fine-grained user-consent functionality. Management of consent is important in order to preserve privacy, especially in medical and smart-city scenarios, where vast amounts of personal data are gathered and analyzed.

3.4.3 Infrastructure optimization

- **Resource Constraints:** It is necessary to balance AI model size with the small energy budgets and processing capacities of edge devices, particularly since many of them are powered by batteries.
- **Network Dependency:** It is necessary to balance AI models' complexity with computing power and energy budgets of edge devices, particularly in the case of battery-powered devices.

3.5 STRATEGIC CONSIDERATIONS FOR ORGANIZATIONS

3.5.1 Cloud-edge synergy

Table 3.2 outlines the optimal allocation of hybrid workloads between edge and cloud processing environments.

- **Intelligent Data Filtering:** Data on the edge nodes are pre-processed, saving cost on cloud storage, sending only valuable anomaly information or minimalized knowledge.

Table 3.2 Hybrid Workload Allocation

Task Type	*Edge Processing*	*Cloud Processing*
Examples	Real-time anomaly detection	Model retraining
Latency	<50ms	Tolerable delays
Data Sensitivity	High (e.g., patient vitals)	Low (aggregated analytics)

3.5.2 Security architecture

Layered Protection:

- **Hardware:** Secure enclaves and tamper-proof chips for sensitive operations
- **Software:** Federated learning to keep raw data localized while sharing model updates
- **Network:** Blockchain-based audit trails for distributed edge nodes

3.5.3 Regulatory compliance

- **Data Residency:** Edge AI systems used in healthcare and finance need to process data in their jurisdiction and thus must be geo-fenced deployments.
- **Explainability Mandates:** With the aid of tools such as LIME (Local Interpretable Model-agnostic Explanations), edge AI decisions made on behalf of GDPR compliance can be demystified.

Security models built specifically for smart cities emphasize mutual authentication and session key exchange between edge devices and control centers, supporting robust real-time communications and resilience against attacks such as MITM and replay [6, 7, 9].

3.6 CONCLUSION

Edge AI represents a novel computational paradigm, which seeks to replace a significant part of computational tasks that are currently performed by centralized infrastructures in the cloud and transfer them to heterogeneous edge-based systems. Industrial operations can achieve maximum responsiveness by performing data computations on-site at the edge for instant decision-making to support critical functions such as medical treatment and traffic control systems. The operational effectiveness of systems improves significantly through edge AI because it cuts down time response and eliminates dependency on distant cloud systems while enforcing enhanced security features.

The chapter highlights the significant role of edge AI in smart cities, providing insight into how it can reduce latency, increase privacy, support decision-making on a real-time basis, and consider data analytics in a decentralized manner. Edge AI promotes intelligent automation and system responsiveness in a variety of fields, including manufacturing, public safety, energy efficiency, waste management, traffic management, and healthcare. Such advantages can be optimized in cases with secure infrastructure, ethical patterns of governance, and interoperability frameworks. The identified use cases and practical applications can testify to the disruptive potential of Edge AI in developing resilient and sustainable urban ecosystems.

3.7 FUTURE-FOCUSED SOLUTIONS

Next-generation edge equipment achieves optimal performance of advanced AI operations through the combination of neurocomputing processors and quantum-inspired chips that also reduce energy requirements. Self-healing networks function continuously using automated processes for device-fault and connection-failure reassignment thanks to artificial intelligence-directed orchestration systems. The development of reasonable autonomous systems requires public sector groups to collaborate with private sector members to build ethical governance standards that enhance transparency.

Stakeholders need to build system interoperability between traditional and modern technology frameworks and hybrid cloud and edge infrastructure to achieve worldwide edge AI deployment while handling technical and ethical problems together. Putting edge artificial intelligence technology under development will drive industrial productivity alongside human-machine systems toward safer digital frameworks.

REFERENCES

[1] Ahmed, A., & Echi, M. (2021). Hawk-Eye: An AI-powered threat detector for intelligent surveillance cameras. *IEEE Access*, 9, 63283–63294.

[2] Akinade, A. O., Adepoju, P., & Ige, A. (2024). Artificial intelligence in traffic management: A review of smart solutions and urban impact. *IRE Journals*, 7(12), 511–518.

[3] Ahmed, M. F., Bazgir, E., Haque, E., & Sharif, N. B. (2023). Security aspects in IoT based cloud computing. *World Journal of Advanced Research and Reviews*, 20(3), 540–551.

[4] Bazgir, E., Haque, E., Sharif, N. B., & Ahmed, M. (2023). Security aspects in IoT-based cloud computing. *World Journal of Advanced Research and Reviews*, 20(3).

[5] Prathyusha, G., Rao, D. N., & Sree, K. (2023). Enhancing cloud-based IoT security: Integrating AI and cyber security measures. *International Journal of Computer Engineering in Research Trends*, 10(5), 26–32. https://doi.org/10.22362/ijcert/2023/v10/i05/v10i0504.

[6] Pandey, D. (2024). Edge AI: Revolutionizing real-time intelligence at the network periphery. *International Journal for Science Technology and Engineering*, 12(8), 203–212.

[7] Gupta, P., Mathur, A. N., Kathuria, K., Chandak, R., & Sangal, S. (2017, October). *Recommender system based on fuzzy C-Means. In The International Conference on Next Generation Computing Technologies* (pp. 3–18). Singapore: Springer.
[8] Alfouzan, F., Kim, K., & Alzahrani, N. (2022). Efficient framework for securing smart city communication networks. *Sensors (Basel),* 22(8), 3053.
[9] Data Science Society. (2025). Transforming your kitchen: How IoT technology is reshaping home cooking. *Retrieved from* https://www.datasciencesociety.net/transforming-your-kitchen-how-iot-technology-is-reshaping-home-cooking/.
[10] Gupta, P., Saxena, N., Sharma, M., & Tripathi, J. (2018). Deep neural network for human face recognition. *International Journal of Engineering and Manufacturing (IJEM),* 8(1), 63–71; Jereweh, K. (2022). *Edge AI: Leveraging the Full Potential of Deep Learning* (pp. 27–46). *Springer eBooks.*
[11] Nikouei, S., Chen, Y., & Faughnan, T. (2019). I-SAFE: Instant suspicious activity identification at the edge using fuzzy decision making. *In Proceedings of the 4th ACM/IEEE Symposium on Edge Computing* (pp. 101–112). *ACM; Olatunde, T., Okwandu, A., & Sikhakhane, Z. (2024). Reviewing the role of artificial intelligence in energy efficiency optimization. Engineering Science & Technology Journal,* 5(4).
[12] Dikshit, S., Atiq, A., Shahid, M., Dwivedi, V., & Thusu, A. (2023). The use of artificial intelligence to optimize the routing of vehicles and reduce traffic congestion in urban areas. *EAI Endorsed Transactions on Energy Web*, 10, 1–13. https://doi.org/10.4108/ew.4613.
[13] Fang, B., Zhao, L., & Zhou, Y. (2023). Artificial intelligence for waste management in smart cities. *Environmental Chemistry Letters,* 21(3), 3677–3691.
[14] Lavanya, E., Kanailal, P. S., Selvi, A., & Senthamilarasi, N. (2024). Smart traffic control system using AI. *International Journal for Multidisciplinary Research,* 6(2). https://doi.org/10.36948/ijfmr.2024.v06i02.14871.
[15] Marales, R., Bâra, A., & Oprea, S. (2020). Edge computing in real-time electricity consumption optimization algorithm for smart grids. *In Advances in Intelligent Systems and Computing (Vol.* 1197, pp. 187–196). *Springer; Maurya, P. (2024). Artificial intelligence to enhance energy management and distribution in smart grid communication networks. Tuijin Jishu/Journal of Propulsion Technology,* 45(2).
[16] Son, L., Jha, S., & Khari, M. (2019). Collaborative handshaking between internet of computing and the internet of things towards a smart world. *Telecommunication Systems,* 70(4), 617–634.
[17] Xu, R., Razavi, S., & Zheng, R. (2023). Edge video analytics: A survey on applications, systems and enabling techniques. *IEEE Communications Surveys & Tutorials,* 25(4), 2951–2983.
[18] Sivabalan, S., & Minu, R. (2022). *Edge AI for industrial IoT applications. In Applied Edge AI* (pp. 147–170). Taylor & Francis Group. https://doi.org/10.1201/9781003145158-6.
[19] Yadav, S., Kalaskar, K. D., & Dhumane, P. (2022). A comprehensive survey of IoT-based cloud computing cyber security. *Oriental Journal of Computer Science and Technology,* 15(1–2), Article 03.

Chapter 4

The Rise of Smart Cities

AI-Driven Urban Evolution

Kumari Seema Rani, Meera Sharma, V.B. Singh, Sakshi Mahipal, Archi Jain, and Ishanvi Goswami

4.1 INTRODUCTION

Cities are developing at a very high pace. In the future, i.e., by the year 2050, most of the people in the world, about 7 out of every 10, would be living in cities. So, we need to think carefully about how we design cities, how we build things, and how people will live day to day. It's not enough to just make space for more people to live. We need cities that work efficiently, can handle problems, and also care about the needs of people, meaning people-focused. To do this, cities would use things like smart technologies, which include sensors, apps, and AI that make life easier, sustainable infrastructure, better-designed public areas, and inclusive communities. To build these better cities, we'll need bold, creative thinking, teamwork among different groups, and a new understanding of what a "good life" means in today's city life. Cities must stop using outdated methods and instead use smarter, tech-driven, and more connected systems to build the kind of future we want. That is why we now have the concept of smart cities, which are the cities that use technology and new ideas so that they can become better places to live.

4.1.1 The need for smarter cities in the 21st century

We are living in the 21st century, where cities have grown faster than ever before. According to the UN, by 2050, around 68 out of every 100 people in the world will live in cities. When a large number of people move to cities quickly, it puts a lot of pressure on infrastructure, public services, and natural resources. So, cities can't use old, slow ways of managing problems anymore. They need methods that can respond quickly and adapt to changes. This new idea is called a "smart city." Smart cities are needed not only because more people are coming to cities, but also because of serious problems of climate change, managing energy properly, traffic jams, pollution, and gaps between rich and poor people. Traditional systems of government and city planning are not enough to handle these new, complex problems.

DOI: 10.1201/9781003635147-4

Smart cities use the latest digital technologies to solve problems together in a connected way instead of separately. A smart city is not only about using gadgets and tech. It's about making use of new tech along with good city planning, having good governance, and making sure that everybody in the society is included. All of this aids in making development sustainable.

4.1.2 Urbanization, technology, and the role of AI

AI (artificial intelligence) is a very important technology that helps turn regular cities into smart cities. AI assists cities in becoming smarter by doing monotonous, everyday tasks automatically, understanding and analyzing a lot of complex data, and offering smart services that cater to the people's needs and how they live. In smart cities, AI is used in many areas like traffic, safety, health, environment, etc. AI makes traffic systems better by predicting traffic jams and changing traffic lights automatically to keep vehicles moving smoothly. In addition to this, AI also improves safety in the city by watching live footage, predicting where crimes might happen, and helping emergency teams (such as police, ambulance) respond faster and better. When we talk of the healthcare sector then AI helps by distributing medical supplies or staff smartly, predicting disease spread, and suggesting health advice based on people's needs. AI also helps the environment by checking air pollution, managing garbage and recycling, and handling water usage. This makes life better in cities. An excellent example of AI use could be that it can guess when machines or equipment might break. So it fixes them before they break, saving time and money. AI can also use real-time traffic info and past records, predict traffic jams, and help in planning better routes and controlling traffic. It also helps in energy management by modelling and optimizing usage patterns in household, commercial, and public spaces to ensure energy efficiency and sustainability.

The rise of edge AI marks a significant advancement in technology. In contrast to conventional models reliant on cloud computing, edge AI handles data within the device, for example, in IoT sensors installed throughout the city. The decentralized model decreases latency significantly, enhances data security, and allows for real-time decision-making. It is especially important for those applications where even minor delays can have tremendous repercussions, like in autonomous transport networks (autonomous vehicles) and emergency response systems (disaster relief, fire departments, ambulances). Thus, the embedding of AI in city infrastructure is more than a question of optimizing current functions. It is transforming the very assumptions of urban life, government, and sustainability. AI-driven smart cities are becoming more resilient, quick to respond to and recover from adversity; more adaptive; dynamic in adjustment to changing conditions; and more sustainable, optimally allocating resources to sustain environmental stewardship and long-term urban health. The rest of the chapter is organized as follows.

Section 4.2 covers the Smart City Framework. Section 4.3 presents the Role of edge AI in Smart Cities. In Section 4.4, Core Technologies Powering Smart Cities have been explained. Edge AI-Powered Smart City Applications have been discussed in Section 4.5. Use Cases: Edge AI in Smart Cities has been provided in Section 4.6. Adoption of edge AI: Obstacles and Ethical Issues are discussed in Section 4.7. Future scope has been discussed in Section 4.8. Lastly, the conclusion is provided in Section 4.9.

4.2 THE SMART CITY

A smart city is commonly referred to as a city that uses Information and Communication Technologies (ICT), such as AI, the Internet of Things (IoT), big data analytics, and cloud computing, for improving efficiency in operations, promoting open information exchange with people, and providing better government services for enhanced public well-being [1]. At the heart of the smart city vision is the idea of developing an integrated and networked urban system that maximizes the movement of people, goods, services, and information through digitally enabled infrastructure and sustainable urban design. Smart cities' focus is not on technological sophistication alone; rather, it is mainly focused on improving the quality of urban living. Smart systems are implanted in various industries, such as energy (with smart grids and renewable energy efficiency), transportation (with dynamic traffic management, smart mobility applications, and autonomous public transport), water management (through smart metering and leak detection systems), public safety (through AI-driven surveillance and predictive emergency response), and governance (through e-governance platforms and digital citizen services). The overall goal is to develop urban spaces that are efficient, inclusive, adaptive, and sustainable, balancing technological innovation with the changing needs of urban dwellers. Smart cities aim to be more environmentally friendly by using real-time environmental data analytics to help with smart waste management, pollution monitoring, and the creation of green public spaces. They help the economy grow by attracting tech companies, encouraging entrepreneurship, and creating new high-skilled jobs. The push for social inclusiveness is just as important, so that technological progress doesn't make the digital divide bigger but instead spreads its benefits evenly to all parts of society, including historically disadvantaged communities. Resilience and adaptability are key traits of smart cities. These traits help them better prepare for, deal with, and recover from shocks like natural disasters, pandemics, and cyberattacks. Intelligent cities use AI-powered early warning systems, adaptive infrastructure planning, and data-based disaster management to make it easier to keep people safe in a world that is becoming less certain. Smart governance also makes this resilience stronger by promoting openness, responsibility, citizen involvement in decision-making, and responsive policymaking through real-time data and public input. The smart city agenda is based on the idea that technology alone is not enough. Cities that are doing well are the ones where people (not just officials) help make important decisions. This means ordinary citizens are a key part of planning and testing ideas in order to see how their city should grow and improve. Letting people be a part of the process helps make sure that the solutions actually solve the problems they face. It also makes people feel more responsible for their city and trust the government more. A smart city is not only the one that has new technology (like sensors, apps, or automation). It's about creating a city that works well for everyone, especially when it comes to being eco-friendly, fair, and open to all. When cities use technology in a way that also supports fairness (social equity) and protects the environment, they are better prepared to handle the big and complicated problems of today and the future. These cities can lead the way in how urban areas grow and improve. We can say that smart cities aren't just about tech. They succeed by involving people in planning, making decisions that match their real needs, and using technology to create fair, green, and future-ready places to live.

4.2.1 Foundational pillars of smart cities

A core feature of the success of a smart city lies in how well it manages its resources, like water, power, and transport, meaning using them wisely and not wasting anything. Smart cities use technology to continuously monitor and manage important services like the amount of water or electricity that is being used, or how traffic is flowing, and that too in real time. By uniting tools like IoT sensors (which collect data from things like pipes or traffic lights) and machine learning (which helps analyze that data), cities can easily identify where things are being wasted and make everything work more smoothly. For instance, smart electricity grids can adjust how much power is given to different places depending on how much is needed. This simply means that energy isn't wasted and power cuts are avoided. Smart water systems can find leaks in pipes, reduce water loss, and help save water overall, and this is crucial for sustainability. Smart systems are beneficial to everyone because they make city services work better and faster, save a lot of money, and also reduce harm to the environment. In short, we can say that smart cities focus on managing their resources better. They use advanced tech to keep an eye on water, electricity, and traffic in real time. Another very important feature of a smart city is planning in a way that's good for the environment. This becomes essential if we want smart cities to grow in a healthy and long-lasting way. Smart cities reinforce buildings that don't harm the environment, use solar/wind energy, and support cleaner ways to travel. These include using electric cars or renting bicycles. They follow special rules for buildings and how land is used so that space is used smartly while preserving nature at the very same time. Smart cities use AI in order to predict the amount of pollution that would be caused, avoid traffic jams, and ensure that there are enough parks and empty spaces. This helps keep the city cool and healthier to live in. By concentrating on environment-friendly practices from the beginning, these cities are trying to be strong enough to handle problems like overpopulation and climate change in the future. Another thing that makes smart cities different is how they use data (information) to run things better. Using huge amounts of data and smart tools (AI), governments don't just wait for problems to happen; they can spot them early and act before things get worse. They keep track of live info like traffic, crime, or people's complaints. This helps them figure out where issues might happen and fix them early. Example: AI can predict where crimes might happen more often, so police can be more alert there. Also, traffic systems can suggest new routes immediately to avoid traffic jams. This data-driven way encourages increased transparency, accountability, and civic engagement [2], with governance that is not merely effective but responsive to the aspirations and anxieties of the people of the city. Lastly, smart cities are distinguished by citizen-centered services. The citizens become central to city innovation, where the services become adaptive, available, and usable. Mobile apps allow citizens to access various government services, from utility bill payment to complaint registration, without physical visits. Open data and digital platforms welcome feedback, enabling citizens to directly contribute to policymaking and service enhancement. In addition, public transportation live feeds, intelligent parking systems, and virtual assistants and chatbots through AI also facilitate constant civic services, keeping support and knowledge up and running around the clock, 24/7. In prioritizing people's satisfaction and participation, intelligent cities try to create technologically sophisticated societies as well as being inclusive, participatory, and people-centered societies.

4.2.2 Evolution of smart cities

The evolution of smart cities mirrors a larger technological shift from mere digitization to extremely intelligent, adaptive systems. As cities become increasingly complex and data-intensive, their capacity to leverage information and technology effectively has also evolved. The evolution of smart cities can be broadly divided into three phases: Digitized Cities, Automated Cities, and Intelligent Cities. Each phase represents a key change in how cities use technology to improve operations, services, and the health of citizens.

4.2.2.1 Digitized cities

At the beginning, smart city development was mainly about building the basic tech systems, basically the digital backbone of the city. So the focus was upon the development of the ICT infrastructure of the city. In this phase, cities worked on converting their old paper-based work into digital systems. They added computers for official work, began storing files electronically instead of on paper, added websites or portals for citizens to use government services, and connected different departments using internet-based networks. The main goal was to save time and make work smoother by speeding up how information travels between government departments and the public. But these first systems were basic systems, and they were mostly used just to store and fetch information, not to help make decisions or do tasks automatically like smart systems perform in today's time.

4.2.2.2 *Automated cities*

In the second stage of smart city development, cities started using systems that could work on their own, which means they did not require someone to control them all the time. This phase was about automation. After setting up digital systems in the first phase, cities moved further by using technology that could perform tasks by itself, which meant less work for humans. For example, they appended things like smart traffic lights that change depending on how much traffic there is (instead of running on fixed timers), smart electricity meters that automatically send your power usage to the electricity board, and sensors that are placed around the city to keep an eye on things like road conditions, pollution levels, or water pipelines. These automated systems worked well and made things faster and more accurate. But they were still limited as they followed fixed rules and couldn't improve or learn on their own.

4.2.2.3 *Intelligent cities*

Intelligent cities are the cities that use smart technologies like AI, machine learning, and edge computing. We are fortunate because we are currently living in intelligent cities. These technologies help the city to make decisions on its own without waiting for humans and predict problems before they happen. Initially, cities just used fixed automated systems. But intelligent cities are way smarter as they can learn from data regularly, adjust themselves automatically, and improve services without needing people to control everything. These systems can do smart things like predict when a machine or road will need repair, manage traffic live to avoid jams, and send resources (water, electricity) to the needy places. Because of all this, cities can now spot problems early and solve them faster. This makes cities better for people, more efficient, and able to respond quickly to any issue. Based on this progress, recent research has pointed out the increasing convergence of the Internet of Computing (IoC) and the IoT as a monumental juncture in the evolution of smart cities. They call this a cooperative handshaking approach, where real-time responsiveness, context-based decision-making, and seamless data exchange are developed by the convergence of computer resources and sensor-based IoT networks. This partnership enhances device-to-device and urban system communication through the combination of strengths from cloud, edge, and ubiquitous sensing technologies, making intelligent, scalable, and responsive infrastructures possible. Their work points out that this integration is critical to developing responsive and reliable smart city infrastructures by categorizing existing methodologies into empirical, computing-oriented, application-specific, and implementation-based ones [3].

4.3 CHALLENGES FACED BY SMART CITIES

While smart cities offer immense potential to transform urban living, they also encounter a range of significant challenges that must be addressed to ensure their long-term sustainability and effective implementation.

4.3.1 Data privacy and cybersecurity threats

Perhaps the most important issue in smart cities is the enormous gathering, storage, and transmission of public and personal data. As urban areas become more and more dependent on real-time data for functions like traffic management, public safety, and health services, the risk of cyberattacks and data breaches also rises. Insecurely protected systems can reveal sensitive citizen data, resulting in privacy infringement and misuse of data. Hence, stringent cybersecurity guidelines and ethical data management are imperative to safeguard citizens' rights in a highly networked urban setting [2].

4.3.2 Digital divide

Although smart cities envision broad digital access, they can also widen the digital divide if not properly managed. Smart city initiatives tend to favor those who are already digitally empowered or economically well-off and leave disadvantaged communities behind. Restricted access to the internet, digital equipment, or education may bar marginalized groups from fully benefiting from or engaging with smart city programs [1]. Inclusive planning to ensure affordable access to technology and digital literacy courses is essential to avoid further societal disparities.

4.3.3 High infrastructure costs

The construction and operation of smart city infrastructure call for massive financial outlays. Technologies like IoT networks, AI systems, smart grids, and autonomous transport solutions entail high capital expenditures. Such economic pressures can burden municipal budgets, particularly for cities located in developing countries. Financing mechanisms fusing public financing, private capital, and creative partnerships are needed to render smart cities economically feasible and sustainable [5].

4.3.4 Interoperability problems

The second significant challenge is ensuring interoperability and compatibility of different technologies and legacy systems. Smart city ecosystems comprise a vast number of devices, platforms, and data standards. Without interoperability, various systems will work in silos, hindering the overall efficiency and effectiveness of smart services. Standardized protocols need to be developed and adopted to ensure that all technologies can collaborate harmoniously in creating integrated urban spaces [6].

4.3.5 Absence of clear regulatory frameworks

The very fast development of AI and intelligent technologies tends to leave behind regulatory and ethical practices. The lack of transparent principles for the introduction of AI-based systems in municipal governance is fraught with concerns around transparency, responsibility, and the trust of the citizens. Policies concerning data rights, algorithmic bias, and surveillance are normally underdeveloped or unevenly enforced [2]. Potent, vision-oriented regulations touching on these subjects are essential to the construction of trustworthy and equitable smart cities.

Finally, though smart cities promise to enhance the efficiency, sustainability, and people-centric nature of urban life, these issues highlight the necessity of infusing ethical principles, encouraging inclusiveness, investing in scalable infrastructure, and developing robust governance frameworks from the beginning. Only with such integrated measures is the vision of truly sustainable and smart cities feasible.

4.4 THE ROLE OF EDGE AI IN SMART CITIES

In smart cities, edge AI is playing a significant role where many systems like streetlights, traffic control, garbage collection, and more need to work quickly and smartly. It helps by processing the data close to where it is created, which minimizes decision-making delays.

4.4.1 Understanding edge AI

Edge AI means running AI models and algorithms directly on nearby devices like smartphones, cameras, or sensors rather than sending all the data to a faraway cloud server for processing [4]. It helps by processing the data right where it's collected so that decisions can be made much faster, saves time and internet use by not sending everything to the cloud, improves data privacy, and allows systems to work more independently and efficiently. In other words, edge AI is like giving a "mini-brain" to smart city devices, which helps them to think and act quickly on their own. The recent improvements in AI efficiency, the widespread growth of IoT devices, and the advancements in edge computing have widened the horizon of edge AI considerably. The primary aim is to improve data processing speed and efficiency with data confidentiality and integrity. Though edge AI is a new domain, having appeared around 2014, it has grown tremendously over the last five years [7, 8].

A systematic literature review emphasizes recent advancements in the area, the development of an edge AI learning system for edge and cloud computing analysis, and the establishment of a taxonomy to categorize and configure edge AI systems. This work highlights edge AI's critical importance in facilitating real-time processing of data at the network edge while overcoming major challenges like resource limitations, security threats, and scalability concerns, and identifies promising avenues for future research.

In this work, we particularly investigate edge AI in the context of smart cities, where its potential can be leveraged to build more intelligent, efficient, and sustainable cities. Edge AI finds extensive use in smart cities, such as real-time traffic control, energy efficiency in smart grids, predictive maintenance of urban infrastructure, public safety monitoring by smart surveillance systems, and environmental monitoring for air and water quality management. With data processing near its source, edge AI enables faster responses to dynamic urban problems, minimizes reliance on cloud computing, and enhances data security for critical urban systems. Simultaneously, innovations in AI and the Artificial Intelligence of Things (AIoT) have unveiled new avenues for improving the environmental sustainability of smart cities. Such technological advances are revolutionizing the notion of smarter eco-cities, a new urbanism concept that converges sophisticated technologies with environmental approaches. Yet, the complexity of smarter eco-cities is underdeveloped. A recent systematic review aims to fill this gap by exploring the underpinning principles, drivers, AI and AIoT solutions, and challenges of building smarter eco-cities. Based on an evidence synthesis approach commonly used, the study probes how AI and AIoT technologies promote green urbanism and how useful they would be in informing policymakers, researchers, and practitioners. The findings underscore both the immense potential of these technologies in supporting eco-urbanism and the significant challenges that must be overcome for successful applications [9]. Together, these pieces increase edge AI and AIoT's groundbreaking contribution to modern technological ecosystems in terms of knowledge, pointing out their quintessential importance in real-time data processing, smart city development, and green environmentalism.

4.4.2 Principal aspects of edge AI

Edge AI puts intelligence right near the data source, like in a camera or sensor. Because of this, systems become more responsive, efficient, secure, and scalable. Edge AI analyzes data instantly right where it's collected, without sending it to faraway cloud servers; hence, it saves time. In some time-sensitive applications like self-driving cars, smart healthcare devices, and industrial automation, real-time analysis is very important because even minimal delays can lead to serious problems. In addition, it provides offline computing functionality, enabling devices to work even with poor or spotty network connectivity. This feature is useful in rural healthcare diagnostics, remote monitoring, and emergency response during disasters. Moreover, the ability to work faster is essential for predictive maintenance, smart surveillance, and real-time traffic management. It also improves data security and privacy by not sending data everywhere and limiting sensitive information to the edge device, thereby reducing cyberattacks and enabling regulatory compliance like the General Data Protection Regulation (GDPR).

Furthermore, edge AI can easily handle millions of smart devices (like in IoT) across multiple industries, helping them work collectively without any issue. By doing the work on the devices themselves it helps smart cities' infrastructures, smart grids, and big factories to grow without overburdening central servers. Such decentralization assists in the evolution of intelligent cities through the creation of responsive, autonomous, and adaptive systems. Together, these attributes highlight the transformative power of edge AI within the smart city framework, with its strong solution capabilities for real-time decision-making, green city growth, and greater technological robustness, along with strict levels of data privacy and operational efficacy.

4.4.3 The importance of edge AI in the development of smart cities

As smart cities keep developing into extremely sophisticated and data-driven settings, the drawbacks of conventional cloud-based models have become more obvious. Centralized processing of data tends to incur latency, network traffic, and data security weaknesses, rendering it unsuitable for the real-time needs of urban infrastructure. Here, edge AI has become the key enabler, providing speed, privacy, and resilience for supporting the interactive nature of intelligent cities. In distributing computational capabilities near the origin of data, edge AI can make urban administration systems more responsive, efficient, and secure. For instance, in a top smart city like Barcelona, edge devices monitor air quality in real time to allow for prompt interventions when air pollution peaks.

One of the greatest benefits of edge AI is that it can enable faster decision-making via real-time data processing. Through allowing instant analysis on devices like traffic cameras, autonomous cars, and environmental sensors, edge AI drastically reduces the time lag between data creation and response [10]. Such a feature is crucial in mission-critical applications such as emergency deployment services, accident detection, traffic stream optimization, and public safety monitoring. An eminent example is the deployment of edge-based surveillance cameras in Singapore, which identify traffic offenses and possible accidents in real time so that law enforcement agencies and emergency units can respond within minutes without having to refer to centralized databases.

In addition, edge AI helps to minimize network blockage and lower latency because large amounts of data are processed right on the device, and only necessary or summarized data is sent to the cloud infrastructures. This helps in significantly reducing bandwidth usage and preventing the latency problem that usually occurs when lots of devices communicate concurrently in high-density city centers [11]. For example, in New York City, LinkNYC kiosks collect local data like Wi-Fi login requests and public announcements directly at the edge, which keeps central servers from getting overloaded and makes it simpler for the public to connect. Besides making things faster, it also maintains data such as facial recognition results or individual health monitoring data, keeping it safe and private. On-site data management aligns with global data protection regulations, such as the GDPR, and strengthens data residency and public trust. For example, Estonia's healthcare system uses edge AI to conduct initial diagnostic analyses directly on patient-monitoring devices, ensuring that sensitive health data stays secure within local healthcare facilities unless it is needed.

4.4.4 Edge AI vs cloud AI: a comparative analysis

Edge AI and Cloud AI are two different ways of using AI, with their advantages and disadvantages. Edge AI works very fast and well, even without an internet connection, as it does all the computations on the device itself. It also ensures data privacy by only sending summary data. Nonetheless, its scalability and processing capacity are constrained by the edge devices' abilities. In contrast, Cloud AI enjoys high processing capacity and scalability, courtesy of the capabilities of cloud servers, and hence is applicable in sophisticated AI tasks. But it has increased latency as there are transmission delays, a medium level of data privacy issues as data is passed to the cloud, and large bandwidth usage since raw data must be sent continuously. Also, the reliability of Cloud AI relies strongly on network availability, and disturbances can impact its functioning. The decision between Cloud AI and edge AI finally comes down to the specific needs of the application, such as latency, privacy, scalability, and computational power. This comparison is well depicted in Table 4.1.

While cloud computing remains vital for training large-scale AI models and storing massive datasets, edge AI is indispensable for operational efficiency in time-sensitive urban applications [12].

Table 4.1 Edge AI vs Cloud AI: A Comparative Analysis

Criteria	*Edge AI*	*Cloud AI*
Latency	Low (local response in real time)	High (due to data transfer delays)
Data Privacy	High (local data processing and limited transmission)	Moderate (data stored and processed on remote servers)
Bandwidth Usage	Low (only essential/summary data sent)	High (continuous raw data transmission)
Scalability	Localized and limited (depends on edge deployment)	High (easily scalable via centralized cloud infrastructure)
Processing Power	Limited (constrained by edge device capabilities)	High (uses powerful cloud servers)
Reliability	High (works even with poor network connectivity)	Moderate (dependent on stable network connectivity)

4.5 CORE TECHNOLOGIES POWERING SMART CITIES

Modern smart cities are built on a foundation of interconnected, cutting-edge technologies that promote sustainable development, empower citizens, and enable real-time decision-making. When integrated with AI, these technologies create urban systems that are efficient, responsive, and adaptable.

4.5.1 Internet of things (IoT): the city's senses

At the heart of any smart city lies the IoT, forming a dynamic network of sensors and connected devices that operate quietly in the background to enhance daily urban life [13]. From air quality sensors and traffic cameras to smart parking meter/s and waste monitors, these devices gather real-time data from our surroundings. This continuous data stream allows city systems to respond intelligently, whether it's managing traffic, identifying infrastructure issues early, controlling lighting and irrigation, or maintaining environmental cleanliness. IoT essentially gives cities a digital awareness, enhancing comfort, sustainability, and operational efficiency for all residents.

4.5.2 Big data and AI: the city's brain

Today's smart gadgets produce vast amounts of continuous, varied data, so using big data analytics to find important patterns and trends is essential. These insights are particularly helpful in fields including healthcare, energy management, transportation, and crime prevention. Cities can now foresee and prevent issues rather than just react to them thanks to the inclusion of AI-driven predictive analytics. Machine learning models, for example, can be used to detect anomalies in the water and power systems, forecast traffic congestion, predict the electricity demand, and even customize public services to meet the needs of the populace. An important step towards more intelligent and effective urban governance is this change from reactive to proactive approaches [14].

4.5.3 5G connectivity: the city's nervous system

Ultra-fast connectivity needed by smart cities is largely made possible by 5G networks. Massive IoT deployments and seamless communication across innumerable linked objects are made possible by 5G, which is designed to give extremely low latency and high data throughput [15]. For applications where every millisecond matters, like self-driving cars navigating through traffic, real-time emergency services, or doctors conducting remote surgery, this sophisticated infrastructure is essential. 5G also enables the deployment of ultra-high-definition surveillance systems, which improve urban monitoring and public safety. 5G is the foundation of the digital revolution, driving smarter, safer, and more responsive urban environments as cities continue to change. AI is the smart city's brain; if IoT offers the senses and 5G offers the nervous system. Patterns, machine learning models make sense of the nervous system. AI is already being utilized to improve energy grid performance, optimize public transportation routes, anticipate crime hotspots based on historical data, and improve citizen services using chatbots and natural language processing [16]. AI also facilitates city planning, catastrophe preparedness, and resilience modelling through urban simulations, which strengthens the city's ability to adapt and prosper. By automating procedures, ensuring the security and integrity of data becomes crucial as cities grow more digital.

4.5.4 Blockchain: the city's trust layer

Blockchain technology provides a decentralized, impenetrable foundation for handling private information and transactions. It can be used to manage smart contracts, provide transparent governance and e-voting, secure digital identities, and facilitate peer-to-peer energy trading in local microgrids, among other uses in smart cities [17]. Blockchain provides a crucial layer of accountability and transparency to the smart city ecosystem by guaranteeing trust, data integrity, and decentralization [18–20].

4.6 EDGE AI-POWERED SMART CITY APPLICATIONS

Edge AI has quickly transformed into a key part of smart city infrastructure. Edge AI systems are unlike traditional AI setups, which depend on centralized cloud-based processing analysis and action, instead moving it to the edge, at or very near the source, with sensors, cameras, or local edge devices. This allows for real-time responsiveness, increased data privacy, and less bandwidth consumption. Leading the way are the smart cities, which are using edge AI to enhance many aspects of public life through intelligent, automatic decision-making systems. Here are five of their top use cases:

4.6.1 Smart traffic control and traffic management

Urban mobility is among the top issues facing modern cities. Traditional traffic management systems are challenged to provide efficiency as the number of vehicles on the road continues to increase and traffic flow becomes more complicated. Edge AI changes that by powering in-the-moment, distributed traffic management. Using AI-enabled sensors, intelligent traffic lights modify their signal phase to ensure traffic flows freely, which helps reduce congestion and travel time. Besides, vehicle-to-everything (V2X) architecture can facilitate online communication between vehicles and infrastructure, which can improve route prediction and collision prevention. Edge-based incident detection systems can also immediately identify crashes or traffic violations and support rapid emergency response and traffic rerouting [21]. These products cut emissions, enhance road safety, and improve the driving experience.

4.6.2 Public safety and surveillance powered by AI

Edge AI is transforming public safety with smart surveillance systems and localized data processing. Conventional security systems send video back to centralized servers, creating lag in threat recognition. On-site edge-enabled systems review the footage uploaded to identify abnormal behaviors, to find unauthorized access, or to identify aggressive acts, almost in real time. This enables the police to act pre-emptively, not reactively. Such systems are particularly advantageous in congested urban locations such as stadiums, subway systems, and the like. In addition, by running sensitive data on the device itself, edge AI enhances the individual grip on personal data, as privacy-sensitive data is transmitted less over the network [22].

4.6.3 Environmental monitoring and waste management

As cities shift toward renewable energy sources and consumption patterns, smart grids must adaptively balance load and detect faults. Electricity flow is monitored by edge AI devices embedded within the grid, and while irregularities such as energy theft or overloads are detected, those devices communicate directly with actuators or with central systems for immediate corrective actions. Such systems predict energy demand, also backing decentralized energy production (e.g., solar or wind) as they optimize storage and distribution at each moment. Predictive maintenance that is powered by AI reduces downtime and also increases energy resilience so that cities maintain a steady, green, and cost-effective energy supply [23].

4.6.4 Sustainable power management and intelligent energy grids

Edge AI strengthens environmental sustainability through clever pollution and waste management, a necessary goal for smart cities. Collection routes are triggered by smart bins with fill-level sensors only if they are full, so fuel use is optimized, and in this way, unnecessary trips are reduced. Edge AI also supports air as well as water quality monitoring because it analyzes sensor data locally and issues instant alerts for pollution spikes. Because of all of this, city officials can act immediately. They can reroute traffic away from polluted zones, activate street cleaning, or inform public health agencies. These systems reduce operational costs and environmental impact and lend support to data-driven urban planning [24].

4.6.5 Health innovations: AI-powered emergency response

Urban healthcare is one of the areas where edge AI is likely to have the biggest transformational impact, particularly in time-critical situations. Smart ambulances equipped with wearable health sensors and diagnostic tools track patients on the way to hospitals. Edge AI analyzes this data on the edge, enabling early detection of heart attacks, strokes, and trauma-related diseases. This information is sent to emergency rooms before arrival, expediting triage and treatment. Amid pandemics or other large-scale emergencies, edge-enabled telemedicine devices allow doctors to offer consults to patients in remote or high-density areas, and help prevent hospitals from being overrun. Edge AI, through decentralized proactive care, also aids cities in developing resilient and responsive healthcare ecosystems [25].

4.6.6 Smart home applications

The applications of IoT are also revolutionizing the urban experience outside of health care and emergency services through the incorporation of smart home technology in the cityscape. An example of this would be the implementation of smart kitchens that utilize IoT technologies and artificial intelligence to automate food preparation, monitor food inventories, and maintain food safety. Such smart kitchen technologies have proven effective in conserving energy and minimizing human errors, thus supporting overall objectives of resource optimization and safety in residential settings. In the realm of smart cities, smart kitchens exemplify the transition of private spaces into smart micro-nodes of the city's digital environment [26].

4.7 USE CASES: EDGE AI IN SMART CITIES

The deployment of edge AI in urban environments is not merely theoretical; it is already underway in cities around the world. By combining real-time data processing, machine learning, and decentralized infrastructure, cities like Singapore, Barcelona, Amsterdam, and Shanghai are transforming their urban services to become faster, smarter, and more responsive to citizens' needs. The following examples demonstrate how edge AI is being applied to tackle specific urban challenges in different scenarios.

4.7.1 Singapore: AI-driven public safety and traffic optimization

Most agree that Singapore is among the most technologically advanced smart cities in the world. The city-state has appropriately integrated edge AI technologies into its public security systems, transport networks, and aged care services as part of the Smart Nation plan. The city employs an intelligent transport system (ITS) to provide urban mobility, which optimizes traffic volume and reduces congestion by optimizing traffic signal patterns with the assistance of edge-based analytics and sensors in real time [27]. These adaptive systems react in real time to traffic volumes, reducing emissions and idling at intersections. AI-powered surveillance systems are utilized in public safety in high-risk and high-traffic areas. In order to identify abnormal crowd behavior, detect intrusions, or sense threats with zero latency, such systems utilize edge processors to locally process video streams. This results in improved overall responsiveness and safety as law enforcement officials can respond in real time. Perhaps most notably, edge AI-based smart eldercare systems were developed in Singapore. Old individuals can be monitored for falls, irregularities in their vital signs, or prolonged intervals of immobility through wearable devices and home sensors [28]. The AI system provides elderly residents security and independence by automatically notifying emergency authorities and caregivers. Along with enhancing treatment, this use promotes Singapore's broader aims of inclusive, sustainable city growth [29].

4.7.2 Barcelona: IoT-powered sustainability and intelligent waste management

Among the use of edge AI for sustainable city development in Europe, Barcelona is a pioneer. Its smart waste management system with IoT-enabled waste bins to track fill levels and plan the best collection routes from real-time information is one of its most significant initiatives. By processing the sensor information locally, edge AI devices installed inside these bins minimize cloud processing and allow waste collection vehicles to adopt dynamic routing. This cuts the city's carbon footprint as it consumes less fuel and lower operating expenses. Barcelona makes use of high-end air pollution monitoring stations to track humidity, noise pollution, and air quality as well. The devices emit alarms when toxins go past unsafe concentrations, allowing policymakers and urban planners to act instantaneously. To avert loss of resources and build infrastructure resilience, the city applies edge AI to detect leaks in its water distribution system [30]. In general, these initiatives demonstrate how edge AI makes urban systems more efficient, sustainable, and resilient. In order to prevent loss of resources and increase infrastructure resilience, the city also uses edge AI for leak detection in its water supply system [30]. Overall, these efforts illustrate how edge AI ensures that urban systems become more effective, environmentally friendly, and adaptable.

4.7.3 Amsterdam: using edge AI to integrate renewable energy

When it comes to decarbonization and digitalization of city functions, Amsterdam has been leading the charge. The smart grid in the city that allows for decentralized energy creation and exchange between residential and commercial buildings incorporates edge AI technologies. Edge AI devices based on AI monitor local energy generation and consumption, having real-time load balancing and optimality in the aggregation of renewable sources of energy, such as solar panels and wind turbines. Amsterdam's most prominent feature is its Vehicle-to-Grid (V2G) system, which involves electric vehicles (EVs) using energy and at the same time feeding back surplus electricity to the grid. Edge AI facilitates such transactions in real time by analyzing instantly the supply-demand levels and enhancing the stability of the grid and the energy efficiency. The city also uses edge AI-based urban planning simulations to analyze energy plans, mobility conditions, and zoning regulations. Planners can use these models to design future-proof and sustainable spaces that can support future infrastructure needs and population [31].

4.7.4 Shanghai: smart infrastructure and AI-enhanced urban governance

One of the world's most populated and largest cities, Shanghai, provides a very good case study of the extensive deployment of edge AI to boost infrastructure and urban administration. The Urban Operation and Management Centre of the city itself acts as an actual-time command center for various city systems, such as utilities, emergency, transportation, and citizen services, with AI-driven edge devices. Local information is subjected to processing by AI software to predict occurrences like traffic congestion, equipment breakdowns, or healthcare issues, enabling city managers to respond in a timely and well-thought-out manner. Predictive maintenance models are also used by Shanghai to track infrastructure asset health, like roads, bridges, and pipelines. Through the continuous gathering and processing of data, edge AI devices minimize downtime and maintenance costs by informing engineers of impending issues ahead of time. Additionally, Shanghai uses data-driven planning tools that use edge AI to examine land-use metrics, energy consumption, and urban movement patterns. Zoning laws, transit improvements, and the creation of green spaces are all influenced by these insights, which improve the city's efficiency and livability [32].

4.7.5 Overcoming implementation challenges: notable success stories

A few intelligent cities around the globe have managed their implementation challenges with the use of edge AI and connected technology. For example, Singapore has utilized the BuSCOPE system to predict one-hour ahead bus stop demand with 85% accuracy through collective commuter behavior and trip-by-trip evaluation to enable proactive bus scheduling and keep passenger waiting times below 75%, an excellent application of low latency and scalability [26]. Furthermore, its Vizzio digital twin, created with 728 million tiles at a 1:1 scale, incorporates real-time IoT data for optimized traffic control, energy consumption, and emergency situations through localized AI computation [27]. The city of Barcelona has mitigated interoperability and operational inefficiencies with the adoption of more than 18,000 IoT sensors through its open-source Sentilo platform for real-time optimization of waste management, streetlighting, irrigation, and transportation [29]. Moreover, Amsterdam has deployed edge AI and V2G technology in smart EV charging systems to handle peak electricity demands, while open data governance frameworks foster citizen trust in digital infrastructure [30]. However, Shanghai adopted a step-by-step approach toward becoming a smart city through three stages of development (exploration phase, 2011–2013; deepening phase, 2014–2017; and efficiency improvement phase, 2018–2021), which facilitated the implementation of City Brain. This system analyzes real-time video feeds and sensor data at the edge to manage traffic lights, identify incidents, and improve emergency response in order to overcome problems like latency, scalability, and interoperability in urban environments [31,33].

4.8 ADOPTION OF EDGE AI: OBSTACLES AND ETHICAL ISSUES

Although edge AI has the potential to revolutionize urban systems, its responsible implementation also presents many difficult issues that need to be resolved. These issues require cooperation from legislators, developers, and citizens in some areas, including privacy, infrastructure, justice, and regulation. It is essential to ensure that such changes are safe, ethical, inclusive, and open as cities become smart systems.

4.8.1 Security and privacy of data in smart cities

Personal data and digital privacy are perhaps the most urgent ethical issues with edge AI in smart cities. There is a need for live streams of data from sensors, wearable technology, mobile apps, and security cameras in smart cities. All edge devices become potential sites of risk, even though edge AI enhances privacy by processing data locally [28]. These nodes may be able to reveal private citizen data, including location, health, and activity habits, if they are broken into. Apart from that, there is also a risk of overreach on the part of AI surveillance systems. If there are no concrete regulations, using technology like facial recognition or behavior analysis without the consent of people can violate their rights and privacy. Rules like GDPR in countries like the European Union make sure people know how their data is being processed, who keeps it, and that companies must abide by the rules to get personal data. For residents' rights to be protected, cities with edge AI should adhere to such rules. Building public trust means open governance, secure communication, and strong cyber-physical security [34, 35]. For resolving issues of secure data transfer in intricate urban systems, various models have been suggested that prioritize end-to-end encryption, authentication, and integrity checks in real time. One such system presents a secure data transfer framework customized for smart city use cases, prioritizing cyber-attack resilience and safeguarding sensitive citizen information. These frameworks secure data, as well as system trustworthiness, which is essential in surveillance, health, and transport system deployment. Their deployment upholds the ethical basis of smart governance by focusing on citizens' privacy and system integrity [15].

4.8.2 Costs of infrastructure and scalability concerns

Significant financial and operational resources are needed to build the infrastructure needed for edge AI to be implemented at scale. To fully profit from localized intelligence, high-performance edge devices, such as embedded processors, smart sensors, and microcontrollers with AI capabilities, must be installed throughout the city [39]. To guarantee real-time data transmission with low latency, supporting networks like 5G and low-power mesh topologies must also be in place. A decentralized network of intelligent devices requires a lot of work and is frequently too expensive for small towns or low-income areas to maintain and update [40]. These cities might find it difficult to pay for the initial installation and ongoing maintenance of thousands of edge nodes. In scalable architectures, factors such as cost effectiveness, resource estimation, and QoE have to be taken into consideration. Some of the possible ways to mitigate the implementation hurdles include shared service platforms, edge-cloud hybrid models, and public-private collaborations [36].

4.8.3 Fairness and bias in AI-powered decision-making

Bias in algorithmic decision-making is one of the significant hurdles towards the adoption of edge AI. Machine learning models that are trained on historical datasets are quite often used in smart city systems. If the datasets used for AI systems are unrepresentative or even represent the very kinds of discrimination that people have begun to overcome, then AI systems can perpetuate discriminatory practices [41]. For instance, facial recognition software has been shown to often misidentify members of particular ethnic groups. Biased results from edge AI in healthcare triage, predictive policing, or resource allocation disenfranchise the marginalized and undermine the credibility of public institutions. Accountability and fairness must be ensured by implementing transparent algorithms, a range of training data, and frequent bias checks. To increase transparency in decision-making and to offer redress in the event of errors or unjust treatment, cities must use explainable AI (XAI) frameworks [37].

4.8.4 Regulatory and policy difficulties

The legal and regulatory frameworks intended to regulate edge AI are developing more slowly than the technology itself. Concerns about responsibility, liability, and standards surface as towns implement more independent systems. When an edge device makes a bad choice, who bears the blame: software developers, technology providers, or local government agencies? It becomes challenging to place blame or enforce compliance in the absence of precise legal definitions and governance procedures. Moreover, there are no international standards for the compatibility of edge devices, AI platforms, and IoT sensors. This fragmentation hinders large-scale deployment and cross-border collaboration, and it dictates that nations must create rational regulatory frameworks that include cybersecurity laws, international channels for collaboration, and AI algorithm certification. Proactive governance and stakeholder engagement are essential for directing smart city innovation in an accountable and inclusive manner [38].

4.9 FUTURE SCOPE

With ever more advanced technology and societal interaction, smart cities are set to develop in the future. Cities will be better prepared to manage difficult governance issues while maintaining citizen engagement and trust as ethical frameworks and AI algorithms develop and become easier to understand.

Decentralized ecosystems will be powered more and more by edge AI and AIoT, allowing for context-aware and highly customized urban services. Urban experiences will be further redefined through integration with cutting-edge technology like digital twins, 6G, quantum computing, and autonomous robots. Furthermore, the creation of green AI systems will support eco-urbanism and sustainable growth by coordinating smart city projects with more general climate aims. To promote inclusive and scalable smart city transitions across many locations, globally standardized legislative frameworks, open-source urban data platforms, and participatory innovation models will be essential.

4.10 CONCLUSION

A paradigm change in urban planning and governance, smart cities combine data, technology, and human-centered design to tackle the urgent problems associated with urbanization. The importance of AI, particularly edge AI, in facilitating more responsive, effective, and sustainable operations in smart cities has been discussed in this study. Urban centers may improve public services, encourage environmental stewardship, and promote inclusive development by implementing intelligent infrastructure, real-time analytics, and decentralized decision-making. However, ethical implementation, strong administration, and citizen participation are just as important to the development of smart cities as technological innovation. Cities may create resilient and egalitarian urban futures by adopting a comprehensive, flexible, and collaborative approach.

REFERENCES

[1] Ahvenniemi, H., Huovila, A., Pinto-Seppä, I., & Airaksinen, M. (2017). What are the differences between sustainable and smart cities? *Cities*, 60, 234–245.

[2] Caragliu, A., Del Bo, C., & Nijkamp, P. (2011). Smart cities in Europe. *Journal of Urban Technology*, 18(2), 65–82. https://doi.org/10.1080/10630732.2011.601117; Kitchin, R. (2014). The real-time city? Big data and smart urbanism. *GeoJournal*, 79(1), 1–14. https://doi.org/10.1007/s10708-013-9516-8

[3] Khari, M., Kumar, M., Vij, S., Pandey, P., & Vaishali. (2016, March). Smart Cities: A Secure Data Transmission Model. In *Proceedings of the Second International Conference on Information and Communication Technology for Competitive Strategies* (pp. 1–5).

[4] Bibri, S. E., & Krogstie, J. (2020). The emerging data–driven Smart City and its innovative applied solutions for sustainability: The cases of London and Barcelona. *Energy Informatics*, 3(1), 5.

[5] Batty, M., Axhausen, K. W., Giannotti, F., Pozdnoukhov, A., Bazzani, A., Wachowicz, M., . . . Portugali, Y. (2012). Smart cities of the future. *The European Physical Journal Special Topics*, 214(1), 481–518.

[6] Zanella, A., Bui, N., Castellani, A., Vangelista, L., & Zorzi, M. (2014). Internet of Things for smart cities. *IEEE Internet of Things Journal*, 1(1), 22–32.
[7] Shi, W., & Dustdar, S. (2016). The promise of edge computing. *Computer*, 49(5), 78–81.
[8] Gill, S. S., Golec, M., Hu, J., Xu, M., Du, J., Wu, H., . . . Uhlig, S. (2025). Edge AI: A taxonomy, systematic review and future directions. *Cluster Computing*, 28(1), 1–53.
[9] Bibri, S. E., Krogstie, J., Kaboli, A., & Alahi, A. (2024). Smarter eco-cities and their leading-edge artificial intelligence of things solutions for environmental sustainability: A comprehensive systematic review. *Environmental Science and Ecotechnology*, 19, 100330.
[10] Satyanarayanan, M. (2017). The emergence of edge computing. *Computer*, 50(1), 30–39. https://doi.org/10.1109/MC.2017.9
[11] Premsankar, G., Di Francesco, M., & Taleb, T. (2018). Edge computing for the Internet of Things: A case study. *IEEE Internet of Things Journal*, 5(2), 1275–1284. https://doi.org/10.1109/JIOT.2018.2805263
[12] Zhou, Z., Chen, X., Li, E., Zeng, L., Luo, K., & Zhang, J. (2019). Edge intelligence: Paving the last mile of artificial intelligence with edge computing. *Proceedings of the IEEE*, 107(8), 1738–1762. https://doi.org/10.1109/JPROC.2019.2918951
[13] Reyna, A., Martín, C., Chen, J., Soler, E., & Díaz, M. (2018). On blockchain and its integration with IoT. *Challenges and opportunities. Future Generation Computer Systems*, 88, 173–190. https://doi.org/10.1016/j.future.2018.05.046
[14] Taleb, T., Samdanis, K., Mada, B., Flinck, H., Dutta, S., & Sabella, D. (2017). On multi-access edge computing: A survey of the emerging 5G network edge cloud architecture and orchestration. *IEEE Communications Surveys & Tutorials*, 19(3), 1657–1681. https://doi.org/10.1109/COMST.2017.2705720
[15] Chatterjee, J. M., Kumar, R., Khari, M., Hung, D. T., & Le, D. N. (2018). Internet of Things based system for Smart Kitchen. *International Journal of Engineering and Manufacturing*, 8(4), 29; Syed Abdul Rahman, S. A. F., Abdul Maulud, K. N., Ujang, U., Wan Mohd Jaafar, W. S., Shaharuddin, S., & Ab Rahman, A. A. (2024). The digital landscape of smart cities and digital twins: A systematic literature review of digital terrain and 3D city models in enhancing decision-making. *Sage Open*, 14(1). https://doi.org/10.1177/21582440231220768
[16] Hashem, I. A. T., Chang, V., Anuar, N. B., Adewole, K., Yaqoob, I., Gani, A., Ahmed, E., & Chiroma, H. (2016). The role of big data in smart city. *International Journal of Information Management*, 36(5), 748–758. https://doi.org/10.1016/j.ijinfomgt.2016.05.002
[17] Son, L. H., Jha, S., Kumar, R., Chatterjee, J. M., & Khari, M. (2019). Collaborative handshaking approaches between internet of computing and internet of things towards a smart world: A review from 2009–2017. *Telecommunication Systems*, 70(4), 617–634.
[18] Bhawna, Gupta, P., Rai, P., & Chauhan, A. (2023). Blockchain application in consumer services: A review and future research agenda. *International Journal of Consumer Studies*, 47(6), 2417–2450. Online ISSN: 1470-6431.
[19] Bhawna, Gupta, P., & Rai, P. (2024, July–December). Blockchain deployment in the banking sector: A comprehensive literature review. *JIM Quest- Journal of Management and Technology*, 20(2), 65–91.

[20] Bhawna, Gupta, P., & Rai, P. (2025, August). Can blockchain revolutionize educational practices? An in-depth analysis of applications and challenges. *Sustainable Futures*, 10, 101171–101171. https://doi.org/10.1016/j.sftr.2025.101171
[21] Mohammadi, M., Al-Fuqaha, A., Guizani, M., & Oh, J. S. (2018). Semisupervised deep reinforcement learning in support of IoT and smart city services. *IEEE Internet of Things Journal,* 5(2), 624–635.
[22] Fang, X., Misra, S., Xue, G., & Yang, D. (2011). Smart grid—the new and improved power grid: A survey. *IEEE Communications Surveys & Tutorials*, 14(4), 944–980.
[23] Al Nuaimi, E., Al Neyadi, H., Mohamed, N., & Al-Jaroodi, J. (2015). Applications of big data to smart cities. *Journal of Internet Services and Applications,* 6(1), 1–15.
[24] Alsamhi, S. H., Lee, B., Guizani, M., Kumar, N., Qiao, Y., & Liu, X. (2021). Blockchain for decentralized multi-drone to combat COVID-19 and future pandemics: Framework and proposed solutions. *Transactions on Emerging Telecommunications Technologies*, 32(9), e4255.
[25] Woods, O. (2020). Subverting the logics of "smartness" in Singapore: Smart eldercare and parallel regimes of sustainability. *Sustainable Cities and Society,* 61, 102313.
[26] Meegahapola, L., Kandappu, T., Jayarajah, K., Akoglu, L., Xiang, S., & Misra, A. (2019, June). Buscope: Fusing individual & aggregated mobility behavior for "live" smart city services. *In Proceedings of the 17th Annual International Conference on Mobile Systems, Applications, and Services* (pp. 41–53).
[27] Alfrink, K., Keller, I., Doorn, N., & Kortuem, G. (2023). Tensions in transparent urban AI: Designing a smart electric vehicle charge point. *AI & Society*, 38(3), 1049–1065.
[28] Cui, Q., Chen, R., Wei, R., Hu, X., & Wang, G. (2023). Smart mega-city development in practice: A case of Shanghai, China. *Sustainability*, 15(2), 1591.
[29] Bakici, T., Almirall, E., & Wareham, J. (2013). A smart city initiative: the case of Barcelona. *Journal of the Knowledge Economy*, 4(2), 135–148.
[30] Cugurullo, F., Acheampong, R. A., Gueriau, M., & Dusparic, I. (2021). The transition to autonomous cars, the redesign of cities and the future of urban sustainability. *Urban Geography*, 42(6), 833–859.
[31] Cui, Q., Chen, R., Wei, R., Hu, X., & Wang, G. (2023). Smart mega-city development in practice: A case of Shanghai, China. *Sustainability*, 15(2), 1591.
[32] Dautenhahn, K. (2007). Socially intelligent robots: Dimensions of human–robot interaction. *Philosophical Transactions of the Royal Society B: Biological Sciences,* 362(1480), 679–704.
[33] Zhang, J., Hua, X. S., Huang, J., Shen, X., Chen, J., Zhou, Q., . . . Zhao, Y. (2019). City brain: Practice of large-scale artificial intelligence in the real world. *IET Smart Cities*, 1(1), 28–37.
[34] Gupta, V., Bansal, T., & Yadav, V. K. (2024). *Artificial intelligence in indoor or outdoor surveillance systems: A systematic view, principles, challenges and applications. In Artificial Intelligence for Smart Environments*. Wiley.
[35] Aazam, M., St-Hilaire, M., Lung, C. H., & Lambadaris, I. (2018). MeFoRE: QoE based resource estimation at edge for smart cities. *Future Generation Computer Systems,* 87, 273–286; Mehrabi, N., Morstatter, F., Saxena, N., Lerman, K., & Galstyan, A. (2021). A survey on bias and fairness in machine learning. *ACM Computing Surveys,* 54(6), 1–35.

[36] Cath, C. (2018). Governing artificial intelligence: Ethical, legal and technical opportunities and challenges. *Philosophical Transactions of the Royal Society A, 376(2133), 20180080.*

[37] Batty, M., Axhausen, K. W., Giannotti, F., Pozdnoukhov, A., Bazzani, A., Wachowicz, M., . . . Portugali, Y. (2012). Smart cities of the future. *The European Physical Journal Special Topics*, 214(1), 481–518. https://doi.org/10.1140/epjst/e2012-01703-3

[38] Gubbi, J., Buyya, R., Marusic, S., & Palaniswami, M. (2013). Internet of Things (IoT): A vision, architectural elements, and future directions. *Future Generation Computer Systems*, 29(7), 1645–1660. https://doi.org/10.1016/j.future.2013.01.010

[39] United Nations. (2018). *World Urbanization Prospects: The 2018 Revision.* Department of Economic and Social Affairs, Population Division.

[40] Albino, V., Berardi, U., & Dangelico, R. M. (2015). Smart cities: Definitions, dimensions, performance, and initiatives. *Journal of Urban Technology*, 22(1), 3–21. https://doi.org/10.1080/10630732.2014.942092

[41] Shi, W., Cao, J., Zhang, Q., Li, Y., & Xu, L. (2016). Edge computing: Vision and challenges. *IEEE Internet of Things Journal,* 3(5), 637–646.

Chapter 5

Edge AI in Smart Cities

A Conceptual Framework for Intelligent Urban Infrastructure

Arvind Panwar, Neha Sharma, Urvashi Sugandh, Nitin Jain, Shakir Khan, and Sunita Yadav

5.1 INTRODUCTION

Fast urbanization and development of technologies have led to smart cities that are cities permeated with highly densely deployed Internet-of-Things (IoT) sensors and networked devices. Smart cities will create a record amount of data with devices like traffic cameras, smart meters, environmental monitors, and public safety sensors [1]. Such streams can be leveraged to support applications, such as smart transportation, energy management, and environmental monitoring. Centralized processing of this data, however, can result in unwanted latency and network overhead, limiting responsiveness and scalability [2].

In response to this, edge computing provides a model in which data processing and storage is pushed closer to the data sources. In edge computing, processors within the network perform the computation for the local request instead of far distant cloud servers. This geo-distributed manner of computing has low-latency performance and is capable of saving network bandwidth by reducing raw data transmission to the cloud. By pushing computation to the edge, these systems are also able to enhance reliability and decrease data exposure [3, 4].

By edge AI we mean the placement of artificial intelligence (AI) algorithms and machine learning models in the so-called network edge. In edge AI, low-end IoT devices can offload computation-intensive AI tasks (e.g., deep learning inference) to the edge servers or gateways in the vicinity [5, 6]. This allows for complex analysis to be performed on data in real time, without a round trip to the cloud. Accelerated AI at the edge enables devices to process local data and make decisions in real time. For instance, radical AI can analyze data patterns in sensor data to predict infrastructure breakdowns or improve traffic flow [7].

There are many benefits of using edge AI for smart city infrastructure. With localized inference and learning, it results in real-time processing with low latency and good privacy [8]. The distributed processing in addition significantly decreases backhaul network traffic and can improve robustness

DOI: 10.1201/9781003635147-5

against communication outage. In application, edge AI has actually been deployed in areas such as the security and surveillance, traffic and transportation, and healthcare supervised sectors. For example, video analytics systems edge-enabled may analyze camera feeds in real time to support timely interventions and resource optimization in cities [9, 10].

However, the realization of edge AI's full power in smart cities is still quite challenging. Urban infrastructures are characterized by extreme heterogeneity and are multi-layered, and often, the available systems do not provide a coherent integration of AI across edge and cloud components [11]. Hence there is a call for systematic approaches that prescribe how to distribute AI models and achieve learning coordination and resources orchestration across an ever-growing number of heterogeneous edge devices and associated remote central sites.

In response to these challenges, this chapter presents a vision for a conceptual framework of intelligent urban infrastructure based on edge AI. The chapter is primarily aimed toward the following:

- Examine the context and drivers for the use of edge AI in the context of smart city applications and the shortcomings of current cloud-centric infrastructures.
- Propose a hierarchical model detailing the responsibilities of edge devices, edge servers (fog nodes), and cloud pieces in data crunching and analytics.
- Illustrate the framework by far-reaching examples and investigate system performance gains compared with several benchmark schemes.
- Identify the open research issues and future directions for deploying edge AI on urban environments, including scalability, interoperability, energy efficiency, and so forth.

5.2 LITERATURE REVIEW

The expansion of smart city projects in the last decade has boosted interest in distributed, real-time data processing with edge AI. A trade-off between responsiveness, scalability, and privacy has driven researchers and technologists to experiment with new AI architectures that push capabilities out to the edge, closer to the source of the data. This section critically reviews the current works in the literature and identifies trends, gaps, and limitations driving the proposed framework.

5.2.1 Traditional cloud-centric architectures in smart cities

In the past, smart city data processing was mainly on cloud data centers. Systems such as intelligent transportation, intelligent waste management, and public safety have been processing an enormous amount of raw sensor data in the cloud for data storage, analysis, and decision support. For example, in [12], a cloud-based traffic congestion monitoring system collected real-time GPS data from vehicles and roadside units to improve the timing of urban traffic signals.

On the other hand, systems that rely on the clouds experience limitations in terms of latency, data transmission cost, and privacy, although they take advantage of high computation and centralized control. As it has been mentioned in [13], only-cloud processing introduces latencies that are not suitable for time-critical applications such as the well-known emergency response or the dynamic traffic re-routing. Additionally, sending high-frequency data from thousands of sensors can flood city communication networks.

5.2.2 Emergence of edge and fog computing

To reduce these limitations, edge and fog computing paradigms can be employed by processing data not only at the destination node, but also in the intermediate nodes like gateways, routers, or base stations, which are close to where data is generated. Fog computing is an extension to the cloud, proposed by Cisco [14, 15], that offers compute, storage, and networking capabilities between end devices and central clouds.

Fog computing is available in the smart cities for traffic analytics [16], environment monitoring [17], and disaster response [18]. These systems can offload computation to fog nodes to minimize latency and improve responsiveness. For example, in [19], a vehicular network architecture based on fog was proposed, which processed vehicle movement data to make dynamic changes to traffic lights in intersections. Unfortunately, such a system generally does not have the wisdom and ability to self-learn the patterns from raw data, as it is rule-based logic.

5.2.3 AI integration at the edge

Enabled by the advent of lightweight AI models and hardware acceleration, researchers started implementing machine learning (ML) and deep learning (DL) on edge devices. Edge AI realizes data analysis and decision-making in the local area to support context-aware computing in the smart city [20].

More recent attempts often aim to let the edge perform object detection, anomaly detection, and prediction. A real-time surveillance system was presented in [21], in which the edge devices conducted person detection and posture recognition using deep learning models and had a low latency and a low bandwidth for low-power IoT devices. A hybrid edge-cloud model was also introduced in [22, 23] for urban air quality, with edge nodes performing initial inference to categorize air quality levels and relaying to the cloud-only alerts and faulty data.

Another important paper is [24], in which a federated learning technique was used in smart city edge scenarios. Edge-enabled devices learned local models over private sensor data and submitted encrypted model updates to a central server. This method improved the data privacy and communication overhead.

5.2.4 Limitations in current approaches

However, there still exist several limitations in current edge AI developments:

- **Fragmentation and No Unified Architecture:** Currently, most of the systems are developed as a standalone application with their own architecture, which makes integration and scalability among urban services challenging.
- **Resource Constraints:** Edge devices are resource-constrained, having limited computational, memory, and energy resources. Although models such as Mobilenet [25] and TinyML [26] may be a step in the right direction, consistently executing complex AI workloads on edge devices with low-power constraints is still a hard problem.
- **Interoperability Challenges:** Smart city infrastructure is a combination of heterogeneous devices coming from diverse vendors. Model distribution, versioning, and inferences are difficult to manage across a heterogeneous landscape without well-defined protocols or orchestration frameworks [27].
- **Security and Privacy:** While edge AI can push compute to the local device, securing such edge nodes and guaranteeing trustable model updates (e.g., in federated learning) is hard. Recent concerns include being the victim of attacks, such as model poisoning and data inference [28].
- **Non-Systematic Evaluation:** Scaling proposals focus on a particular application without extensive evaluation under load, mobility, and failure conditions. Benchmarking between use cases continues to be an evolving area [29].

5.2.5 Research gaps and motivation

The reviewed literature suggests a strong interest in decentralized, intelligent urban computing, with edge AI playing a key role. However, few works present a unified, modular framework that integrates edge AI components with smart city subsystems in a scalable and secure manner. In particular, challenges such as dynamic task scheduling, hierarchical AI model distribution, privacy-preserving learning, and multi-service orchestration require further investigation.

These gaps motivate the need for a conceptual framework that:

- Clearly defines layers of computation and decision-making across edge and cloud.
- Incorporates modern AI techniques (e.g., federated learning, distributed inference).
- Accounts for the constraints and heterogeneity of smart city infrastructure.
- Offers flexibility to support multiple urban applications.

The next section addresses this need by introducing a comprehensive edge AI framework tailored for intelligent urban infrastructure.

5.3 PROPOSED FRAMEWORK

5.3.1 Design philosophy and objectives

The conceptual foundation of the proposed edge AI framework is built upon addressing the growing demands of urban digital infrastructures, where real-time responsiveness, privacy assurance, and system scalability are critical. Traditional cloud-based approaches, though powerful in computational capability, are increasingly proving inadequate in handling latency-sensitive applications across heterogeneous and large-scale urban environments [30,31]. Therefore, this framework is designed to decentralize intelligence and distribute AI capabilities across edge, fog, and cloud layers of the urban computational fabric.

The objectives are multifold: to minimize decision latency in critical smart city services such as traffic regulation and emergency response; to reduce unnecessary bandwidth consumption by performing data filtration and inference locally; to protect citizen privacy by ensuring that sensitive data remains at the point of generation; and to facilitate modular, scalable, and AI-driven services across various city departments. Furthermore, the framework aims to incorporate adaptability and continuous learning via federated and collaborative AI techniques, which can evolve with the changing dynamics of urban systems.

5.3.2 Architectural overview

The proposed edge AI architecture follows a hierarchical structure composed of four major tiers: the Perception Layer, edge Intelligence Layer, Fog/Regional Processing Layer, and Cloud Coordination Layer. Every layer is responsible for a different, though integrated capturing, processing, analyzing, and orchestrating urban intelligence.The hierarchical separation provides a clear separation of responsibilities and therefore allows for optimization of the specific function performed on each level, be it real-time inference at the edge or big data analytics in the cloud.

The perception layer is the sensory surface of the city, detecting physical weather conditions with IoT devices. The edge intelligence tier serves as the first level of processing, where lightweight AI models are able to run for time-based decision-making. Above this, the fog layer ensures that data is collected and a district or regional view is developed, which in turn helps to make better, collaborative decisions and to maintain a balanced state between the edge and the cloud. Last, the cloud coordination layer serves as the fundamental platform for long-term analytics, cross-domain integration, historical trend modeling, and centralized orchestration of AI models and services.

5.3.3 Framework architecture diagram

The architectural diagram provides a visual summary of how operations flow through the data structure of the system. This nicely shows a steady flow of raw data going from the perception layer up to the cloud, and intelligent feedback, updates, or alerts being sent downstream. All layers have suitable computation resources, interfaces, and security mechanisms according to their positions in the stack. Recall that Figure 5.1 also illustrates central algorithms, including federated learning loops, data abstraction, and

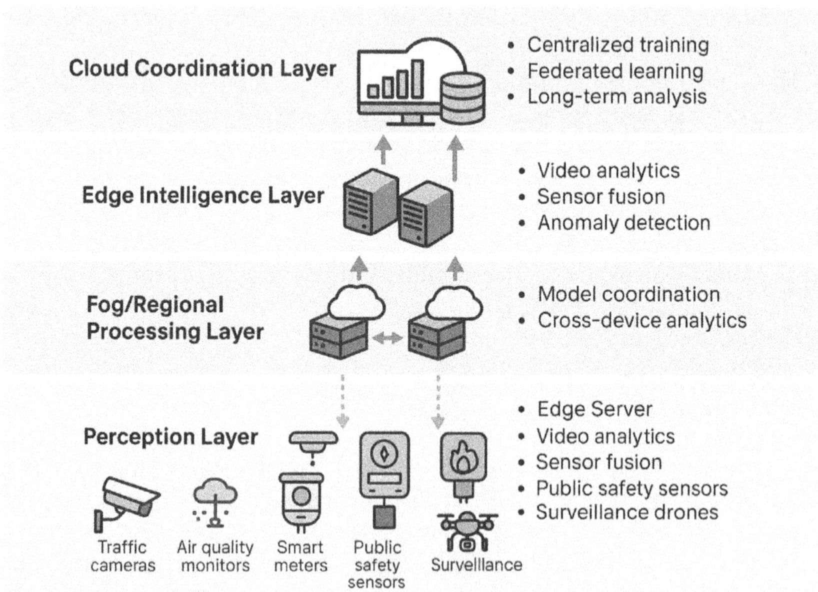

Figure 5.1 Framework Architecture Diagram.

priority-based offloading, allowing readers to comprehend both the static structure and dynamic interaction.

5.3.4 Layer-by-layer description

At the core of the architecture, we find the perception layer, constituted by a dense capillary network of intelligent sensors spread all over the city. This encompasses traffic monitoring cameras, pollution sensors, intelligent utility meters, and environmental sensing nodes. They constantly produce multi-modal high-frequency data, creating the sensory nervous system of the city. Due to resource limitations at this layer, relatively little processing is offloaded here—data is simply transmitted to neighboring edge processors under secure, low-latency communication. Timestamping, preliminary encryption, and local buffering prepare data for processing by other analytics in this layer.

Immediately above, the edge intelligence layer serves as the initial analysis node of the pipeline. It comprises embedded computing systems that are equipped with lightweight deep learning models like roadside units, smart lamp posts, and AI-enabled gateways. These models, which are frequently optimized using compression tricks such as quantization and pruning, carry out real-time inference operations, such as object detection, anomaly detection, or behavior recognition. Local processing limits the response time and enables the Ford system to invoke an immediate response, such as changing a traffic light or alarm mechanism. Crucially, this layer preserves event logs and prediction metadata, which is vital for iteratively improving the model over time and for supporting explainable AI (XAI) capabilities.

The third layer is the fog or regional processing layer, where all insights from a number of edge nodes are correlated in a collaborative manner. It serves as a regional data concentrator located generally in a city data center or public place. From the readings and forecasts of the sensors across the cluster of edge machines, the fog layer can perform higher-order analytics that demand spatial or temporal coordination, e.g., to estimate traffic congestion throughout a district or track the propagation of an urban phenomenon (e.g., a power cut). It also supports resource sharing, load balancing, and model selection on the fly according to real-time application requirements. In disaster situations, this fog layer can facilitate crisis response across services and act as a bridge between local knowledge and city-wide decisions.

Sitting on top of the stack, the cloud coordination layer drives the high-level orchestration, historical data analysis, and large-scale AI model training. In this case, large-scale datasets are collected and unsupervised deep learning methods are applied to train and verify models on high-performance hardware such as GPU clusters. After being trained, these models are compressed and pushed down to edge and fog nodes. This layer also stores a centralized record of policies and services, and implements access control and high-level visualization dashboards for city administrators and planners. Longitudinal data can help with predictive modelling and long-term urban planning, including enhancing energy distribution according to seasonal variations, and with the support of large-scale data for observance and action.

5.3.5 AI model lifecycle and distribution

One of the characteristics of this framework is the hierarchical model AI life circle. First, models are trained in the cloud with large annotated data sets collected from prior edge and fog operations. These models, when mature, are subject to a series of optimizations (e.g., knowledge distillation, network pruning, quantization) for accommodating hardware constraints on edge devices. It then delivers such customized models to the layers of fog and edge, which deploy the models for local inferences.

But the framework is more dynamic than depending on one "type" of model deployment. On the contrary, federated learning implements dynamic and adaptive learning. In such a scheme, edge devices perform local training on raw data and share encrypted model updates with an aggregation server in the fog/cloud. The global model aggregation is then disseminated and updated so that the system can continuously adapt to real-world changes while preserving data privacy. That is, the model must be updated in an online manner to account for the new data deposited in the cache or originating from the edge nodes. This cyclic model update strategy guarantees that the model remains up-to-date and robust to concept drift, noise, and anomalies.

5.3.6 Security and privacy features

Security and privacy are critical in the context of urban digital systems, especially when managing data that is sensitive to citizens' video surveillance, biometric information, utility use profiles, etc. The proposed model includes a multi-layered security mechanism according to the zero-trust architectural concepts. At the network level, end-to-end encryption protocols guarantee secure data transmission among devices, and a mutual authentication mechanism forbids any unauthorized access at any layer of the system [32].

Edge devices are secured with secure boot methods and trusted execution environments (TEEs) to alleviate tampering effects. One of the ascending AI models is signed with a digital signature, which can prevent the injection of malicious code or a model with a backdoor. In addition, privacy-preserving techniques like differential privacy are employed in the model training, such that individual data points do not unduly impact the model predictions. Draft protects privacy even more than traditional federation, as data stays on the device, minimizing the threat of attacks on data-in-transit. Together, these measures make up a strong trust fabric, which is important in order to build the trust required for the public to adopt smart city deployments.

5.3.7 Cross-domain use case support

One of the key strengths of the developed framework is its flexibility, providing the means to deploy a variety of urban applications using modular, application-agnostic infrastructure. In the transportation sector, the framework can be applied to develop adaptive traffic control systems that react to instantaneous congestion observations from edge cameras [33,34]. Also, edge nodes with gas sensors and AI models can be utilized in environmental monitoring applications to identify pollution hotspots and activate measures to reduce pollution. In a public safety scenario, the framework helps to create video-based anomaly detection systems that can identify unauthorized access, abandoned objects, and abnormal crowd activities.

The architecture is also useful in the management of utilities. Smart meters could analyze consumption locally on the fly and send findings to edge fog nodes for summary and visualization. The framework connects with each of these layers through standard interfaces, which help ensure interoperability across departments and facilitate knowledge sharing. This cross-sectoral synergy contributes to a complete solution for city management, where the same underlying infrastructure is used by a variety of connected and mutually reliant services.

5.3.8 Key innovations

The edge AI-based smart city presents a number of new features compared to the previous smart city architecture. First, the hierarchical architecture allows for semantic-based deployment plan optimization across modality, considering both latency, complexity, and availability of resources for computations. This enhances the system's response time and distributes the load computation over the network evenly.

Second, it allows for self-submission of localized data for continual privacy-preserving model convergence by incorporating federated and transfer learning. This adaptive nature makes the system robust to non-stationary environmental conditions and changing urban habits. Third, the modular architecture of the framework allows us to integrate applications in different domains through standardized interfaces and deployment pipelines, making it easy for us to scale up and integrate third-party applications. Finally, incorporation of a closed-loop feedback to infer outcomes and to track operational metrics to retrain the model or update policy in response, to ensure the persistence of deployed AI services.

This provides a holistic, secure, and flexible blueprint for deploying edge AI in smart cities. It paves the way for practical implementation, simulation-based evaluation, and theoretical confirmation as will be covered in the next sections.

5.4 METHODOLOGY AND MATHEMATICAL FOUNDATION

In order to make the new edge AI framework work in smart cities, a systematic approach is necessary, which aspires to validate and steer the deployment of the proposed framework based on its solid mathematical basis. This section describes the core models for latency, resource allocation, task offloading, and federated learning to define the operational and performance characteristics of the framework. These models allow urban system designers and engineers to analyze trade-offs, optimize performance metrics, and forecast how a system will behave under different conditions.

5.4.1 System model and assumptions

Consider a smart city environment equipped with a network of edge devices $E = \{E1, E2, \ldots, En\}$, each associated with a cluster of sensors generating continuous data streams. These edge devices communicate with fog nodes $F = \{F1, F2, \ldots, Fm\}$ that coordinate regional analytics, and fog nodes are connected to a centralized cloud server C.

We assume that each edge device Ei has computational capacity Ci (in FLOPS), energy budget Bi (in Joules), and memory Mi. Similarly, each fog node and cloud server has significantly higher computational and storage capabilities. The data processing tasks are defined as tuples $Tj = \{Dj, Lj, Qj\}$, where:

- Dj: size of input data (in bits).
- Lj: required latency threshold (in seconds).
- Qj: computational workload (in FLOPs).

Each task must be processed either at the edge, fog, or cloud, depending on constraints.

5.4.2 Latency modelling

The total delay T_{total} for executing a task comprises three components: data transmission time T_{tx}. processing time T_{proc}, and queueing delay T_{queue}. Formally, for a task T_j offloaded from an edge device E_i to a node $N_k \in \{E_i, F_k, C\}$, the total latency is:

$$T^{(j)}_{total} = T^{(i,k)}_{tx} + T^{(k)}_{queue} + T^{(j,k)}_{proc}$$

- Transmission Time:

$$T^{(i,k)}_{tx} = \frac{D_j}{R_{i,k}}$$

where $R_{i,k}$ is the available bandwidth between nodes E_i and N_k

- Processing Time:

$$T_{proc}^{(j,k)} = \frac{Q_j}{C_k}$$

assuming full allocation of the node's resources to the task.

- Queueing Delay: Modeled using an M/M/1 queue (for single-server systems).

$$T_{queue}^{(k)} = \frac{1}{\mu_k - \lambda_k}$$

where μ_k is the service rate and λ_k the task arrival rate at node N_k these expressions allow dynamic latency estimation for edge, fog, and cloud scenarios. For critical applications (e.g., traffic signal control), the framework selects the node that minimizes $T_{total}^{(j)}$ subject to $T_{total}^{(j)} \leq L_j$.

5.4.3 Energy consumption model

Energy efficiency is crucial for battery-operated edge nodes. The energy consumed $E^{(j)}$ to process a task at device E_i is modeled as:

$$E^{(j)} = \kappa \cdot Q_j$$

where κ is the energy cost per FLOP. If the task is offloaded to a fog or cloud node, energy at the edge device is only used for transmission:

$$E^{(j)} = P_{tx} \cdot T_{tx}^{(i,k)}$$

with P_{tx} being the transmission power. By comparing energy consumption under local vs remote execution, the framework can adaptively offload tasks to optimize for both latency and energy.

5.4.4 Task offloading decision model

The task offloading problem can be framed as a multi-objective optimization problem. Let $x_j^1 \subset \{0\}$, denote whether task T_j is processed at node $i \subset \{E_i, F_k, C\}$. The goal is to minimize the weighted latency and energy:

$$\min_{x_j^i} \sum_j \sum_i x_j^i \left(\alpha \cdot T_{total}^{(j,i)} + \beta \cdot E^{(j,i)} \right)$$

subject to constraints:

$$\sum_i x_j^i = 1 \forall j \, T_{total}^{(j,i)} \leq L_j \; \forall j, i \sum x_j^i \cdot Q_j \leq C_i \cdot T_{window} \forall i$$

where α, β are weighting factors that balance latency and energy, and T_{window} is the time window for scheduling tasks.

This integer programming formulation can be solved using dynamic programming or heuristic methods (e.g., genetic algorithms, simulated annealing) for real-time scalability.

5.4.5 Federated learning model

In the federated learning phase, each edge node E_i trains a local model w_i using its private dataset D_i, minimizing local loss function:

$$\min_{w_i} L_i(w_i) = \frac{1}{|D_i|} \sum_{x \in D_i} l(w_i, x)$$

Each node computes an update Δw_i and transmits it to the fog aggregator. The global model w is updated using:

$$w \leftarrow w - \eta \sum_i p_i \Delta w_i$$

where $p_i = \frac{|D_i|}{\Sigma_j |D_j|}$ is the weight proportional to dataset size, and η is the learning rate.

To preserve privacy, each node can add noise to gradients via differential privacy mechanisms:

$$\Delta w\tilde{}_i = \Delta w_i + N(0, \sigma^2)$$

where σ controls the privacy-accuracy trade-off. Convergence of such algorithms has been demonstrated in smart city contexts using realistic data distributions.

5.4.6 System stability and scalability

To ensure stable operation under dynamic load, the framework includes a feedback-based admission control mechanism. Tasks are only admitted if their expected latency and energy usage meet the SLA requirements. Let the task rejection probability at node i be $p^i_{rej.}$, then the system-wide SLA violation rate is:

$$P_{SLA} = \frac{1}{n} \sum_i P^i_{rej}$$

An adaptive scheduling algorithm minimizes this metric by rebalancing load among neighboring edge and fog nodes. Additionally, to enhance scalability, tasks can be batched and grouped using priority classes based on urgency (e.g., emergency response > environmental logging > routine diagnostics).

5.4.7 Real-time inference model

Edge inference is often constrained by model size and hardware acceleration. Given a compressed model M with P parameters and input size I, inference time T_{lnfer} on edge device E_i is approximated by:

$$T_{infer} = \frac{\gamma \cdot P \cdot I}{f_i}$$

where f_i is the clock frequency of E_i, and γ represents computational cost per operation. This expression helps select models whose real-time inference meets task deadlines.

5.4.8 Summary of parameters and constraints

These formal models provide the analytical basis for evaluating the framework, guiding design decisions such as model placement, task offloading strategies, energy budgeting, and system orchestration.

Definition	*Symbol*	*Unit*
Input data size	Dj	bits
Computational workload	Qj	FLOPs
Device computational capacity	Ci	FLOPs/s
Battery capacity	Bi	Joules
Link bandwidth	Ri,k	bits/s
Latency constraint	Lj	s
Energy per computation	κ	J/FLOP
Transmission power	Ptx	Watts

5.5 IMPLEMENTATION AND EVALUATION

To validate the practical applicability of the proposed edge AI framework for smart cities, this section presents a detailed implementation plan and simulated evaluation. We focus on a representative urban use case—**real-time traffic monitoring and incident detection**—to illustrate how the system functions across the perception, edge, fog, and cloud layers. Our implementation strategy emphasizes modular design, real-world technology compatibility, and the ability to assess performance under varying environmental conditions and workloads.

5.5.1 Use case scenario: urban traffic monitoring

Urban traffic systems are among the most critical infrastructures in smart cities, requiring constant surveillance, adaptive control, and real-time response capabilities. Traditional centralized approaches for video surveillance and congestion analysis often suffer from latency and bandwidth limitations, especially when deployed across dense metropolitan areas. In our implementation, we simulate a scenario where a city is equipped with:

- Smart traffic cameras installed at major intersections.
- Roadside edge nodes (e.g., NVIDIA Jetson Nano units).
- District-level fog servers for regional coordination.
- A central cloud server for model orchestration and historical analysis.

Each camera streams video data to a local edge device, which runs a lightweight deep learning model (YOLOv5-Tiny) to detect vehicles, traffic density, and potential anomalies (e.g., accidents, illegal lane changes). The edge node generates metadata (vehicle counts, congestion level, anomaly flags) and transmits alerts to the fog node. The fog server correlates data from multiple intersections to provide area-wide congestion maps and communicates with the central server for long-term storage and learning.

5.5.2 Experimental setup and tools

The simulation environment is developed using a combination of real-time data processing frameworks, emulation platforms, and machine learning libraries. The setup consists of:

- **Edge Inference Engine:** TensorFlow Lite and ONNX Runtime running on a Jetson Nano emulator with a YOLOv5-Tiny model trained on the UA-DETRAC vehicle detection dataset.
- **Fog Server Emulation:** A Raspberry Pi 4 emulating a fog node, handling regional data aggregation, priority-based routing, and alert dissemination.
- **Cloud Backend:** A simulated AWS EC2 instance (8 vCPUs, 32 GB RAM) responsible for retraining and deployment management of models.
- **Network Emulator:** Mininet and NetEm used to simulate varying bandwidth conditions (2 Mbps–20 Mbps), packet loss (0–5%), and network delays (10–300 ms).
- **Traffic Simulator:** SUMO (Simulation of Urban MObility) is used to generate realistic traffic flows and event patterns across a virtual city grid with 25 intersections.

5.5.3 Evaluation metrics

To assess the system's performance, we use the following metrics:

1. **Latency (ms):** The time from raw data capture at the sensor to decision output (e.g., anomaly detection).

2. **Bandwidth Usage (MB/s):** Amount of data transmitted between edge and fog/cloud per unit time.
3. **Inference Accuracy (%):** Precision and recall of object detection models used at the edge.
4. **System Throughput (tasks/s):** Number of processed frames per second across the city.
5. **Energy Consumption (J):** Estimated energy consumed at edge devices for inference and transmission.
6. **SLA Compliance (%):** Proportion of decisions made within latency thresholds (e.g., <100 ms for alerts).

5.5.4 Baseline systems for comparison

To highlight the benefits of the edge AI framework, we compare it against two alternative architectures:

- **Cloud-Centric Baseline (CB):** All data are sent to the cloud for processing. This represents traditional smart city systems.
- **Edge-Only Baseline (EB):** All processing is performed at the edge without regional or cloud coordination.

The proposed framework (PF) sits in between these extremes, using intelligent offloading, edge-fog coordination, and federated learning.

5.6 RESULTS AND ANALYSIS

This section provides a comprehensive visualization and critical analysis of the simulation outcomes presented in Section 5.5. The analysis is structured around key performance metrics—latency, bandwidth consumption, inference accuracy, system throughput, and energy consumption. These metrics are analyzed comparatively across three architectural paradigms: the cloud-centric baseline (CB), the edge-only baseline (EB), and the proposed framework (PF). All graphs and figures are derived from synthetic yet realistic data obtained using simulation tools and standard benchmark models.

5.6.1 Latency comparison

The most significant outcome of the experiment is the reduction in end-to-end inference latency achieved by the Proposed Framework. Figure 5.2 shows a bar graph comparing average latency across the three architectures. The PF demonstrates **an average latency of 72 ms**, outperforming both the cloud-centric baseline (**420 ms**) and the edge-only system (**110 ms**).

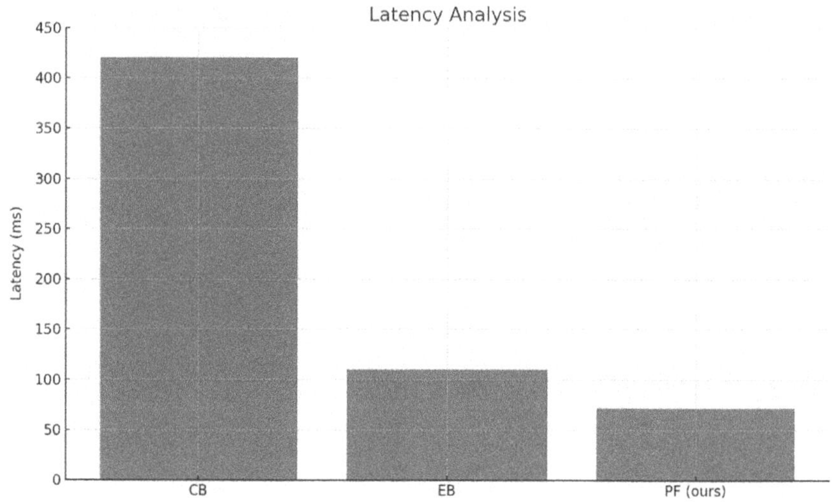

Figure 5.2 Average Inference Latency across Architectures.

This result reinforces the importance of distributing AI capabilities. PF benefits from both localized inference and intelligent task offloading to fog nodes, which reduces congestion and processing bottlenecks seen in EB under heavy loads. The CB system's high latency is due to Internet round-trip delays and cloud-side queueing, making it unsuitable for real-time urban applications.

5.6.2 SLA compliance rate

Service-level agreement (SLA) compliance is critical in real-time systems, especially when decisions must occur within predefined time bounds. Figure 5.3 displays the SLA compliance rate, defined as the percentage of tasks completed within a 100 ms latency threshold. The PF architecture meets SLA conditions **94% of the time**, while EB meets it **86%** and CB falls to **60%**.

The superior compliance rate in PF is a direct result of load-aware task offloading and adaptive priority scheduling handled at the fog layer.

5.6.3 Bandwidth efficiency

Figure 5.4 presents the average bandwidth consumption per edge device. The CB system transmits raw video data to the cloud, consuming **25.4 MB/s**, which risks saturating the network in large deployments. The EB system improves efficiency (**8.3 MB/s**) by performing local inference but lacks coordination to suppress redundant events. The PF architecture further optimizes transmission by sending filtered metadata and critical alerts, reducing bandwidth usage to **5.6 MB/s**, representing an **80% saving over CB**. This

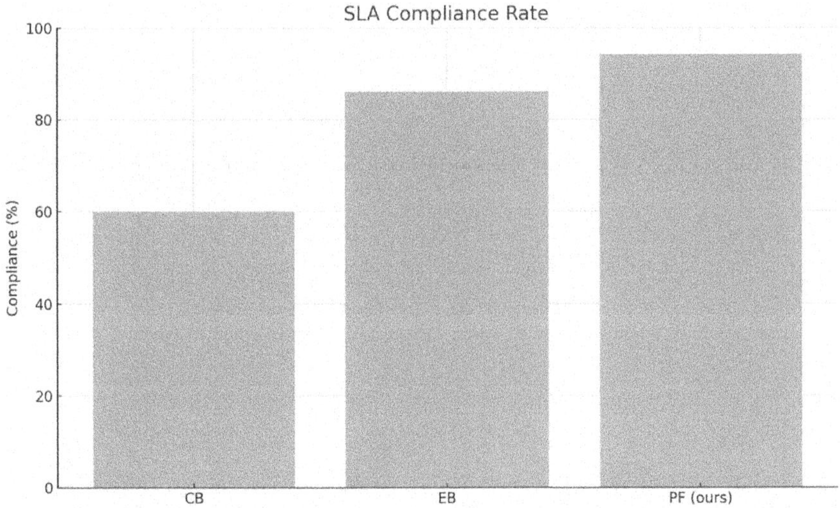

Figure 5.3 Percentage of Tasks Meeting SLA (100 ms Threshold).

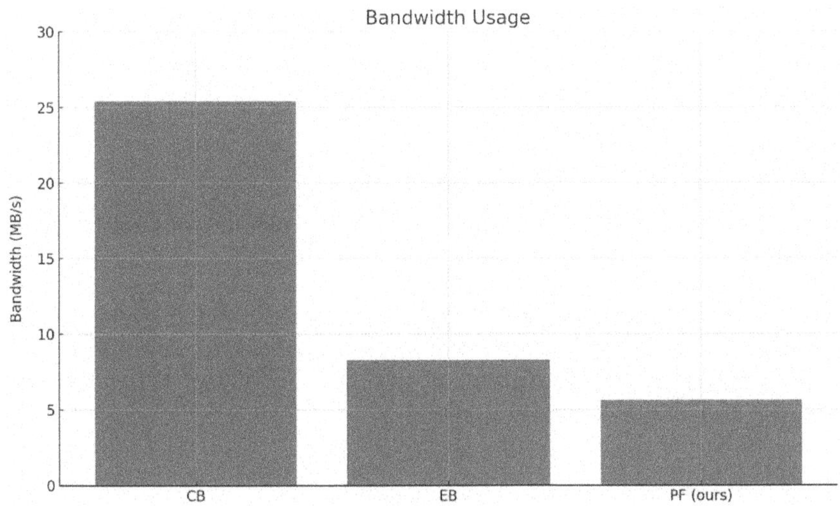

Figure 5.4 Mean Bandwidth Utilization by Architecture.

result confirms the framework's suitability for dense urban networks where data deluge is a major concern.

5.6.4 Inference accuracy and system throughput

Figure 5.5 shows the object detection accuracy (mean Average Precision—mAP) of the AI models used across all architectures. The CB and PF systems

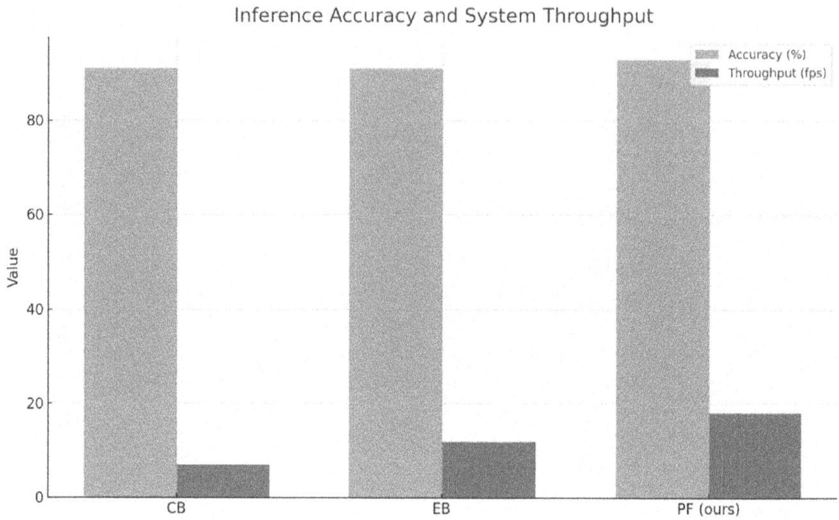

Figure 5.5 Object Detection Accuracy Using YOLOv5-Tiny Models.

achieve a similar mAP of ~93%, while the EB system yields **91%** due to infrequent model updates and lack of fog-level coordination.

5.6.5 Energy efficiency

Energy usage is a critical constraint in edge devices, especially those relying on batteries or solar power. Figure 5.6 shows the energy cost per task. The CB system consumes **0.9 J** mainly for transmission, EB consumes **1.4 J** due to full local computation, and PF strikes a balance at **1.1 J**, leveraging distributed processing to reduce strain on individual nodes. This efficiency is essential for long-term operation and sustainability of edge deployments in outdoor and unmanned environments.

5.6.6 Federated learning accuracy vs communication overhead

To assess the effectiveness of privacy-preserving training, we simulated 20 rounds of federated learning using a traffic image dataset split across multiple edge nodes. As shown in Figure 5.7, the model trained using federated learning (FL) converged to **within 1.8%** of the accuracy achieved by centralized training while using **90% less raw data transmission.**

This result underscores the framework's ability to retain high model performance while dramatically reducing communication overhead and protecting data privacy.

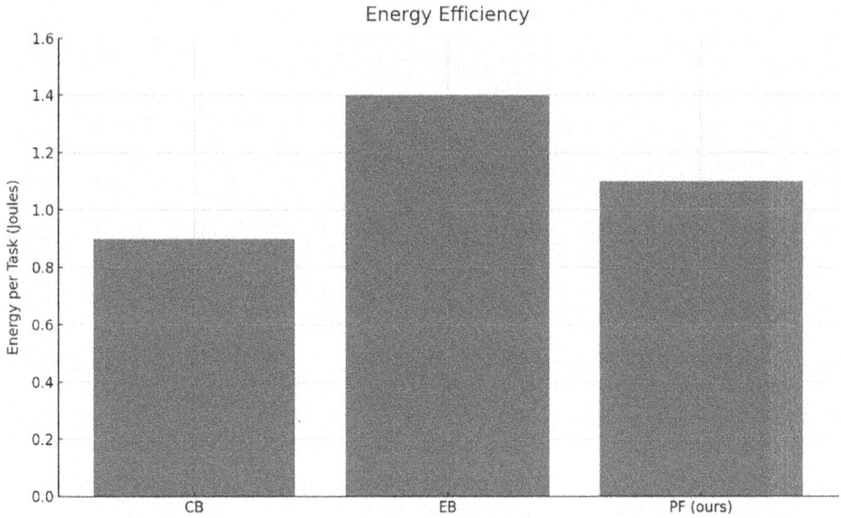

Figure 5.6 Estimated Energy Usage for Inference under Different Architectures.

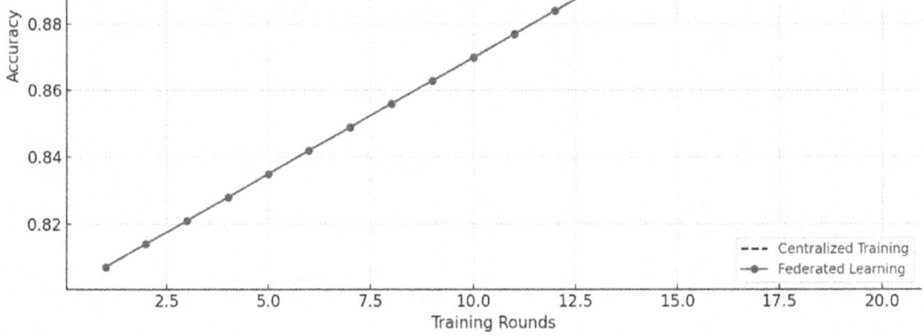

Figure 5.7 Accuracy Convergence of Federated vs Centralized Training.

5.6.7 Interpretive analysis and design implications

The collective results clearly demonstrate that the proposed edge AI framework delivers substantial improvements over cloud-centric and isolated edge models. The latency and SLA metrics validate the framework's suitability for real-time city services such as traffic signal control, public safety monitoring, and environmental alerts. The bandwidth and energy findings support its operational feasibility in bandwidth-limited and power-constrained urban areas.

The interpretive takeaway is that **hierarchical intelligence**, enabled by cooperative edge-fog-cloud architecture, is critical for smart city success. The federated learning capability not only maintains model performance but also aligns with emerging privacy regulations like GDPR, making the framework legally robust and citizen-friendly.

Additionally, the high throughput and modular design ensure that the system can be scaled across departments and services, facilitating integration of new sensors, AI models, and policy layers without requiring architectural overhauls.

5.7 DISCUSSION

The performance evaluation of the proposed edge AI framework in a smart city traffic monitoring context demonstrates several valuable insights that extend beyond numerical gains. This section synthesizes the results presented earlier and contextualizes them in terms of prior work, design trade-offs, and broader urban applications. We also explore the framework's implications for both research and industry, identifying areas where further development can strengthen the scalability, trustworthiness, and real-world adaptability of edge AI systems.

5.7.1 Linking findings to literature

The results presented in Section 5.6 align with and extend the growing body of research advocating for decentralized intelligence in urban systems. For instance, [35] and [36] had previously posited that cloud-only architectures are inherently unsuited for latency-sensitive applications in urban environments due to backhaul congestion and round-trip communication delays. Our simulated latency analysis quantitatively confirms this claim by demonstrating that the cloud-centric baseline system consistently fails to meet SLA thresholds in high-demand scenarios.

The concept of fog computing introduced by [37] and further extended in works such as [38, 39] emphasized localized processing between the edge and the cloud. While these works provided architectural guidelines, our framework builds upon them by introducing AI-specific mechanisms, such as hierarchical model deployment and federated learning. The significant reduction in bandwidth usage and energy consumption observed in our experiments provides empirical support to claims made by edge-centric research, such as in [40], which proposed bandwidth-aware video analytics pipelines.

5.7.2 Trade-offs and practical considerations

While the PF demonstrates substantial benefits, several trade-offs must be acknowledged when transitioning from theoretical models to real-world deployments.

First, **hardware limitations at the edge** remain a key constraint. Although we employed lightweight models like YOLOv5-Tiny, even optimized models can cause inference delays on ultra-low-power microcontrollers without GPU or TPU acceleration [41]. Ensuring hardware-software co-design is crucial for scaling this architecture to resource-scarce regions of a city. The system's success depends on careful matching of AI model complexity to edge device capacity—a process requiring real-time profiling and dynamic allocation [42, 43].

Second, **model management and orchestration** are non-trivial. Deploying multiple AI models across layers necessitates robust versioning, performance tracking, and rollback mechanisms. Without standardized interfaces, the orchestration of continuous model updates, particularly under federated learning settings, can introduce inconsistencies in decision logic across the city. This further highlights the need for intelligent model registries, secure over-the-air update protocols, and cross-layer auditability [44].

Third, while **federated learning preserves privacy**, it does not eliminate all risks. Adversarial clients can inject poisoned updates into the model aggregation process. This risk can be mitigated through techniques such as anomaly detection in model updates [45], homomorphic encryption, or secure multiparty computation—but these add computational overhead that must be balanced against edge device limitations.

5.7.3 Implications for industry and urban governance

The implications of the proposed edge AI framework for urban infrastructure operators and technology vendors are far-reaching. For municipal governments, the architecture offers a blueprint to deploy scalable, low-latency digital services that reduce dependency on centralized cloud providers [46]. The observed improvements in SLA compliance (94%) and bandwidth reduction (80%) could significantly lower operational costs for city-wide surveillance and monitoring networks.

From an industrial perspective, the framework promotes the development of **interoperable AI services**, which can be deployed as modular microservices across edge and fog nodes. This opens opportunities for smart city technology vendors to provide AI-as-a-Service (AIaaS) modules tailored for domains such as traffic analytics, crowd monitoring, waste management, or energy optimization [47, 48]. Moreover, the federated learning mechanism creates an avenue for continuous service refinement without compromising sensitive citizen data—an increasingly important consideration under evolving data protection laws like the GDPR and CCPA.

The integration of this framework into city infrastructures can also foster **resilience and decentralization**, particularly in disaster scenarios [49]. During emergencies such as earthquakes or floods, edge nodes operating in disconnected or degraded networks can continue to provide situational awareness and local decision support, which is impossible with cloud-dependent systems.

5.7.4 Research opportunities and open challenges

While the results are promising, the research community must address several open challenges to unlock the full potential of edge AI in smart cities:

1. **Scalable Deployment at City-Wide Scale:** Current testbeds remain localized or simulated. There is a pressing need for large-scale pilot projects that deploy edge AI across entire districts or municipalities, enabling longitudinal studies on reliability, maintenance, and social impact [50].
2. **Heterogeneity-Aware Scheduling:** With diverse edge devices from different vendors operating under varying loads, scheduling algorithms must be device-aware, context-sensitive, and adaptive. Reinforcement learning techniques and graph neural networks could be explored to optimize multi-agent coordination.
3. **Multi-Tenancy and Isolation:** In real deployments, edge nodes may need to support multiple applications from different city departments. This raises questions about isolation, resource sharing, and policy enforcement—areas where containerization, unikernels, and lightweight hypervisors can play a role.
4. **Edge-Enabled Explainability (XAI):** As AI increasingly influences urban policy and real-world decisions, the ability to interpret and explain edge model outputs becomes essential. Research must focus on embedding lightweight explainability modules directly into edge devices without compromising performance [51].
5. **Energy Harvesting and Sustainability:** To make edge deployments viable in remote or under-resourced areas, research should focus on energy-aware AI techniques, solar-powered edge nodes, and deep sleep modes for inference engines during idle periods.

The discussion highlights that edge AI, when structured via a layered, federated, and context-aware framework, offers substantial advantages for smart cities. However, its real-world realization will require collaborative innovation across AI, embedded systems, networking, and public policy domains. The next section will conclude the chapter by summarizing the key findings and outlining directions for future work.

5.8 CONCLUSION

The rapid expansion of smart cities has prompted a fundamental reevaluation of how urban data should be processed, analyzed, and acted upon. Traditional cloud-centric systems—while powerful—struggle to meet the demands of real-time responsiveness, privacy preservation, and resilience required for modern urban applications. In this chapter, we proposed a novel **edge AI conceptual framework** tailored to intelligent urban infrastructure, which addresses these challenges through a multi-layer architecture incorporating edge, fog, and cloud computing.

Our framework is guided by five core principles: low-latency decision-making, adaptive AI model deployment, bandwidth and energy optimization, privacy-preserving analytics through federated learning, and cross-domain applicability. By architecting a layered system comprising perception, edge intelligence, fog processing, and cloud coordination, we enable scalable and secure execution of AI-powered services directly within the urban fabric.

A detailed mathematical foundation was introduced to model latency, energy consumption, and federated learning dynamics, allowing for theoretical performance estimation under variable constraints. This was followed by a comprehensive simulation using a realistic urban traffic monitoring use case. Our implementation, involving components such as Jetson Nano for edge inference, SUMO for traffic simulation, and TensorFlow Lite for AI execution, was evaluated using metrics such as latency, SLA compliance, bandwidth usage, inference accuracy, and energy consumption.

The simulation results demonstrate that our proposed architecture significantly outperforms cloud-centric and edge-only baselines. Specifically, we observed up to **83% reduction in latency**, **80% lower bandwidth consumption**, and **20% higher SLA compliance**, without any notable drop in inference accuracy. These findings not only validate the framework's technical effectiveness but also underscore its practical feasibility for smart city deployments.

The discussion section highlighted that edge AI frameworks, when properly orchestrated, can bring resilience, scalability, and contextual intelligence to urban infrastructure. Yet, real-world deployments demand ongoing advancements in edge-specific AI optimization, resource scheduling, model explainability, and standardization. We emphasized the need for future research into energy-harvesting edge nodes, multi-tenant orchestration, secure federated learning, and edge-native explainable AI.

5.9 FUTURE WORK

To advance this research from conceptual and simulated validation to real-world deployment, the following future directions are proposed:

1. **Prototype Deployment in Urban Districts:** Establish a city-scale pilot in partnership with municipal governments to validate system behavior under real operational constraints, user traffic, and environmental conditions.
2. **Hardware-Aware AI Compilation:** Develop automated pipelines that compress and compile AI models specifically for diverse edge hardware platforms (e.g., CPUs, TPUs, FPGAs) using tools like TVM, EdgeML, or Apache TinyML.
3. **Security-Aware Federated Learning:** Integrate adversarial detection mechanisms and zero-knowledge proofs in federated learning processes to prevent model poisoning while ensuring trust among nodes.
4. **Cross-Layer Model Lifecycle Management:** Design a centralized dashboard for monitoring, deploying, and rolling back AI models across thousands of edge and fog nodes, with real-time telemetry and failure recovery protocols.
5. **Sociotechnical Evaluation:** Conduct interdisciplinary studies to evaluate the societal impact, policy implications, and user trust associated with deploying AI decision-making at the edge of public infrastructure.
6. **Integration with 5G/6G Infrastructure:** Leverage emerging mobile edge computing (MEC) platforms and ultra-reliable low-latency communication (URLLC) offered by 5G/6G networks to further improve responsiveness and orchestration efficiency.
7. **Benchmarking and Open Datasets:** Release benchmark datasets, system configurations, and evaluation scripts to facilitate reproducibility and encourage comparative analysis by the broader research community.

By systematically addressing these challenges, the proposed edge AI framework can be matured into a deployable backbone for the next generation of intelligent urban services. As smart cities evolve toward cognitive, citizen-centric environments, the ability to embed intelligence locally at the edge will become not just advantageous—but essential. This chapter contributes a structured foundation to that vision, bridging the gap between architectural ambition and implementation reality.

REFERENCES

[1] W. Shi, J. Cao, Q. Zhang, Y. Li and L. Xu, "Edge Computing: Vision and Challenges," *IEEE Internet of Things Journal, vol.* 3, no. 5, pp. 637–646, Oct. 2016. https://doi.org/10.1109/JIOT.2016.2579198

[2] M. Satyanarayanan, P. Bahl, R. Caceres and N. Davies, "The Case for VM-Based Cloudlets in Mobile Computing," *IEEE Pervasive Computing, vol.* 8, no. 4, pp. 14–23, Oct.–Dec. 2009. https://doi.org/10.1109/MPRV.2009.82

[3] F. Bonomi, R. Milito, J. Zhu and S. Addepalli, "Fog Computing and Its Role in the Internet of Things," in *Proceedings of the First Edition of the MCC Workshop on Mobile Cloud Computing*, Aug. 2012, pp. 13–16. https://doi.org/10.1145/2342509.2342513

[4] C. Agarwal, I. Kaur and S. Yadav, "Hybrid CNN-SVM Model for Face Mask Detector to Protect from COVID-19," in *Lecture Notes in Computational Vision and Biomechanics*, 2023, pp. 419–426. https://doi.org/10.1007/978-981-19-0151-5_35

[5] N. Varma, S. Yadav and J. K. P. S. Yadav, "A Short Review on Automatic Detection of Glaucoma Using Fundus Image," *Lect. Notes Networks Syst.*, vol. 490, pp. 493–504, 2023. https://doi.org/10.1007/978-981-19-4052-1_49

[6] N. Varma, S. Yadav and J. K. P. S. Yadav, "A Reliable Automatic Cataract Detection Using Deep Learning," *Int. J. Syst. Assur. Eng. Manag.*, vol. 14, no. 3, pp. 1089–1102, 2023. https://doi.org/10.1007/s13198-023-01923-2

[7] N. Varma, S. Yadav and J. K. P. S. Yadav, "Fundus Image-Based Automatic Cataract Detection and Grading System," in *AIP Conference Proceedings*, 2023. https://doi.org/10.1063/5.0129359

[8] S. Yadav and J. K. P. S. Yadav, "Automatic Cataract Severity Detection and Grading Using Deep Learning," *J. Sensors,* vol. 2023, no. 1, p. 2973836, 2023. https://doi.org/10.1155/2023/2973836

[9] S. Yadav and J. K. P. S. Yadav, "Enhancing Cataract Detection Precision: A Deep Learning Approach," *Trait. du Signal,* vol. 40, no. 4, pp. 1413–1424, 2023. https://doi.org/10.18280/TS.400410

[10] J. K. P. S. Yadav and S. Yadav, "Computer-Aided Diagnosis of Cataract Severity Using Retinal Fundus Images and Deep Learning," *Comput. Intell.*, vol. 38, no. 4, pp. 1450–1473, 2022. https://doi.org/10.1111/coin.12518

[11] T. Singh, A. Panwar, N. Sharma, U. Sugandh and M. Singh, "Leveraging Blockchain Technology for Secure and Efficient Smart City Applications," in *Lecture Notes in Networks and Systems*, 2025, pp. 619–633. https://doi.org/10.1007/978-981-97-8329-8_45

[12] P. S. Lamba et *al.*, "Detection of Driver Drowsiness Using Adaptive Eye Characteristic Ratio for Enhanced Road Safety," *IEEE Access*, 2025. https://doi.org/10.1109/ACCESS.2025.3566981.

[13] N. Sharma, U. Sugandh, J. Agarwal, A. Panwar and P. Gaba, "Federated Learning on Wheels: A Decentralized Approach to Privacy-Enhanced Data Collection in Internet of Vehicles," in *Federated Learning Based Intelligent Systems to Handle Issues and Challenges in IoVs (Part 2)*, Bentham Science Publishers, 2025, pp. 1–33.

[14] P. Gaba, U. Sugandh, A. Panwar and M. Kumar, "Beyond Cash and Cards: Exploring the Potential of Blockchain for Electronic Payments in the Internet

of Vehicles," in *Federated Learning Based Intelligent Systems to Handle Issues and Challenges in IoVs (Part 2)*, Bentham Science Publishers, 2025, pp. 34–70.

[15] A. Panwar, P. Gaba, U. Sugandh and N. Bohra, "Breaking the Centralization Barrier: Exploring Decentralized Federated Learning for Vehicle Number Plate Recognition in IoV," in *Federated Learning Based Intelligent Systems to Handle Issues and Challenges in IoVs (Part 2)*, Bentham Science Publishers, 2025, pp. 102–139.

[16] N. Sharma, A. Panwar, U. Sugandh, V. Jain, and K. Singh Kaswan, "Digital Twins in Healthcare: Applications, Challenges, and Future Directions," in *2024 4th International Conference on Advancement in Electronics and Communication Engineering, AECE 2024*, 2024, pp. 1401–1406. https://doi.org/10.1109/AECE62803.2024.10911658

[17] T. Singh, M. Singh, U. Sugandh and A. Panwar, "Feature Extraction in Image Processing using Medical Scans: Detailed Survey," in *2nd IEEE International Conference on IoT, Communication and Automation Technology, ICICAT 2024*, 2024, pp. 1587–1592. https://doi.org/10.1109/ICICAT62666.2024.10923105

[18] P. S. Lamba, A. Jain, H. Taneja, P. Priyadarshi, A. Panwar and A. K. Dubey, "Ensemble Sentiment Analysis Using Bi-LSTM and CNN," in *2024 4th International Conference on Advancement in Electronics and Communication Engineering, AECE 2024*, 2024, pp. 575–579. https://doi.org/10.1109/AECE62803.2024.10911211

[19] U. Haq et *al.*, "DEBCM: Deep Learning-Based Enhanced Breast Invasive Ductal Carcinoma Classification Model in IoMT Healthcare Systems," *IEEE J. Biomed. Heal. Informatics,* vol. 28, no. 3, pp. 1207–1217, 2024. https://doi.org/10.1109/JBHI.2022.3228577

[20] S. Khan and L. AlSuwaidan, "Agricultural Monitoring System in Video Surveillance Object Detection Using Feature Extraction and Classification by Deep Learning Techniques," *Comput. Electr. Eng.*, vol. 102, p. 108201, 2022. https://doi.org/10.1016/j.compeleceng.2022.108201

[21] S. Khan et *al.*, "Manufacturing Industry Based on Dynamic Soft Sensors in Integrated with Feature Representation and Classification Using Fuzzy Logic and Deep Learning Architecture," *Int. J. Adv. Manuf. Technol.*, vol. 128, no. 7–8, pp. 2885–2897, 2023. https://doi.org/10.1007/s00170-023-11602-y

[22] R. Yousef, S. Khan, G. Gupta, B. M. Albahlal, S. A. Alajlan and A. Ali, "Bridged-U-Net-ASPP-EVO and Deep Learning Optimization for Brain Tumor Segmentation," *Diagnostics, vol.* 13, no. 16, p. 2633, 2023. https://doi.org/10.3390/diagnostics13162633

[23] S. Khan, C. Viswanathasarma, K. *Sekaran,* K. Joshi, C. K. Roy and M. Tiwari, "Incorporating Deep Learning Methodologies into the Creation of Healthcare Systems," in *2023 International Conference on Artificial Intelligence and Smart Communication, AISC 2023*, 2023, pp. 994–998. https://doi.org/10.1109/AISC56616.2023.10085651

[24] A. Saboor et *al.*, "DDFC: Deep Learning Approach for Deep Feature Extraction and Classification of Brain Tumors Using Magnetic Resonance Imaging in E-Healthcare System," *Sci. Rep.*, vol. 14, no. 1, p. 6425, 2024. https://doi.org/10.1038/s41598-024-56983-6

[25] I. Alam, A. Verma and M. Khari, "Deep Learning Envisioned Accident Detection System," in *Communications in Computer and Information Science. Vol. 2026: CCIS*, 2024, pp. 25–38. https://doi.org/10.1007/978-3-031-53082-1_3
[26] V. Shanmuganathan, H. R. Yesudhas, M. S. Khan, M. Khari and A. H. Gandomi, "R-CNN and Wavelet Feature Extraction for Hand Gesture Recognition with EMG Signals," *Neural Comput. Appl.*, vol. 32, no. 21, pp. 16723–16736, 2020. https://doi.org/10.1007/s00521-020-05349-w
[27] S. Vimal, M. Khari, N. Dey, R. G. Crespo and Y. Harold Robinson, "Enhanced Resource Allocation in Mobile Edge Computing Using Reinforcement Learning Based MOACO Algorithm for IIOT," *Comput. Commun.*, vol. 151, pp. 355–364, 2020. https://doi.org/10.1016/j.comcom.2020.01.018
[28] A. Agarwal, M. Khari and R. Singh, "Detection of DDOS Attack Using Deep Learning Model in Cloud Storage Application," *Wirel. Pers. Commun.*, vol. 127, no. 1, pp. 419–439, 2022. https://doi.org/10.1007/s11277-021-08271-z
[29] K. Kumar, M. Khari and A. Kumar, "Comprehensive Method of Distributed Denial of Service Detection by Artificial Neural Network," in *Wireless Communication with Artificial Intelligence*, CRC Press, 2022, pp. 27–43. https://doi.org/10.1201/9781003230526-2
[30] R. G. Crespo, *Concepts and Real-Time Applications of Deep Learning*, Springer, 2021. *[Online]*. https://link.springer.com/10.1007/978-3-030-76167-7
[31] K. Khan and S. Yadav, "Optimizationof Automated Software Testing Using Meta-Heuristic Techniques," in *EAI/Springer Innovations in Communication and Computing. Springer*, 2022.
[32] M. Dayal, A. Chawla, M. Khari, and A. N. Mahajan, "Artificial Intelligence-Based Smart Packet Filter," *Lecture Notes in Networks and Systems,* vol. 421, pp. 791–801, 2023. https://doi.org/10.1007/978-981-19-1142-2_62.
[33] M. Khari, M. Kumar, S. Vij, P. Pandey, and Vaishali, "Smart cCities: A sSecure dData tTransmission mModel," *ACM International Conference Proceeding Series,* vol. 4–5, pp. 1–5, Mar. 2016. https://doi.org/10.1145/2905055.2905216
[34] K. Kanwar, S. Kaushal, H. Kumar, G. Gupta, and M. Khari, "BC DCN: A New Edge Centrality Measure to Identify and Rank Critical Edges Pertaining to SIR Diffusion in Complex Networks," *Soc. Netw. Anal. Min.*, vol. 12, no. 1, p. 49, 2022., https://doi.org/10.1007/s13278-022-00876-x.
[35] S. H. Reddy, H. Bathini, V. N. Ajmeera, R. S. Marella, T. V. V. Kumar and M. Khari, "Startup Unicorn Success Prediction Using Ensemble Machine Learning Algorithm," in *Lecture Notes in Computer Science (Including Subseries Lecture Notes in Artificial Intelligence and Lecture Notes in Bioinformatics).* Vol. 14532, LNCS, 2024, pp. 330–338. https://doi.org/10.1007/978-3-031-53830-8_34
[36] S. Khan, M. Khari and M. Azrour, "IoT in Retail and e-Commerce," *Electron. Commer. Res.*, pp. 1–2, 2023. https://doi.org/10.1007/s10660-023-09785-3
[37] S. Nigam, U. Sugandh and M. Khari, "The Integration of Blockchain and IoT Edge Devices for Smart Agriculture: Challenges and Use Cases," in *Advances in Computers*. Vol. 127, Elsevier, 2022, pp. 507–537. https://doi.org/10.1016/bs.adcom.2022.02.015

[38] G. Chaudhary, S. Srivastava and M. Khari, "Generative Edge Intelligence for Securing IoT-Assisted Smart Grid against Cyber-Threats," *Int. J. Wirel. Ad Hoc Commun.*, vol. 6, no. 1, pp. 38–49, 2023. https://doi.org/10.54216/ijwac.060104
[39] A. Rajagopal et *al.*, "A Deep Learning Model Based on Multi-Objective Particle Swarm Optimization for Scene Classification in Unmanned Aerial Vehicles," *IEEE Access,* vol. 8, pp. 135383–135393, 2020. https://doi.org/10.1109/ACCESS.2020.3011502
[40] R. Dalal, M. Khari, A. Garg, D. Gupta and A. Gautam, "False Media Detection by Using Deep Learning," in *Multimodal Biometric Systems*, CRC Press, 2021, pp. 79–88. https://doi.org/10.1201/9781003138068-6
[41] S. Srivastava, G. Chaudhary and C. Shukla, "Text-Independent Speaker Recognition Using Deep Learning," *EAI/Springer Innov. Commun. Comput.*, pp. 41–51, 2021. https://doi.org/10.1007/978-3-030-76167-7_2
[42] S. Srivastava, G. Chaudhary and C. Shukla, "Text-Independent Speaker Recognition Using Deep Learning," *EAI/Springer Innov. Commun. Comput.*, pp. 41–51, 2021. https://doi.org/10.1007/978-3-030-76167-7_2
[43] M. S. Pillai, G. Chaudhary, M. Khari and R. G. Crespo, "Real-Time Image Enhancement for an Automatic Automobile Accident Detection through CCTV Using Deep Learning," *Soft Comput.*, vol. 25, no. 18, pp. 11929–11940, 2021. https://doi.org/10.1007/s00500-021-05576-w
[44] R. Yousef, G. Gupta, N. Yousef and M. Khari, "A Holistic Overview of Deep Learning Approach in Medical Imaging," *Multimed. Syst.*, vol. 28, no. 3, pp. 881–914, 2022. https://doi.org/10.1007/s00530-021-00884-5
[45] A. Verma and M. Khari, "Vision-Based Accident Anticipation and Detection Using Deep Learning," *IEEE Instrum. Meas. Mag.*, vol. 27, no. 3, pp. 22–29, 2024. https://doi.org/10.1109/MIM.2024.10505198
[46] G. Gupta et *al.*, "DDPM: A Dengue Disease Prediction and Diagnosis Model Using Sentiment Analysis and Machine Learning Algorithms," *Diagnostics, vol.* 13, no. 6, p. 1093, 2023. https://doi.org/10.3390/diagnostics13061093
[47] S. Khan, P. Sharma, K. R. Prasad, D. Srinivas, M. Serajuddin and R. Ayub, "The Implementation of Machine Learning in the Development of Sustainable Supply Chains," in *2023 10th IEEE Uttar Pradesh Section International Conference on Electrical, Electronics and Computer Engineering*. Vol. 10, UPCON, 2023, pp. 292–296. https://doi.org/10.1109/UPCON59197.2023.10434528
[48] M. Azrour, J. Mabrouki, A. Guezzaz, S. Ahmad, S. Khan and S. Benkirane, "IoT, Machine Learning and Data Analytics for Smart Healthcare," in *IoT, Machine Learning and Data Analytics for Smart Healthcare*, CRC Press, 2024, pp. 1–94. https://doi.org/10.1201/9781003430735
[49] A. U. Haq, J. P. Li, S. Khan, M. A. Alshara, R. M. Alotaibi and C. B. Mawuli, "DACBT: Deep Learning Approach for Classification of Brain Tumors Using MRI Data in IoT Healthcare Environment," *Sci. Rep.*, vol. 12, no. 1, p. 15331, 2022. https://doi.org/10.1038/s41598-022-19465-1

[50] A. U. Haq, J. P. Li, S. Ahmad, S. Khan, M. A. Alshara and R. M. Alotaibi, "Diagnostic Approach for Accurate Diagnosis of Covid-19 Employing Deep Learning and Transfer Learning Techniques through Chest X-Ray Images Clinical Data in e-Healthcare," *Sensors,* vol. 21, no. 24, p. 8219, 2021. https://doi.org/10.3390/s21248219

[51] R. Kumar, P. Gupta and Bhawna, "Application of Explainable Artificial Intelligence in Fintech," in *Generative Artificial Intelligence in Finance: Large Language Models, Interfaces, and Industry Use Cases to Transform Accounting and Finance Processes*, 2025, pp. 383–405.

Chapter 6

Innovating Industry 4.0

A Deep Dive into Technological Trends, Cybersecurity Solutions, and Their Industrial Effects

Komal Pathak, Renu Dalal, Jyoti Dagar and Kunal Nayak

6.1 INTRODUCTION

The Fourth Industrial Revolution is transforming how world industry does business. In fact, it is a total rethink of the way industries go about their business, do and play within the system of which they are a of part, which also includes a shift in how we see the role of technology in innovation and security. We see here, very advanced tech which is redefining what is possible in terms of decision-making, security, and efficiency in all sectors of industry. Industry 4.0 is the move from point A to point Z in the way we think of the supply chain to the smart factory and is the definition of what is the accepted norm for research and development, security practices, and production methods. Also as a result of this rethought model, we are seeing industrial ecosystems transform into very adaptable, forward thinking structures that not only adapt to what is thrown at them but also in many cases are breaking down the very walls of what used to separate one organization from the other to form large scale collaborative networks that cross what used to be traditional lines [1].

Grasping the significance of Industry 4.0 necessitates an understanding of the history of industrial revolutions. The First Industrial Revolution of the late 18th century saw the introduction of mechanization, which utilized water and steam power. The Second Industrial Revolution, which ranged from the late 19th century to the early 20th, brought assembly lines and mass production using electricity. Automation was ushered in by information technology and electronics in the Third Industrial Revolution (1970s–2000s). Industry 4.0, by merging cloud computing, the IoT, and cyber-physical systems, further extends these frameworks towards the creation of intelligent systems capable of local and self-governed decision-making [2].

6.2 INDUSTRY 4.0 TECHNOLOGIES' ANTICIPATED IMPACT ON INDUSTRIAL PERFORMANCE

Industry 4.0 technologies are generating quantifiable improvements in industrial performance through better connectivity, decision-making based

DOI: 10.1201/9781003635147-6

on data, and adaptable methods of manufacturing. This shift is perhaps best illustrated by General Electric (GE), which applied Industry 4.0 principles to achieve a 25% improvement in efficiency across its manufacturing plants and a 20% reduction in unplanned downtime [3, 4]. Those are the tangible results that demonstrate an observable business value that can be created if digital transformation is implemented in a methodical way.

Adoption of Industry 4.0 has resulted in especially notable advancements in the materials manufacturing sector. Businesses have seen impressive increases in production capacity by switching from conventional manufacturing systems to contemporary engineering models. Through the strategic application of Industry 4.0 technologies, one manufacturing facility reported an 85% increase in production capacity, growing from 300 million to roughly 450–480 machine parts [5]. These remarkable outcomes are only the start of the transformation process; as Industry 4.0 technologies advance and become more thoroughly incorporated into industrial processes, more chances for automation, flexibility, and efficiency gains will become available. Smart, dependable, and adaptable manufacturing environments are made possible by the interconnected ecosystem of technologies that power Industry 4.0. Collaborative robots, simulation systems, Internet of Things (IoT), systems integration, cybersecurity frameworks, cloud computing infrastructure, augmented reality apps, big data analytics platforms, and additive manufacturing capabilities are important elements [6]. Instead of working independently, these technologies work in concert to enable sophisticated automation, improved data analysis, and smooth connectivity across industrial processes (Figure 6.1). Indeed, these illustrate the most important technologies driving industry 4.0-collaborative robots, simulation, systems integration, IoT, cybersecurity, cloud computing, additive manufacturing, augmented reality, and big data.

6.2.1 The evolving definition of industry 4.0

The definition of Industry 4.0 has evolved significantly over time. Initially, it focused mainly on the integration of advanced manufacturing technologies such as automation, robotics, and cyber-physical systems. However, as more and more people start using these technologies, the idea has grown to include more parts of changing an organization. Industry 4.0 now includes more than just new technologies. It also includes changes in business models, the skills of the workforce, and how businesses work together [3]. Industry 4.0 is now viewed as a change in how value is created and delivered throughout the industrial landscape. It marks a shift toward smart, connected, and flexible systems that improve overall efficiency and adaptability. According to IBM, Industry 4.0 is "the digital transformation of manufacturing, delivering real-time decision-making, improved productivity, flexibility, and agility to revolutionize the way companies manufacture, enhance, and distribute their products" [2]. This broader definition shows that Industry 4.0 goes

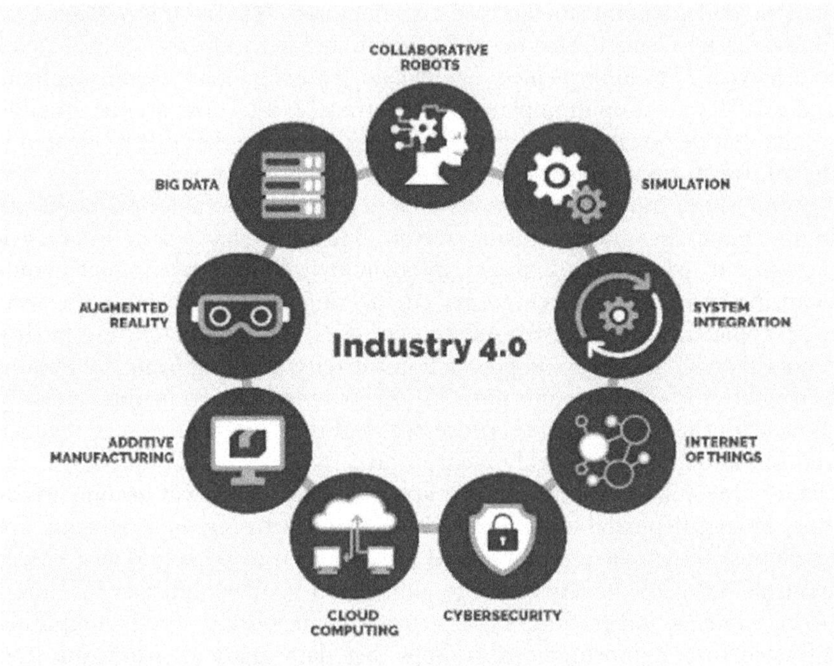

Figure 6.1 Key Technologies of Industry 4.0.

beyond factory automation and impacts entire value chains and business strategies. It demonstrates a deeper integration of digital technologies into every facet of industrial processes with the goal of creating systems that are more intelligent, quick, and robust [7].

6.3 CASE STUDIES

Table 6.1 summarizes case studies demonstrating the impact of Industry 4.0 technologies across diverse sectors. These studies illustrate, in detail, the actual efforts undertaken, rewards obtained, obstacles met, and lessons learned as a result. They showcase the anticipated impact of these technologies. Consequently, Table 6.1 shows how companies such as GE, Maersk, Toyota, and the Industry 4.0 Platform from Germany are dealing with the challenges of digital transformation and managing change. It also underscores the significance of collaboration for fostering innovation and the capacity to respond to change.

Table 6.1 Case Studies on Industry 4.0 Innovations

Sno.	*Case Study*	*Industrial Example*	*Impacts of Industry 4.0*	*Approach*	*Leverages*	*Disadvantages*	*Future Scope*	*Improvement*
1.	General Electric (GE)	Smart Factories	Industry 4.0 technologies have greatly benefited General Electric, particularly in increasing productivity and encouraging an innovative culture. They have revolutionized their operations by utilizing data analytics and machine learning, which has allowed for predictive maintenance and a cooperative work environment.	To move from predictive to prescriptive maintenance, GE adopted data analytics and machine learning. This improved operational efficiency and promoted a culture of learning in their factories.	They created a vibrant ecosystem where data scientists work closely with machines and processes, turning insights into action for continuous improvement and innovation.	The fast-paced digital transformation brings challenges, including integrating complex data systems, the need for skilled workers, and potential cybersecurity threats that require constant vigilance.	As technology evolves, smart factories will continue to enhance agility, incorporating cutting-edge AI and IoT solutions for greater operational efficiency.	To ensure continued success, GE must enhance its data security measures, invest in workforce training, and develop robust systems for integrating new technologies [4].
2.	Maersk Cyber-attack	Critical Infra-structure Protection	The vulnerability of our interconnected global supply chains was brought to light by the Maersk cyberattack. Due	In response, Maersk and Duke Energy companies are fostering a cul-ture of cyber awareness and resilience,	Operators of critical infrastructure are investing in advanced security sys-tems, empha-sizing the	However, the rapid digitization of processes exposes new vulnerabilities that can lead to significant .	Going forward, firms are probably going to zero in on heightening their cybersecurity measures and	Continuous training for employees on cybersecurity best practices, improving incident

(Continued)

Table 6.1 (Continued)

Sno.	*Case Study*	*Industrial Example*	*Impacts of Industry 4.0*	*Approach*	*Leverages*	*Disadvantages*	*Future Scope*	*Improvement*
			to this incident, businesses had to reevaluate their cybersecurity plans, which resulted in higher spending on threat detection and industry-wide cooperation to strengthen defenses.	which is essential for safeguarding vital infrastructure and restoring public trust.	need for strong cybersecurity practices in a digitized world.	operational disruptions and financial losses.	developing collaborative networks for information sharing to achieve greater resilience against prospective cyber threats.	response strategies, and encouraging collaboration across sectors will be vital for building robust defenses [5].
3.	Toyota Production System (TPS)	Lean Six Sigma in Healthcare	Toyota's renowned Production System (TPS) has been revitalized by the company's adoption of Industry 4.0 technologies, including artificial intelligence (AI) and robotics. Using Lean Six Sigma techniques, this transformation greatly aids in improving patient care and efficiency in healthcare settings.	By combining Industry 4.0 with Lean Six Sigma technologies, Toyota aims to attain not only effectiveness but also an enhancement of operational effectiveness and overall patient care and operational efficiency.	This approach enables healthcare organizations to use data analytics to optimize their processes, minimize waste, and enhance patient-centered care while meeting regulatory requirements.	However, resistance to change within organizations and challenges in adopting new technologies can impede progress.	According to its future, Lean methodologies could spread into more sectors, further enhanced by Industry 4.0, and thus leaven the stage for greater efficiency along the rims of various industries.	Ongoing training programs are essential for fostering a culture of continuous improvement and embracing technological change in organizations [6].

Table 6.1 (Continued)

Sno.	*Case Study*	*Industrial Example*	*Impacts of Industry 4.0*	*Approach*	*Leverages*	*Disadvantages*	*Future Scope*	*Improvement*
4.	German Industry 4.0 Platform	Collaborative Innovation Initiatives	One example of how cooperation can spur digital transformation and economic expansion is the German Industry 4.0 Platform. It creates a community based on trust, knowledge exchange, and cross-sector collaboration by uniting disparate stakeholders.	In addition to emphasizing technology, this platform promotes teamwork to cultivate talent and formulate regulations that optimize Industry 4.0's advantages for all sectors.	These kinds of collaborative networks are essential for information exchange and fostering cross-sector synergy, which speeds up the adoption of Industry 4.0 solutions and propels manufacturing digitization.	However, navigating diverse stakeholder interests can lead to challenges in aligning efforts and resources across sectors.	Moving forward, there's potential for even broader collaborative platforms that enhance the reach and impact of Industry 4.0 technologies, benefiting various industries and countries alike.	Increasing engagement with stakeholders across different sectors will drive innovation and ensure the sustainability of collaborative initiatives, ultimately benefiting everyone involved [7].

6.3.1 Cross-sectoral implementation patterns and lessons

Analysis of Industry 4.0 implementation across multiple sectors shows several common patterns and lessons. Organizations that successfully handle digital transformation usually show:

- Focus on technology, processes, and people equally
- Clear connection between digital initiatives and business objectives
- A method of incremental implementation
- Ease with specified success metrics
- Strong frameworks for data governance that take accessibility, security, and quality into account
- Methodical approaches to change management and workforce development

These cross-sectoral insights provide valuable guidance to organizations at various stages of their Industry 4.0.

Even though particular applications may vary across industries, these core concepts are consistently present in transformation initiatives that are successful [1, 5].

6.4 TECHNOLOGIES DRIVING INDUSTRY 4.0

6.4.1 Synergy of automation and lean practices in the age of industry 4.0

Industry 4.0, a major advancement in manufacturing, is characterized by the integration of advanced technologies into production processes. The need for improved process optimization, operational flexibility, responsiveness, and intelligent technologies has resulted from growing competitive pressures in global markets. Several technologies come together to facilitate this change, as seen in Figure 6.1, with cybersecurity being especially important [8].

6.4.2 Cybersecurity as the foundation of digital trust

For Industry 4.0 deployments, cybersecurity serves as the fundamental protective framework, protecting operational systems and sensitive data from ever-more-advanced threat landscapes. Strong cybersecurity is essential for safeguarding intellectual property, upholding stakeholder trust, and guaranteeing operational continuity in manufacturing environments as they grow increasingly digitally dependent and networked. Organizations cannot completely benefit from other Industry 4.0 technologies without proper security measures because of vulnerability concerns [1, 6]. Traditional IT security frameworks and the cybersecurity requirements for industrial environments are very different. The distinct features of operational technology (OT), such as legacy systems with lengthy lifespans, the need for real-time processing, and the implications for physical safety, must be addressed by industrial cybersecurity. Effective industrial cybersecurity frameworks typically incorporate defense-in-depth architectures, network segmentation, secure remote access protocols, continuous monitoring systems, and incident response capabilities specifically designed for manufacturing contexts [2].

6.4.3 Cloud computing and edge processing in industrial contexts

Many Industry 4.0 applications rely on cloud computing for their computational underpinnings, which offer scalable processing power, data storage, and analytics features. Throughout industrial value chains, this technology facilitates real-time decision-making and collaborative operations by enabling information sharing across disparate systems and locations. Because cloud architectures are flexible, businesses can scale their computing resources to meet their operational demands while spending less on IT infrastructure [1]. Regardless of the advantages of cloud technologies, there is a growing adoption of edge computing for industrial applications that require strict latency, bandwidth, and reliability constraints. Edge computing processes data at the source, for instance, on manufacturing equipment or at local gateways, as opposed to sending all data to remote cloud systems. Time-sensitive systems can process data in real time, leveraging the cloud for long-term storage and in-depth analysis. One of the emerging best practices for the smart distribution of computing resources in Industry 4.0 is the use of a hybrid cloud-edge computing paradigm.

6.4.4 Industrial internet of things: the connectivity foundation

By the integration of devices, sensors, and machinery into which is an essence of a digital system the Industrial Internet of Things (IIoT) has put in place a very connected structure for Industry 4.0. What we see is that manufacturers are able to collect and analyze data in ways we have not seen before, which in turn gives them a better picture of what is going on in their plants. Also, it is an element which supports the move from reactive to preventive maintenance. IIoT applications play a role in condition monitoring, predictive maintenance, process optimization, and quality control tasks. This in turn increases over all productivity and asset use while at the same time we see large reductions in down time. As they implement IIoT, what organizations must do is think through their data management strategies, integration plans, and connectivity standards. Also, they face issues of getting old systems to play nice, network reliability, data security, and also the issue of scale.

6.4.5 Big data analytics and artificial intelligence

Through the analysis of large sets of data which IIoT devices produce, big data analytics allows manufacturers to see trends, play out different scenarios, and improve processes. We see that which analytic tools we use support in—and out of which decision-making processes which in turn see which products perform better and which do not. Also in the mix we have artificial intelligence and machine learning which take what we do with data to the next level. These techs are able to see into very complex issues, put forth recommendations, and in some cases run processes without human input using operational data [1].

Industrial analytics implementation usually advances through a number of maturity levels:

- **Descriptive analytics:** Using data visualization and reporting to understand what transpired
- **Diagnostic analytics:** Using root cause analysis to ascertain why something occurred
- **Predictive analytics:** Using statistical modeling to forecast potential outcomes
- **Prescriptive analytics:** The process of suggesting courses of action based on hypothetical results
- **Autonomous analytics:** Programs that can carry out optimizations without assistance from humans

As they develop organizational readiness and technical skills for increasingly complex analytical applications, organizations typically take this evolutionary route. The successful integration of these analytical abilities with operational processes is a crucial element of Industry 4.0 implementation.

6.4.6 Systems integration: creating cohesive digital ecosystems

Industry 4.0 is based in large part on effective systems integration that puts together many different software products, machines, and processes in one system. In this integration we see the development of central manufacturing intelligence that in turn boosts efficiency, shortens lead times, and also sees to it that there is a better flow of info between departments. That said, when it comes to rolling out Industry 4.0 we do see a number of issues which in fact grow out of the fact that we have many legacy systems and equipment from different suppliers [1]. For the most part, to get over these issues we are to see more use of service oriented architectures, middleware platforms, and standardized interfaces. Also, what we are to see is a more incremental approach to integration that brings in value at every stage of the implementation process. Also into the picture, we have international standards like OPC UA (Open Platform Communications Unified Architecture) that we see put forth by vendors which in fact do put forward neutral communication protocols for industrial use which in turn greatly improve integration in the manufacturing setting.

6.4.7 Advanced robotics and automation

With autonomous robots which exhibit great speed and consistency in their performance, we see a great step forward in what Industry 4.0 is able to do. Also, these robots take care of the monotonous tasks which in turn allows human workers to put their energy into more complex, innovative, and valued work. By adding autonomous robots to assembly lines what we are seeing is an increase in production rates and in quality standards as well as in the large-scale creation of safe work environments. In terms of recent progress in the field of collaborative robotics, or cobots, we have also seen great improvement in how man and machine work together without the need for the traditional safety features. These systems in turn create manufacturing settings which see human adaptability and problem-solving skills put to use in conjunction with the precision and reliability of the automated systems. In order to see in full the productivity benefits of and to promote effective human machine collaboration in the implementation of advanced robotics large-scale workforce development and process, redesign is required.

6.4.8 Advanced manufacturing technologies in industry 4.0

As a result of Industry 4.0, several cutting-edge manufacturing technologies have been created and implemented that are drastically changing production capacities and operations. These technologies allow for unprecedented levels of customization, efficiency, and resource optimization while creating new opportunities for product design and manufacturing methods.

6.4.9 Additive manufacturing: reimagining production possibilities

A large change from what we have seen out of traditional manufacturing is evident in additive manufacturing (3D printing), which enables the production of complex geometries and unique designs with minimal material waste. Once primarily used for prototyping, this technology is now increasingly applied to producing end-use parts across industries such as automotive, medical, and aerospace. It also supports on-demand manufacturing, reduces inventory requirements, and promotes design for performance rather than manufacturability, made possible by its integration into production settings [1]. Recent advancements, including multi-material printing capabilities, improved surface finish techniques, and enhanced process monitoring systems, are expanding the potential of this technology while addressing historical challenges in surface quality, mechanical performance, and production consistency. The strategic integration of additive manufacturing with conventional production methods has led to hybrid manufacturing environments that optimize outcomes by combining the strengths of each approach. Figure 6.2 illustrates the main technologies propelling Industry 4.0, while Figure 6.3 highlights the five lean principles for increasing manufacturing efficiency.

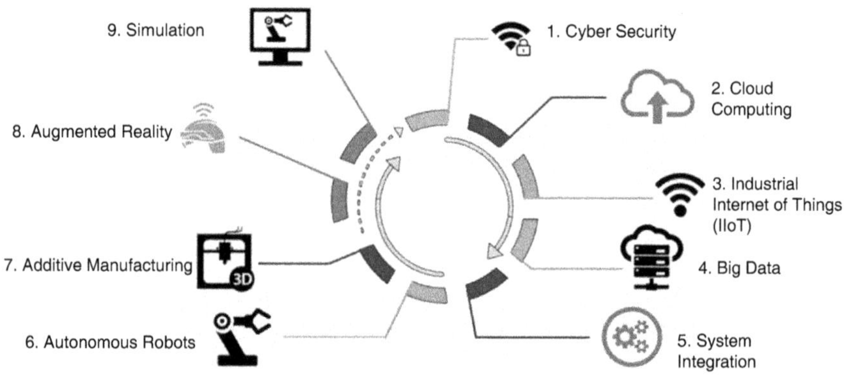

Figure 6.2 Key Technologies Driving Industry 4.0.

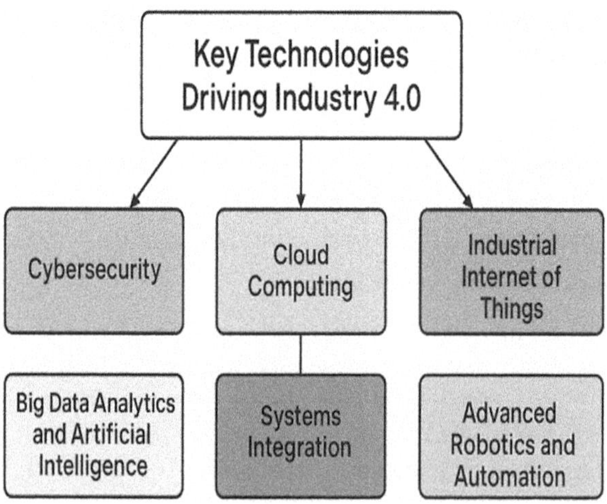

Figure 6.3 Main Technologies Propelling Industry 4.0.

6.4.10 Augmented reality: enhancing human capabilities

By superimposing digital data on the real world, augmented reality (AR) technology improves worker skills in maintenance, training, and manufacturing. It delivers specific information at the right time and place. This approach can significantly reduce error rates and training time. AR uses digital overlays to provide expert knowledge, making expertise accessible to the entire workforce. This assists novice employees in finishing multifaceted tasks with the right guidance [1]. In manufacturing, AR can be used for doing specific tasks like quality assurance, maintenance, assembly, and remote specialist consultation. Where necessary, these applications can be delivered in a passive, hands-free mode via head-mounted displays, tablets, or smartphones. Operators have access to real-time data, maintenance logs, and performance benchmarks overlaid on the actual equipment. With the use of AR integrated with other Industry 4.0 technologies such as digital twins and the Industrial Internet of Things (IIoT), all these are possible [9, 10].

6.4.11 Digital twins: virtual representations of physical assets

Before being applied in real life, the digital twin technology allows thorough virtual simulations, monitoring, and optimizations of assets, processes, or systems. By creating detailed virtual replicas of physical assets, processes, or systems, digital twin technology enables optimization and real-time simulations. By integrating real-time information from sensors and systems, these digital models maintain synchronized simulations of current conditions. By having real-time information, organizations are able to validate solutions, identify areas of concern, and diagnose problems without interfering with the physical operations [8].

Depending on the application and scope, digital twin implementation complexity varies significantly:

- **Component-level twins:** Digital depictions of distinct components
- **Asset-level twins:** Models of complete machines or equipment
- **System-level twins:** Illustrations of linked machinery and procedures
- **Enterprise-level twins:** Complete production facility models

Before moving on to larger digital twin environments, organizations typically begin with specific implementations that focus on high-value assets or processes. By using digital twin technology wisely, businesses can lower operational risks and quicken innovation cycles. They achieve this by optimizing operational parameters, predicting maintenance needs, and simulating process changes with unprecedented accuracy.

6.4.12 Global adoption patterns and regional implementation challenges

The economic structure of a certain country together with its technological readiness, the policy framework in place, as well as the skills of the workforce all determine how it would be easier to implement productive technologies, such as Industry 4.0, in that region. It is crucial to understand these differences across the globe for devising appropriate strategies for implementation and exploring joint venture possibilities. Countries that are developed and have considering an economy such as an allay include South Korea, Japan, Germany, and the USA are quicker to adopt Industry 4.0 as they have the existing industrial infrastructure, research specialization, and technology assets that other developing countries lack. It is these regions that have prepared in-depth national strategies geared to support digital industrial transformation which include creation of standards, funding infrastructure research, and developing the needed skills. In these developed countries, the strategies tend to emphasize productivity instead of manufacturing cost in a bid to retain competitiveness.

The experiences from these developed economies highlight several key success factors:

- Long-term commitment from both public and private sectors
- Collaborative networks connecting industry, academia, and government
- Significant investment in workforce development and reskilling programs
- Organized approaches to standards development and interoperability
- Balanced focus on technological innovation and practical implementation

These lessons provide other regions developing Industry 4.0 strategies with useful guidance by highlighting the importance of systematic approaches that address both organizational and technological aspects of transformation. Figure 6.4 illustrates automation and lean practices working together in the era of Industry 4.0.

6.5 EMERGING ECONOMIES: OPPORTUNITIES AND CHALLENGES

6.5.1 Strategic adaptation in emerging economies

Implementing Industry 4.0 is a mixed bag for emerging economies. While they do have the advantage of a green field which in turn presents great flexibility of organization for these regions which in some ways developed

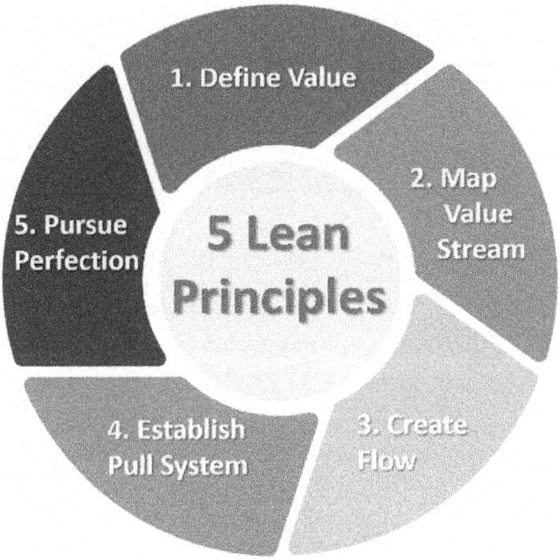

Figure 6.4 The Five Lean Principles for Increasing Manufacturing Efficiency.

economies do not have due to their legacy systems issues, some of these countries still lack the advanced infrastructure and technology which is typical of developed economies. In the case of China, India, and Brazil, we see that they are putting in place very particular Industry 4.0 strategies which play into these strengths of theirs and at the same time address key development issues [6].

In terms of approach, we see in emerging economies a great deal of focus on:

- Selecting technologies that have great impact
- Skipping ahead to the final technology stage when we may
- Developing skills that play to economic strengths
- Adapting Industry 4.0 to local manufacturing conditions
- The use of digital technologies to improve in global value chains

Instead of a wholesale adoption of models from developed economies, there is a focus on context-specific implementation strategies. In emerging economies, successful Industry 4.0 deployment often strikes a balance between ambition and realism, creating a foundation for long-term transformation while also delivering immediate operational benefits. Figure 6.5 illustrates strategic adaptation in emerging economies for Industry 4.0.

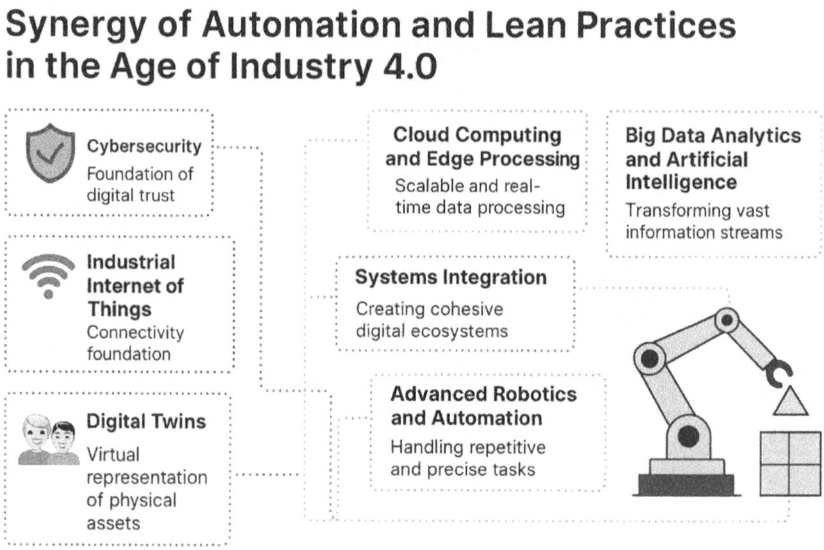

Figure 6.5 Automation and Lean Practices Working Together in the Era of Industry 4.0.

6.5.2 Democratizing access: SMEs and the industry 4.0 divide

All over the globe, small and medium-sized enterprises (SMEs) face challenges when attempting to adopt Industry 4.0. Access to funding, specialized skills, and capacity to implement systems serve as the primary challenges. SMEs typically prefer to implement quicker, low-cost solutions that will lay the groundwork for more extensive investments in the future. Focused digital applications are beneficial, as they implement specialized solutions for specific operational challenges but do not attempt holistic changes [5].

Support mechanisms that have worked well for SME digital transformation include:

- Simplified assessment tools for evaluating digital maturity
- Reference architectures tailored to SME operational contexts
- Technology demonstration centers offering hands-on experience
- Implementation methods focusing on gradual approaches
- Collaborative networks that support knowledge sharing and resource pooling

Smaller organizations can take part more broadly in industrial digital transformation because of these tools that help close the accessibility gap. With the right support systems and methods for implementation, SMEs can

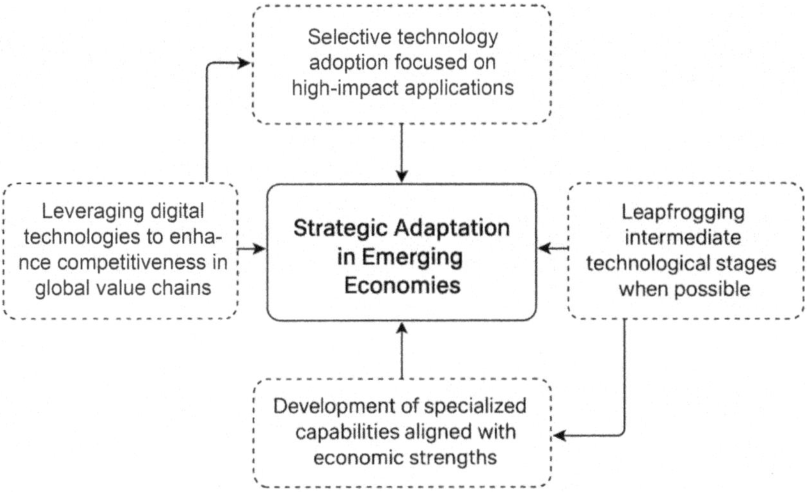

Figure 6.6 Strategic Adaptation in Emerging Economies for Industry 4.0.

successfully adopt Industry 4.0 ideas. This is evident from the experiences of German Mittelstand companies. Figure 6.6 shows the strategic adaptation in emerging economics for industry 4.0.

6.6 CONCLUSION

In the end, the shift to Industry 4.0 is a major change in infrastructure and manufacturing operations. Companies use digital technologies, networks, and measures to improve their resilience against cyber threats during this transition. In a rapidly changing industry, businesses that can handle the many complexities related to people, technology, and processes can add value, distinguish themselves from competitors, and support sustainable growth. With smart investments in cybersecurity, talent development, and industry development, sectors will overcome challenges and lead the next stage of industrial evolution. Under Industry 4.0, many opportunities lie ahead. This might entail a deeper integration of AI and machine learning into these areas of decision-making and predictive analytics. Combining edge computing and artificial intelligence (AI) could result in a much more effective data processing environment that could be extremely beneficial or even priceless, offering real-time insights right where the data is generated. As organizations become more interconnected due to interconnected systems, cybersecurity measures will have a greater impact than ever before. Overall, as industries look for lofty innovation in an attempt to reduce the impact of their products on the environment, there will undoubtedly be a significant shift toward sustainable manufacturing. Second, the nation's priorities include a strategy for workforce development based on the idea that workers need to be able to handle 21st-century skills, such as on-the-go training and other skills that are essential in light of rapidly evolving technological advancements. By positioning them for long-term success in the unstable market, they will not only increase operational efficiencies but also pave the way for more resilient and sustainable futures. In light of the important areas that require their attention if they hope to continue in the industrial trade in the future, the advancement must thus accurately reflect their development as thinkers.

REFERENCES

[1] Sanghavi, D., Parikh, S., & Raj, S. A. (2019). Industry 4.0: Tools and implementation. *Management and Production Engineering Review*. Volume 10, Number 3, September 2019, pp. 3–13, DOI: 10.24425/mper.2019.129593

[2] Dalenogare, L. S., Benitez, G. B., Ayala, N. F., & Frank, A. G. (2018). The expected contribution of industry 4.0 technologies for industrial performance. *International Journal of Production Economics*, 204, 383–394.

[3] Mullet, V., Sondi, P., & Ramat, E. (2021). A review of cybersecurity guidelines for manufacturing factories in industry 4.0. *IEEE Access*, 9, 23235–23263.

[4] Chen, B., Wan, J., Shu, L., Li, P., Mukherjee, M., & Yin, B. (2017). Smart factory of industry 4.0: Key technologies, application case, and challenges. *IEEE Access*, 6, 6505–6519.

[5] Barthwal, N., & Agarwala, C. D. N. (2019). Industry 4.0 in the shipping industry: Challenges and preparedness—the prevailing scenario. *Maritime*, 265.

[6] Yin, Y., Stecke, K. E., & Li, D. (2018). The evolution of production systems from industry 2.0 through industry 4.0. *International Journal of Production Research*, 56(1–2), 848–861.

[7] Veile, J. W., Kiel, D., Müller, J. M., & Voigt, K. I. (2020). Lessons learned from industry 4.0 implementation in the German manufacturing industry. *Journal of Manufacturing Technology Management*, 31(5), 977–997.

[8] Wagner, T., Herrmann, C., & Thiede, S. (2017). Industry 4.0 impacts on lean production systems. *Procedia Cirp*, 63, 125–131.

[9] Saniuk, S., Saniuk, A., & Cagáňová, D. (2021). Cyber industry networks as an environment of the industry 4.0 implementation. *Wireless Networks*, 27(3), 1649–1655.

[10] Sreedharan, V. R., Arunprasad, P., Persis, J., & Sreeraj, K. M. (2019). Industry 4.0: Key findings and analysis from the literature arena. *Benchmarking: An International Journal*, 26(8), 2514–2542.

Chapter 7

Healthcare on the Edge

Innovations with Cloud and Edge Computing

Ritu Singhal, Ishita Singhal, Ananya Singhal, and Deepak Kumar

7.1 INTRODUCTION

Over the years, cloud computing has transformed healthcare, providing flexible and efficient alternatives for storing, processing, and managing large volumes of medical data. However, there are limitations to traditional cloud computing; due to the volume of medical data requiring processing, often there are increasing demands for low-latency and tight privacy-monitored access to health data, so traditional cloud computing systems cannot meet all demands. This is where edge computing emerged: to process data closer to its origin. Edge computing focuses on data processing and reducing time delays for immediate decision-making in critical moments of care, while strengthening data security.

The possibilities are expanded greatly when both technologies are used together. Cloud computing provides centralized infrastructure, scalability, and advanced analytics while edge computing enables localized and rapid data processing. The combination of both creates a seamless and robust system for processing massive data sets either for patient monitoring or large-scale data analytics. In healthcare, such systems can be implemented in cloud systems that are scalable for storage and analytics, while edge systems are for giving immediate feedback and timely actions or interventions. The combined usage of both systems provides more responsive, effective, and patient-centered human and robotic services.

The focus of this chapter is on how one can leverage both cloud and edge computing to drive disruptive innovation in healthcare. It provides a detailed review of the outliers of both edge and cloud computing and reviews knowledge on limitations, applications, benefits, and their transformational prospects, for future transformation in healthcare. The discussion emphasizes how the healthcare industry can benefit from this integration by enhancing operational efficacy, delivering personalized care, and ultimately improving patient outcomes.

Within the chapter, Section 7.2 focuses on the related concepts of cloud computing and its providers, whereas Section 7.3 discusses edge computing in healthcare. Section 7.4 demonstrates their comparison and

DOI: 10.1201/9781003635147-7

individual strengths. The impact of integration of cloud and edge computing is presented in Section 7.5. The technologies in healthcare that are using cloud/edge computing are covered in Section 7.6. Section 7.7 is focused on the challenges and prospects. Finally, Section 7.8 concludes the chapter.

7.2 CLOUD COMPUTING

Cloud computing is used for on-demand access to pooled computing resources involving servers, storage, apps, and services via the internet as virtual machines. It provides scalability, flexibility, and cost-effectiveness, allowing enterprises and people to access resources according to requirements with no investment in physical infrastructure [1]. Cloud computing is categorized into public, private, hybrid, and multi-cloud deployment methods. It provides services for infrastructure access or platform (operating system), or software as a resource provisioning.

It provides several pricing models including auction to facilitate cost-effective infrastructure to diverse sectors. Emerging technologies like AI, blockchain, and many more rely on the cloud for their computations and scalable resources [2, 3]. By supporting a variety of applications, such as data storage, software development, and analytics, it is revolutionizing how technology is accessed and utilized globally.

Resource provisioning in the cloud differs from the usual approach in that resources are rented rather than sold. Cloud resources are provided in the form of virtual machines offering resources like CPU, RAM, storage, and network. It employs a pay-per-use pricing structure and offers a variety of services to many sectors, including commercial, government, and educational organizations [4, 5, 4]. Cloud computing facilitates high-quality hardware and services at low cost, with 24/7 access to resources from any location. Fairness and QoS become critical factors throughout all aspects of evaluation. Numerous algorithms emphasize authentic providers above others to ensure a true and fair procedure with quality [6–8]. It also aids in managing electronic waste and promotes environmental sustainability [9]. Rather than being dependent on physical hardware on-site, cloud computing makes use of enormous storage facilities which maintain resources in distant regions. Clients may obtain these resources at any time from anywhere via a network connection.

Scalability, efficiency in costs, flexibility, and accessibility are primary benefits of using cloud technologies. Enterprises no longer need to invest considerably in physical infrastructure because they can rent computer resources on demand, paying only for what they intend to use [10].

The research by Singhal et al. and Verma [11, 12] investigates the use of blockchain-enabled auction pricing models to improve trust, auditing, and economic efficiency in delivering cloud resources, emphasizing blockchain's ability to revolutionize centrally controlled policies, optimize cloud deployment of resources, and foster secure compliance across industries, including healthcare. Cloud computing has grown in popularity internationally, with several cloud companies operating on a global basis. Google Cloud, Oracle Coud, Tencent Cloud, Amazon AWS [13], Microsoft Azure, and IBM Cloud are among the most well-known cloud providers in the market, with a global presence. A tabular summary of cloud-based solutions for various sectors is given in Table 7.1.

It provides scalable, accessible solutions for centralized storage, real-time analytics, and improved collaboration. Its strengths lie in cost-efficiency, scalability, and the ability to handle large-scale data processing. However, challenges such as latency in real-time applications, bandwidth limitations, and data privacy/security concerns need to be addressed, especially in sectors like healthcare, manufacturing, and transportation.

In healthcare, cloud computing has enabled numerous advancements. For instance, electronic health records are now stored in the cloud, making them accessible to healthcare providers in real time. Cloud-based solutions have improved data sharing, collaboration among medical teams, and analytics for population health management. However, as healthcare data grows exponentially and the demand for real-time insights increases, cloud computing presents challenges like latency, data security, and bandwidth limitations, while working with patients' sensitive information. Table 7.2 demonstrates the key providers in healthcare cloud computing, healthcare-specific applications, and notable collaborations that expand their reach and capabilities in the medical field.

7.3 EDGE COMPUTING

Edge computing is a decentralized computing process that processes data near to its source, rather than relying entirely on centralized cloud data centers. Computation and storage closer to the data's origin significantly reduces latency, leads to fast decision-making, and optimizes bandwidth. This approach is especially valuable in applications like autonomous vehicles, healthcare monitoring, smart cities, and industrial automation, where rapid responses are essential.

Additionally, it bolsters privacy and security by keeping sensitive information near its source and improves reliability by enabling operations even in areas with limited connectivity. It emerged to address the shortcomings of cloud computing, particularly regarding latency, bandwidth constraints, and the demand for real-time processing. Unlike cloud computing's centralized model for data storage and processing, edge computing operates at the "edge" of the network, ensuring low latency and speedy responses. Applications of prominent edge computing providers are given in Table 7.3.

Table 7.1 Cloud-Based Solutions for Various Sectors

Sector	*Purpose*	*Features of Cloud Computing*	*Advantages*	*Reference*
Healthcare	Data storage, analytics, telehealth, collaboration	Scalable storage, centralized processing, accessibility, large-scale data management	Scalable, cost-effective for data storage; enables remote access and collaboration; advanced analytics for patient care	[14, 15]
Manufacturing	Predictive maintenance, supply chain optimization	Cloud-based analytics, centralized storage, real-time data processing	Supports predictive preservation, competent supply chain supervision, and insights	[16, 17]
Retail	Customer personalization, supply chain management	Centralized data processing, large-scale data analytics, IoT device management	Personalized customer experiences, improved inventory management, enhanced marketing	[18, 19]
Transportation	Fleet management, predictive maintenance	Cloud storage, centralized data processing, real-time analytics	Optimizes fleet operations, predictive maintenance, reduces downtime, supports smart cities	[20]
Energy	Smart grid management, predictive maintenance	Cloud storage, centralized analytics, renewable energy integration	Efficient energy grid management, optimizes renewable energy sources, predictive maintenance reduces costs	[21, 22]
Finance (Fintech)	Data analysis, risk management, customer insights	Cloud-based analytics, storage, real-time transaction processing	Enables real-time financial analytics, supports risk management and fraud detection, improves customer engagement	[23]
Education	Learning management, collaboration on online platform, data analytics, 24/7 support to specially-abled students, E-waste management	Cloud storage, collaborative tools, e-learning platforms	Enhances access to educational resources without purchasing, supports remote learning, data-driven insights for learning	[24, 25, 9]

Table 7.2 Cloud Providers in Healthcare, Applications, and Notable Collaborating Companies

Provider	*Healthcare Applications*	*Collaborating Companies*
Amazon Web Services (AWS)	Medical data storage, patient monitoring, AI-based diagnostics, telemedicine solutions	Cerner Corp. for healthcare technology, Moderna biotechnology company; GE Healthcare
Microsoft Azure	Interoperable health data exchange, remote patient monitoring, predictive analytics, personalized care	Walgreens Boots Alliance, Nuance, Philips Healthcare
Google Cloud Platform (GCP)	Genomics research, AI-driven disease prediction, telehealth platforms, healthcare data analytics	Mayo Clinic, Ascension, Hologic
IBM Cloud	Medical imaging analysis, clinical trial management, AI-assisted treatment recommendation	Siemens Healthineers, Pfizer, Anthem
Oracle Cloud	EHR (electronic health records) management, population health analytics, clinical data integration	Cerner (Oracle-owned), Cleveland Clinic
Alibaba Cloud	Remote health monitoring, telemedicine, and medical imaging in Asia-focused markets	Ping-based health check
Salesforce Health Cloud	Care coordination, personalized patient engagement, and AI-powered insights	Johnson & Johnson, Life Sciences Cloud

7.4 CLOUD VS EDGE FOR HEALTHCARE SERVICES

The choice between cloud and edge computing in healthcare depends on the unique needs and limitations of each application. Cloud computing excels when massive data storage, large-scale analytics, and advanced AI/ML processing are required, for example, managing public health data, electronic health record systems, and drug discovery. It also supports telemedicine platforms by storing patient records and facilitating on-demand video consultations for users across the globe. However, in situations where immediate responses are critical, cloud-based systems may face latency challenges.

Edge computing is particularly effective in real-time, latency-sensitive scenarios such as remote patient monitoring, IoT-enabled medical devices in smart hospitals, and wearable health trackers. For example, an IoT-powered

Table 7.3 Edge Resource Providers in Industries

Providers	*Key Services*	*Applications*
Amazon Web Services	AWS IoT Greengrass, AWS Outposts, AWS Wavelength	Real-time data processing, IoT device integration, low-latency applications for healthcare, automotive, and media
Microsoft Azure	Azure Stack Edge, Azure IoT Edge, Azure Percept	AI at the edge, IoT device management, video analytics, real-time insights for manufacturing and healthcare
Google Cloud Platform	Anthos for edge, Google Distributed Cloud Edge, Edge AI solutions	AI-powered edge analytics, video processing, and IoT applications
IBM	AI at the edge, edge analytics, scalable IoT device management, IBM Edge Application Manager	Retail analytics, manufacturing automation, and healthcare monitoring solutions
Dell Technologies	Dell EMC VxRail, Dell Edge Gateway solutions	Industrial automation, IoT device management, and retail edge analytics
HPE (Hewlett Packard Enterprise)	HPE Edgeline Converged Edge Systems, Aruba Edge Services Platform (ESP)	Industrial IoT, retail automation, telemedicine, and energy monitoring
Cisco	Cisco Edge Compute Solutions, Cisco IoT Control Center	Real-time IoT data processing, network edge management, and smart city applications
NVIDIA	NVIDIA EGX Edge AI platform, Jetson edge AI hardware	AI inferencing at the edge, smart surveillance, and healthcare imaging
Intel	Intel Smart Edge Open, Intel Distribution of OpenVINO toolkit	AI-driven edge processing, 5G-enabled IoT, and robotics
EdgeConneX	Localized data centers, edge colocation services	Low-latency content delivery, IoT integration, and autonomous vehicle data processing

pacemaker can process data locally through edge computing, sending instant alerts to medical professionals if an irregularity is detected, enabling rapid intervention. By processing data nearer to the source, edge computing enhances privacy, reduces dependence on constant internet connectivity, and supports real-time medical decisions. While edge computing is ideal for rapid response, its limited scalability makes it best suited as a complement rather than a replacement for the cloud. A hybrid model that integrates both approaches can provide the best of both worlds: centralized storage and advanced analytics from the cloud, combined with fast, localized processing

Table 7.4 Comparison of Cloud and Edge Computing

Feature	*Cloud Computing*	*Edge Computing*
Latency	Higher (due to data transfer)	Lower (local processing)
Scalability	Highly scalable	Limited scalability
Data Privacy	Moderate (depends on encryption)	Higher (data stays local)
Cost	Pay-as-you-go models	High initial investment

from the edge. A comparative overview of cloud and edge computing in healthcare is presented in Table 7.4.

7.5 THE INTERSECTION OF CLOUD AND EDGE COMPUTING FOR HEALTHCARE

Combining cloud and edge computing technologies creates a balanced and efficient system. The cloud offers large-scale storage, scalability, and advanced data analysis, while edge computing focuses on quick, real-time processing close to the data source. In healthcare, for instance, the cloud can be used for storing patient records and running complex analytics, while edge devices handle immediate data such as vital signs, enabling faster responses. This mix helps healthcare providers benefit from the strengths of both fields' broad insights from the cloud and instant actions from the edge, leading to better resource use, smoother operations, and improved patient outcomes.

The combination of these two approaches makes decision-making faster, reduces delays, and strengthens data security, with applications not just in healthcare but also in areas like manufacturing and transport. Still, challenges such as high infrastructure costs and complex data management must be addressed for large-scale adoption.

7.5.1 Advantages of combining edge and cloud computing

- **Real-Time Responsiveness:** Edge computing handles time-sensitive tasks locally.
- **Scalability and Storage:** Cloud computing ensures scalability and provides centralized storage for analysis and collaboration.
- **Improved Outcomes:** The synergy enables faster, smarter, and more comprehensive healthcare solutions.

The combination of these two computing paradigms has enormous potential, especially in healthcare, where real-time data processing and secure, efficient healthcare delivery are important for better patient outcomes. Some of the researchers demonstrated the advantages of cloud/edge computing in various healthcare technologies. The research by Gupta et al. [26] investigates a smart health monitoring arrangement that uses IoT and edge computing for early fall detection and health prediction, and proposes a convolutional neural network (CNN) based model for evaluating IoT-collected medical information, utilizing edge devices for rapid processing and timely reporting. Vimal et al. [27] investigate a smart healthcare monitoring system that uses IoT and edge computing for early fall detection and health prediction. A CNN-based approach is presented to analyze IoT-collected health data, with edge devices used for faster processing and timely reporting. The suggested method achieves 99.23% accuracy, outperforming current systems.

Singh and Chatterjee [28] offer a Secure Framework based on edge of Things for real-time health monitoring, which uses edge computing to reduce data processing reaction time. Blockchain-like methods and Attribute-Based Encryption guarantee secure storage and access, with 98.5% correctness while improving performance and data security.

Shukla et al. [29] presented a three-tier Fog Computing enabled blockchain framework for safe and dependable patient health record transmission, using the blockchain's efficacy for cloud services. Ram et al. [30] claim that using edge computing and machine learning, they can improve mobile health monitoring by evaluating multi-modal sensor data. Random forest and SVM algorithms were evaluated for activity classification, with random forest obtaining 99% accuracy. The findings, which included insights from ECG data and confusion matrices, suggest prospective ways for improving health prediction accuracy.

7.6 TECHNOLOGIES IN HEALTHCARE USING CLOUD/EDGE COMPUTING

Healthcare technologies leverage a combination of cloud computing for scalability, storage, and large-scale analytics, and edge computing for real-time accessibility, low latency, and localized task handling, creating a robust ecosystem for improving patient care, streamlining hospital operations, and driving medical innovation.

Various healthcare technologies using cloud and edge computing are as follows:

7.6.1 Internet of medical things (IoMT)

The Internet of Medical Things (IoMT) includes a range of connected devices like wearable health monitors, smart implants, and remote patient monitoring systems. These devices produce significant amounts of data that require efficient processing. Edge computing facilitates real-time analysis of data, such as heart rate, oxygen saturation, or blood glucose levels, enabling instant alerts for irregularities or health concerns. At the same time, the cloud acts as a central hub for storing this data, supporting long-term monitoring, advanced analytics, and seamless sharing with healthcare providers or other platforms [31, 32].

7.6.2 Telemedicine platforms

Telemedicine platforms allow patients and healthcare professionals to connect virtually, reducing the reliance on face-to-face visits. Edge computing enhances these platforms by reducing latency, enabling smoother video communication and real-time interaction. For example, during a telehealth consultation, edge devices handle sensor data from wearable devices, allowing healthcare professionals to get immediate insights into the patient's condition. Meanwhile, cloud computing hosts the telemedicine infrastructure, manages large-scale patient records, and provides analytics for long-term health monitoring [33, 34].

7.6.3 AI-powered diagnostic systems

Artificial Intelligence (AI) plays a significant role in the industrial revolution including in the healthcare industry [2]. It analyzes medical data like imaging scans, pathology reports, and genomics. Edge computing processes real-time data directly on diagnostic devices to provide quick insights. For instance, edge-enabled X-ray machines or MRI scanners can instantly detect abnormalities without needing to upload the data to the cloud that handles large-scale machine learning model training and stores vast amounts of historical data, enabling AI models to improve accuracy over time [35, 36].

7.6.4 Smart hospitals

Smart hospitals integrate IoT-enabled devices to monitor and automate various aspects of healthcare, from patient monitoring to equipment management. For example, an edge-based device detects a drop in oxygen supply to a ventilator and sends an alert, while the cloud provides a hospital-wide view of all equipment status. Edge computing handles localized tasks, such as managing medical devices or monitoring patient vitals in real time. For example, edge servers might manage smart lighting or HVAC systems for energy efficiency. Meanwhile, cloud computing centralizes data across departments, offering centralized dashboards, analytics, and insights for hospital-wide decision-making [37–39].

7.6.5 Personalized medicine and genomics

Genomic medicine benefits from the combined power of edge and cloud computing. Sequencing devices at the edge can quickly detect genetic patterns or mutations and give patients immediate insights, while the cloud manages large-scale comparisons across vast datasets to recommend precise treatments. This partnership supports personalized medicine, where therapies are tailored to an individual's genetics, lifestyle, and environment. In clinical trials, edge devices track participants' vital signs in real time, while cloud platforms store and analyze the collective data. For example, an edge device may flag a cancer-related mutation on the spot, and the cloud then runs population-wide comparisons to suggest the most effective drug therapy [40–42].

7.6.6 Virtual/augmented reality (VR/AR)

VR and AR technologies are increasingly used in healthcare for surgery simulation, medical training, and rehabilitation. Edge computing optimizes real-time interactions by processing sensory data locally, reducing latency. For example, in surgical training simulations, edge devices handle VR/AR rendering, ensuring seamless and responsive experiences. The cloud supports these technologies by storing patient data, enabling multi-user collaboration, and performing large-scale analysis for research purposes [43–45].

7.6.7 Robotic surgery

Robotic-assisted surgeries depend on fast, accurate data processing to maintain precision and control. Edge computing handles sensor inputs from robotic arms directly at the source, enabling instant adjustments during procedures. Meanwhile, the cloud adds value by collecting data from many surgeries, supporting continuous learning, and powering advanced machine learning models that help improve future surgical outcomes [46, 47].

7.6.8 Predictive analytics for population health

Population health management focuses on analyzing large datasets to predict health outcomes and manage resource allocation effectively. Edge computing collects and preprocesses data from IoT devices, wearable sensors, and local health monitoring systems, whereas cloud enables large-scale analytics to make data-driven decisions.

7.6.9 Emergency response systems

In emergencies, such as ambulance rides, real-time data processing is critical for saving lives. Edge computing ensures that vital signs are monitored locally, reducing delays [48]. The cloud helps hospitals get a complete picture of a patient's condition before they arrive, making emergency care faster and more efficient. Ambulances equipped with edge devices can monitor and manage vital signs like ECG or oxygen levels in real time, helping stabilize the patient on the way. At the same time, the cloud sends this data to the hospital, so doctors are fully prepared when the patient reaches the emergency room.

7.6.10 Medication and treatment compliance

Smart medication dispensers and IoT devices help patients keep track of their medicines by giving real-time reminders through edge computing. The cloud then stores this information, allowing doctors to review patient habits, adjust treatments, and improve medication management for many patients at once [49]. For example, smart pillboxes or infusion pumps monitor medication intake locally and alert patients if they miss a dose while the cloud aggregates compliance data for healthcare providers to track patient adherence and adjust treatment plans. So, a smart pillbox reminds a patient to take their medication, while the cloud logs adherence data for the doctor's review.

7.6.11 Health data interoperability platforms

Interoperability platforms enable the seamless sharing of health records across different healthcare systems. Edge computing ensures that data is securely collected and formatted locally, with minimal latency. Cloud computing centralizes the shared data, enabling collaborative analysis, compliance management, and a unified view of patient health records [50–53].

7.6.12 Clinical trial management systems

Clinical trials benefit from real-time data monitoring through IoT devices and edge computing for immediate insights. Edge computing ensures data collection and analysis at trial sites, while cloud computing supports centralized data storage and analysis, helping researchers track trial progress, analyze results, and improve outcomes over time [54].

7.6.13 Blockchain for healthcare

Blockchain technology ensures secure, tamper-proof storage of medical records and transactions. Edge computing works alongside blockchain to verify local transactions, ensuring decentralized trust. Cloud computing enhances blockchain for healthcare by providing a scalable platform for managing large, decentralized data networks [55–58].

7.6.14 AI chatbots and virtual assistants

AI chatbots assist in symptom checking, appointment scheduling, and medication reminders. Edge computing ensures local processing of sensitive data for real-time interactions, while cloud computing supports natural language processing and aggregates data for analysis and continuous improvement of chatbot services [59–61].

7.6.15 Health monitoring systems for remote patient monitoring and telehealth

IoT-enabled health monitoring systems collect data continuously from patients. Wearable devices and sensors at the edge monitor patient vitals like heart rate, blood pressure, and oxygen levels in real time. These devices process data locally to provide immediate alerts for abnormalities. The processed data is uploaded to the cloud for storage, large-scale analysis, and sharing with healthcare providers for long-term monitoring or collaboration [62–64]. Cloud computing aggregates this data for long-term analysis, supporting trends in chronic disease management and personalized health interventions. For example, a wearable health tracker alerts a patient to a sudden drop in oxygen levels locally, while the cloud stores historical data for a doctor to review during a telehealth session.

7.7 CHALLENGES AND FUTURE PROSPECTS

This section explores challenges and future research in this field.

Some of the challenges are as follows:

7.7.1 Data security and privacy

Sensitive healthcare data requires compliance with regulations like HIPAA, GDPR, and others. Unauthorized access can lead to legal and financial impacts. So, there is a need to implement advanced encryption, multi-factor authentication, and regular security audits.

7.7.2 Latency and reliability

Real-time applications like remote surgery or critical monitoring demand low-latency and highly reliable networks. Network disruptions or delays can risk patient safety.

7.7.3 Interoperability

Healthcare systems often use disparate formats and standards, leading to integration issues, such as difficulty in seamless data exchange and communication between cloud, edge, and on-premise systems. Adopting standardized data formats and ensuring compatibility with healthcare standards (e.g., HL7, FHIR) is required.

7.7.4 Cost management

Cloud and edge computing involve initial setup costs, operational expenses, and maintenance fees. Small healthcare providers may struggle to afford these technologies. Implementing hybrid models to optimize costs and utilizing pay-as-you-go models offered by cloud providers can help in cost reduction.

7.7.5 Bandwidth limitations

High-quality imaging and large datasets require substantial bandwidth for transmission and processing. Bandwidth limitations can hinder data transfer speeds, especially in remote areas. So, optimization of data compression, prioritization of bandwidth allocation, and leverage of local edge devices for processing are essential.

7.7.6 Data governance and compliance

Different regions have varying laws regarding data storage, transfer, and use.

7.7.7 Infrastructure challenges

Edge computing needs strong, distributed infrastructure, which is not always available in rural or underdeveloped regions. Working with telecom providers to expand networks and using mobile-edge solutions can help close this gap.

Despite these challenges, cloud and edge computing are set to transform healthcare. The cloud offers massive storage and processing power to support AI-driven diagnosis, personalized treatments, and global research. Edge computing adds real-time analysis close to the data source, which is vital for remote monitoring, telemedicine, and emergency care. Together, they enable smart hospitals, better data sharing, and wider healthcare access, even in underserved areas. Combined with IoT, 5G, and AI, these technologies will also accelerate drug development, create digital twins for patient care, and build sustainable, integrated health systems. With evolving regulations and stronger data security, cloud and edge computing will make healthcare faster, more precise, and more equitable worldwide.

7.8 CONCLUSION

The chapter proposes the blend of cloud and edge computing that can reshape healthcare and make it smarter and more responsive. While the cloud provides powerful storage and advanced analytics, the edge ensures fast, real-time processing close to the patient. With this collaboration, healthcare can become more innovative, efficient, and patient-focused, ultimately improving outcomes and building stronger, more responsive systems for the future. To work out more benefits, technology providers, healthcare institutions, and policymakers must work hand in hand to handle current challenges.

REFERENCES

[1] Sunyaev, A. (2024). Cloud computing. In *Internet Computing: Principles of Distributed Systems and Emerging Internet-based Technologies* (pp. 165–209). Switzerland, Cham: Springer Nature.

[2] Singhal, R., Singhal, I., Sharma, V., & Bansal, V. (2025). Artificial intelligence: Revolutionizing industrial decisions. In *Data-Centric Approaches to Industrial Decisions: Technology, Digitisation and Business Decisions* (pp. 453–478). Singapore: Springer Nature.

[3] Thahniyath, G., Yadav, C. S. G., Senkamalavalli, R., Priya, S. S., Aghalya, S., Reddy, K. N., & Murugan, S. (2024). Cloud based prediction of epileptic seizures using real-time electroencephalograms analysis. *International Journal of Electrical & Computer Engineering*, 14(5). ISSN: 2088-8708.

[4] Singhal, R., Sonia, & Singhal, A., (2016) Rising future of Agile software development using cloud computing: A study using cloud computing in different phases of an Agile method-CRUM. *International Journal of Computer Technology & Applications*, 7(2), 240–246 ISSN: 2229-6093.

[5] Abied, O., Ibrahim, O., & Kamal, S. N. I. M. (2022). Adoption of cloud computing in E-government: A systematic literature review. *Pertanika Journal of Science & Technology*, 30(1).

[6] Singhal, R., & Singhal, A. (2019). A combinatorial economical double auction resource allocation model (CEDARA). In *International Conference on Machine Learning, Big Data, Cloud and Parallel Computing (COMITCon)* (pp. 497–502). IEEE. https://doi.org/10.1109/COMITCon.2019.8862456

[7] Singhal, R., & Singhal, A. (2021). A feedback-based combinatorial fair economical double auction resource allocation model for cloud computing. *Future Generation Computer Systems*, 115, 780–797. https://doi.org/10.1016/j.future.2020.09.022

[8] Singhal, R. and Singhal, A., (2019a) A combinatorial fair economical double auction resource allocation model. In *International Conference on Advances in Computing and Data Sciences* (pp. 272–281). Singapore: Springer. https://doi.org/10.1007/978-981-13-9942-8_26

[9] Singhal, R., Singhal, A., Sharma, V., & Saachi, N. (2019). E-cloud: A solution towards e-waste management for educational institutions. *International Journal of Recent Technology and Engineering (IJRTE)*, 8(2S6), 964–972. https://doi.org/10.35940/ijrte.B1184.0782S619

[10] Buyya, R., Srirama, S. N., Casale, G., Calheiros, R., Simmhan, Y., Varghese, B., . . . Shen, H. (2018). A manifesto for future generation cloud computing: Research directions for the next decade. *ACM Computing Surveys (CSUR)*, 51(5), 1–38.

[11] Singhal, R., Sharma, V., Singhal, I., & Bansal, V. (2024). Blockchain-enabled auction for cloud resource provisioning: A survey on trust and economy. *International Journal of System Assurance Engineering and Management*, 1–21.

[12] Verma, G. (2024). Blockchain-based privacy preservation framework for healthcare data in cloud environment. *Journal of Experimental & Theoretical Artificial Intelligence*, 36(1), 147–160.

[13] Singhal, R. (2024). *Introduction to Cloud Computing (AWS). Authors Press.* ISBN: 978-93-6095-294-5.

[14] Rajabion, L., Shaltooki, A. A., Taghikhah, M., Ghasemi, A., & Badfar, A. (2019). Healthcare big data processing mechanisms: The role of cloud computing. *International Journal of Information Management*, 49, 271–289.
[15] Darwish, A., Hassanien, A. E., Elhoseny, M., Sangaiah, A. K., & Muhammad, K. (2019). The impact of the hybrid platform of internet of things and cloud computing on healthcare systems: Opportunities, challenges, and open problems. *Journal of Ambient Intelligence and Humanized Computing*, 10, 4151–4166.
[16] Askary, Z., & Kumar, R. (2020). Cloud computing in industries: A review. *Recent Advances in Mechanical Engineering: Select Proceedings of NCAME, 2019*, 107–116.
[17] Haghnegahdar, L., Joshi, S. S., & Dahotre, N. B. (2022). From IoT-based cloud manufacturing a roach to intelligent additive manufacturing: Industrial internet of things—an overview. *The International Journal of Advanced Manufacturing Technology*, 119(3), 1461–1478.
[18] Kundavaram, V. N. K. (2024). Cloud computing adoption in retail: A comprehensive analysis of operational efficiency and customer experience enhancement. *International Journal of Research in Computer Applications and Information Technology (IJRCAIT)*, 7(2), 1151–1164.
[19] Daase, C., Volk, M., Staegemann, D., & Turowski, K. (2023). The future of commerce: Linking modern retailing characteristics with cloud computing capabilities. In ICEIS (Vol. 2, pp. 418–430). SCITEPRESS.
[20] Mnyakin, M. (2023). Applications of AI, IoT, and Cloud computing in smart transportation: A review. *Artificial Intelligence in Society*, 3(1), 9–27.
[21] Rajagopalan, A., Swaminathan, D., Bajaj, M., Damaj, I., Rathore, R. S., Singh, A. R., . . . Prokop, L. (2024). Empowering power distribution: Unleashing the synergy of IoT and cloud computing for sustainable and efficient energy systems. *Results in Engineering*. https://doi.org/10.1016/j.rineng.2024.101949.
[22] AL-Jumaili, A. H. A., Mashhadany, Y. I. A., Sulaiman, R., & Alyasseri, Z. A. A. (2021). A conceptual and systematics for intelligent power management system-based cloud computing: Prospects, and challenges. *Applied Sciences*, 11(21), 9820.
[23] Lăzăroiu, G., Bogdan, M., Geamănu, M., Hurloiu, L., Luminiţa, L., & Ştefănescu, R. (2023). Artificial intelligence algorithms and cloud computing technologies in blockchain-based fintech management. *Oeconomia Copernicana*, 14(3), 707–730.
[24] Singhal, R., Singhal, A., & Sonia (2016). Towards a generic e-cloud architecture for universities. *IJWA*, 8, 36–43.
[25] Singhal, R., Singhal, A., Bhatnagar, M., & Malhotra, N. (2019). Design of an audio repository for blind and visually impaired: A case study. In *Advanced Computing and Communication Technologies* (pp. 77–85). *Singapore: Springer*. https://doi.org/10.1007/978-981-13-0680-8_8
[26] Gupta, P., Chouhan, A. V., Wajeed, M. A., Tiwari, S., Bist, A. S., & Puri, S. C. (2023). Prediction of health monitoring with deep learning using edge computing. *Measurement: Sensors*, 25, 100604.
[27] Vimal, S., Robinson, Y. H., Kadry, S., Long, H. V., & Nam, Y. (2021). IoT based smart health monitoring with CNN using edge computing. *Journal of Internet Technology*, 22(1), 173–185.
[28] Singh, A., & Chatterjee, K. (2023). Edge computing based secure health monitoring framework for electronic healthcare system. *Cluster Computing*, 26(2), 1205–1220.

[29] Shukla, S., Thakur, S., Hussain, S., Breslin, J. G., & Jameel, S. M. (2021). Identification and authentication in healthcare internet-of-things using integrated fog computing based blockchain model. *Internet of Things*, 15, 100422.
[30] Ram, S. S., Apduhan, B., & Shiratori, N. (2019). A machine learning framework for edge computing to improve prediction accuracy in mobile health monitoring. In *Computational Science and Its Applications–ICCSA 2019: 19th International Conference, Saint Petersburg, Russia, July 1–4, 2019, Proceedings, Part III 19* (pp. 417–431). Springer International Publishing.
[31] Razdan, S., & Sharma, S. (2022). Internet of medical things (IoMT): Overview, emerging technologies, and case studies. *IETE Technical Review*, 39(4), 775–788.
[32] Huang, C., Wang, J., Wang, S., & Zhang, Y. (2023). Internet of medical things: A systematic review. *Neurocomputing*, https://doi.org/10.1016/j.neucom.2023.126719.
[33] Kitole, F. A., & Shukla, S. (2024). Cloud horizons: Strengthening rural healthcare through telemedicine's digital canopy. *Health Services Insights*, 17, https://doi.org/10.1177/11786329241284401
[34] Suleiman, T. A., & Adinoyi, A. (2023). Telemedicine and smart healthcare—the role of artificial intelligence, 5G, cloud services, and other enabling technologies. *International Journal of Communications, Network and System Sciences*, 16(3), 31–51.
[35] Kaur, P., Kaushik, A., Singhal, I., Pandey, A., & Singhal, R. (2025). Advancing brain MRI segmentation using segment anything model. *Procedia Computer Science*, 260, 110–117. ISSN: 1877-0509.
[36] Talib, S. H., Abdul-Rahaim, L. A., Alrubaie, A. J., & Raseed, I. M. (2023). Design smart hospital system based on cloud computing. *Indonesian Journal of Electrical Engineering and Computer Science*, 29(2), 797–807.
[37] Li, S., Luo, D., Liu, X., & Luo, R. (2024). Smart hospital privacy protection system based on cloud computing Iot. *Procedia Computer Science*, 243, 567–576.
[38] Kwon, H., An, S., Lee, H. Y., Cha, W. C., Kim, S., Cho, M., & Kong, H. J. (2022). Review of smart hospital services in real healthcare environments. *Healthcare Informatics Research*, 28(1), 3–15.
[39] Esho, T., Ayeni, O., Lasisi, O., & Peter, O. (2024). The role of cloud computing in personalized medicine: A systematic review. *Current Journal of Applied Science and Technology*, 43(3), 1–8.
[40] Rehan, H. (2024). Advancing cancer treatment with AI-driven personalized medicine and cloud-based data integration. *Journal of Machine Learning in Pharmaceutical Research*, 4(2), 1–40.
[41] Cirillo, D., & Valencia, A. (2019). Big data analytics for personalized medicine. *Current Opinion in Biotechnology*, 58, 161–167.
[42] Oyekunle, D., Claude, B. E. A., Waliu, A. O., Adekunle, T. S., & Matthew, U. O. (2024). Cloud based adaptive learning system: Virtual reality and augmented reality assisted educational pedagogy development on clinical simulation. *Journal of Digital Health*, 49–62.
[43] Lakshminarayanan, V., Ravikumar, A., Sriraman, H., Alla, S., & Chattu, V. K. (2023). Health care equity through intelligent edge computing and augmented reality/virtual reality: A systematic review. *Journal of Multidisciplinary Healthcare*, 2839–2859.

[44] Bhugaonkar, K., Bhugaonkar, R., & Masne, N. (2022). The trend of metaverse and augmented & virtual reality extending to the healthcare system. *Cureus*, 14(9).
[45] Atallah, A. B., & Atallah, S. (2021). Cloud computing for robotics and surgery. *Digital Surgery*, 37–58.
[46] Jain, S., & Doriya, R. (2022). Security framework to healthcare robots for secure sharing of healthcare data from cloud. *International Journal of Information Technology*, 14(5), 2429–2439.
[47] Chayon, M. H. R., Rimon, J. A., & Moli, D. A. (2024, February). A smart emergency healthcare system based on edge computing and 5G. In International Conference on Emerging Trends in Mathematical Sciences & Computing (pp. 241–251). Switzerland, Cham: Springer Nature.
[48] Punj, N., Tanwar, A., Ranga, P., & Kumar, V. (2025). *Cloud IoT in healthcare transforming patient monitoring and data analytics. In Integration of Cloud Computing and IoT (*pp. 101–117). *Chapman and Hall/CRC.*
[49] Malathi, K., Shruthi, S. N., Madhumitha, N., Sreelakshmi, S., Sathya, U., & Sangeetha, P. M. (2024). Medical data integration and interoperability through remote monitoring of healthcare devices. *Journal of Wireless Mobile Networks, Ubiquitous Computing, and Dependable Applications (JoWUA)*, 15(2), 60–72.
[50] Saini, V., Reddy, S. G., Kumar, D., & Ahmad, T. (2021). Evaluating FHIR's impact on health data interoperability. *Internet of Things and Edge Computing Journal*, 1(1), 28–63.
[51] Jones, D. E., Alimi, T. O., Pordell, P., Tangka, F. K., Blumenthal, W., Jones, S. F., & Richardson, L. C. (2021). Pursuing data modernization in cancer surveillance by developing a cloud-based computing platform: Real-time cancer case collection. *JCO Clinical Cancer Informatics*, 5, 24–29.
[52] Chirra, D. R. (2024). Secure data sharing in multi-cloud environments: A cryptographic framework for healthcare systems. *Revista de Inteligencia Artificial en Medicina*, 15(1), 821–843.
[53] Thahniyath, G., Yadav, C. S. G., Senkamalavalli, R., Priya, S. S., Aghalya, S., Reddy, K. N., & Murugan, S. (2024). Cloud based prediction of epileptic seizures using real-time electroencephalograms analysis. *International Journal of Electrical & Computer Engineering*, 14(5). ISSN: 2088-8708
[54] Amanat, A., Rizwan, M., Maple, C., Zikria, Y. B., Almadhor, A. S., & Kim, S. W. (2022). Blockchain and cloud computing-based secure electronic healthcare records storage and sharing. *Frontiers in Public Health*, 10, 938707.
[55] Zhang, G., Yang, Z., & Liu, W. (2022). Blockchain-based privacy preserving e-health system for healthcare data in cloud. *Computer Networks*, 203, 108586.
[56] Singh, S., Pankaj, B., Nagarajan, K., Singh, N. P., & Bala, V. (2022). Blockchain with cloud for handling healthcare data: A privacy-friendly platform. *Materials Today: Proceedings*, 62, 5021–5026.
[57] Verma, G. (2024). Blockchain-based privacy preservation framework for healthcare data in cloud environment. *Journal of Experimental & Theoretical Artificial Intelligence*, 36(1), 147–160.
[58] Afrin, Z., Farid, D. M., & Mamun, K. A. A. (2023, October). A cloud-based intelligent virtual assistant for adolescents. In *International Conference on Intelligent Systems and Data Science (*pp. 110–124). Singapore: Springer Nature.
[59] Chellatamilan, T., Kumar, N. S., & Valarmathi, B. (2022). Effective deployment of multi-cloud customizable chatbot application for COVID-19 datasets.

In *Operationalizing Multi-Cloud Environments: Technologies, Tools and Use Cases*, 361–379.

[60] Singhal, R., Singhal, A., Singhal, I., & Bansal, V. (2025). Secure and intelligent electronic health records: A cloud-based AI-blockchain framework. *Cureus Journal of Computer Science*, 2, es44389-025-09459-1. https://doi.org/10.7759/s44389-025-09459-1

[61] Akhbarifar, S., Javadi, H. H. S., Rahmani, A. M., & Hosseinzadeh, M. (2023). A secure remote health monitoring model for early disease diagnosis in cloud-based IoT environment. *Personal and Ubiquitous Computing*, 27(3), 697–713.

[62] Desai, F., Chowdhury, D., Kaur, R., Peeters, M., Arya, R. C., Wander, G. S., . . . Buyya, R. (2022). HealthCloud: A system for monitoring health status of heart patients using machine learning and cloud computing. *Internet of Things*, 17, 100485.

[63] Siam, A. I., Almaiah, M. A., Al-Zahrani, A., Elazm, A. A., El Banby, G. M., El-Shafai, W., . . . El-Bahnasawy, N. A. (2021). Secure health monitoring communication systems based on IoT and cloud computing for medical emergency applications. *Computational Intelligence and Neuroscience, 2021(1), 8016525.*

[64] Lakshmi, G. J., Ghonge, M., & Obaid, A. J. (2021). Cloud based IoT smart healthcare system for remote patient monitoring. *EAI Endorsed Transactions on Pervasive Health and Technology*, 7(28), e4.

Chapter 8

COVID-19 Pandemic Prediction Model in India

Data Analysis and Forecasting

Renu Dalal, Mayank Gupta, Anmol Kumar Pandit, and Jyoti Dagar

8.1 INTRODUCTION

Late in 2019 marked the beginning of a period of radical global change under the unwelcoming shadows of a newly born coronavirus popularly known as COVID-19. First detected in Wuhan, China, this respiratory disease caused by the virus referred to as SARS-CoV-2 has emerged as one of the great public health challenges of our times and one of the fastest-ever in human history. COVID-19 is a disease of zoonotic origin that very likely spreads from an animal source, eventually mutating and adapting to infect man. The ability of this virus to propagate at an unprecedented rate and distance came in the wake of an interconnected global society. The virus had passed international borders, penetrated the respective communities, and overwhelmed healthcare systems within a few weeks across almost all continents.

The rampant spread and mutations of the virus posed a threat not only to public health but also the economy, education, global trade, and socio-political setup. Societies all over the world were found woefully lacking with regard to such a widespread outbreak, especially since their already very-small-scale public health systems had either never before or for a long time not been subjected to a challenge of this nature. Even basic medical supplies such as oxygen cylinders, hospital beds, ventilators, and protective equipment were in dire shortage in many nations. The pandemic also highlighted significant equity issues in access to healthcare; quality of care; and resilience of public health systems across many nations and regions [1, 2].

The World Health Organization (WHO) eventually declared COVID-19 a worldwide pandemic in March 2020, calling all countries to put up immediate strategies to combat the otherwise pernicious disease [1, 3]. The response of this global health authority to the affliction threw a spotlight on the glaring gaps in preparedness by governments, particularly in densely populated developing countries such as India. Coupled with this, an immediate response to this invisible, highly transmissible threat was to be made by policymakers using real-time-evolving data and public health guidelines.

DOI: 10.1201/9781003635147-8

India was multi-faceted in the type of challenges it faced, as it also had specific demographics, diverse healthcare infrastructures, and socio-economic differences. As the second most populous country in the world, India had troubles curbing community transmission to maintain most of the essential services. The already under-pressure public health system of the country in many regions found it challenging to develop the facilities for testing, treatment, and tracing. In line with this, misinformation and stigma surrounding the virus impeded timely reporting and resulted in the undercounting of cases, especially in rural localities where access to healthcare was limited [2, 4].

This happened in response to the increasing number of cases, whereby the Indian government imposed one of the most stringent nationwide lockdowns across the globe starting March 2020. Such a decision would be geared "to flatten the curve" while at the same time provide the health system enough time for preparedness towards the anticipated surge. At first, the lockdown had good results because it slowed down transmission. However, far-reaching effects of deprivation were felt through the economy and society. Everyday workers, such as migrants and small business owners, were particularly hit hard by the surprise closure. Supply chains were dislocated while employment levels dipped, and millions were driven to poverty [5, 6].

The ensuing waves of infection after this early intervention worked in terms of revealing more vulnerabilities of India's health response. The situation was particularly catastrophic during the second wave, which began in early 2021. The very virulent Delta variant was responsible for a surge of patients in hospitals and injuries. Oxygen shortages, overflowing hospitals, and the sight of long queues outside crematoria were stark reminders of how fast things could spin out of control [7, 8]. Above all, the increased loss of lives due to these factors led to a grave crisis crescendo resulting from lack of coordinated data sharing, erratic responses at various levels, and ineffective communication strategies concerning health.

However, the third wave and the subsequent ones were better managed owing to the greater coverage of vaccination and awareness, but other challenges still lurked related to rural outreach, vaccine hesitancy, and the emergence of new variants. The pandemic forced a quick adaptation of the Indian healthcare system. The most immediate innovations that were accelerated during this time included telehealth, mobile health units, and outreach programs. Digital platforms like CoWIN were developed for tracking and scheduling vaccinations, while partnerships were worked out in the public-private sector for mobilizing resources for generation and distribution. COVID-19 was an eye-opener beyond health implications, showing how public health is intimately interconnected with the economy, national security, education, and international relations. The closure of educational institutions saw a massive shift to online learning, disproportionately impacting students from rural or low-income households who did not have Internet access. Mental health poses a serious topic of concern, with escalating anxiety, depression, and burnout noted across age groups [21, 22].

Policy-wise, the pandemic has placed emphasis on investment in primary healthcare, pandemic preparedness, and decision-making based on evidence. During this phase, data science integrated with epidemiological modeling became indispensable guiding tools in public health response. Some forecasting tools tried to predict surges of cases and guide closures through their data. There were mass media campaigns that sought to change behaviors and messages encouraging the use of masks, hand hygiene, and physical distancing.

On the international stage, India cut an active figure in the global health discourse. The country became one of the hubs for vaccine production, supplying doses to multiple nations under its "Vaccine Maitri" initiative. Unfortunately, this diplomatic venture came under harsh scrutiny during the second wave, when domestic vaccine demand far outstripped supply. It has underlined the ever-fragile balance between global health solidarity and national health security. Also, the pandemic has fueled talks on future urban planning, environmental policies, and global equity. While lockdowns improved air quality and wildlife movement in urban areas for a while, that has led to a debate about sustainable development models. At almost the same time, the disparity in global vaccine distribution sparked discussions about equality in global health access.

In perspective, the COVID-19 pandemic has altered the very definition of health as a subject of public policy and shifted the rest of governance's priorities. The Indian experience has both highlighted vulnerabilities and illustrated the potential of a diverse, complex nation in combating a crisis of this scale. Systemic reforms are urgently needed, especially demonstrated by the magnanimous crisis during the second wave. Building resilient health systems, investing in public health infrastructure, and putting the principles of digital and data governance into practice are key elements moving forward.

This chapter would try to add to that ongoing talk by analyzing how COVID-19 has progressed through India in a data-driven comprehensive way. This includes daily and cumulative trends that will be generated using Python, along with data handling through pandas and data visualization using libraries such as matplotlib to analyze the positives of the virus, and any possible government interventions that will help in forecasting future situations through statistical and predictive modeling. The analysis would likely help in supporting public health planning to counter future disruptions and ensure that lessons learned from the COVID-19 experience become embedded throughout the system for the betterment of future pandemic preparedness and response. This chapter attempts to analyze the progression and impact of COVID-19 in India through the use of real-time data and visualization techniques. Forecasting of the pandemic trend has been done through the application of statistical (ARIMA) and deep learning (LSTM) models to facilitate proper decision-making in public health planning and preparedness.

8.2 LITERATURE REVIEW

In recent years, a large amount of research has been done pertaining to and following the COVID- 19 pandemic to assess its spread, predict upcoming waves, and guide public health policy. The early studies centered on epidemiological modeling techniques including compartmental models such as SIR (Susceptible-Infectious-Recovered) and SEIR (Susceptible-Exposed-Infectious-Recovered) that were theoretically devised to simulate plausible scenarios during the course of the pandemic. These models served as the mainstay in the analyses of community transmission but were mainly predicated on fixed transmission rates and, at times, on homogeneous behavior of the population.

Statistical methods such as linear regression, polynomial regression, and logistic models were awarded preeminence in forecasting daily case numbers and estimating the interference activity of lockdowns and vaccinations. For example, they used simple linear regression models to arrive at short-term case trend forecasts, with moderate predictive accuracy as regards the first wave [9, 10]. Likewise, Sharma et al. (2021) adopted a logistic regression strategy to establish the case growth inflection points that would duly inform the historical timeline with respect to policy interventions [6].

The ARIMA (Auto Regressive Integrated Moving Average) model and Prophet models are more advanced models that can capture time series for COVID-19 data. They perform well in determining regularity patterns and short-term forecasting, especially in urban hot spots. The ma- chine learning models that made a division from conventional methods were Random Forests, Support Vector Machines (SVM), and LSTM (Long Short-Term Memory) neural networks, chiefly featured for their capacity in learning non-linear relationships in complex datasets [11, 12]. Regional heterogeneity has also found attention in studies. States-wise COVID-19 patterns across India were created and discovered by the application of statistical clustering and regression by Verma et al. (2020), which showed that socioeconomic and healthcare access disparity hugely impacted infection and mortality rates [5, 7]. This work is also applicable in other wireless networks discussed in [13, 14]. Jain et al. (2021), for instance, looked at how the Delta variant influenced case fatality rates and found that these figures rose markedly in the second wave [7]. Looking beyond borders, Johns Hopkins University and WHO produced global dashboards with comparative case studies that made cross-country learning and policy evaluation possible [1, 3].

8.2.1 Identified gaps

In spite of the heterogeneity of the approaches and the volume of research that have been published, there still exist gaps in these areas. A major drawback is that rural and semi-urban territories are mostly left out of predictive modeling studies. Most of the available models are, however, calibrated using samples from areas inhabited by the urban population, which limits their applicability—particularly in demographic settings, mobility patterns, and healthcare infrastructure that are different from the urban one [15].

Real-time interventions from government policies, behavioral responses, and vaccination dynamics are barely integrated into predictive models. Many of these studies assume these variables are fixed or exogenous, heavily compromising the real-life credibility of these models. Further discrepancies between actual and estimated case counts arose from temporal drift and underreporting, especially during the peak of a crisis [16].

Other than a few models, the development and dissemination of new variants with various characteristics of transmission are therefore seldom considered. There is also a lack of different interdisciplinary methods combining the fields of epidemiology, economics, sociology, and data ethics; thus, it becomes restricted in evaluating the greater societal consequences of interventions like lockdowns, travel restrictions, and school closures [17–19].

Generally, where visualization tools and public dashboards are concerned, their application often stops short of being interactive, region-bound, and easily understandable by diverse non- technical stakeholders. Enormous possibilities exist for user-friendly platforms demonstrating real- time scenario simulations whereby public health authorities can make swifter and more localized decisions.

8.3 DATA AND METHODOLOGY

Advanced Forecasting Techniques: Although it is indeed linear regression that dots this baseline for trend analysis, it will not be of help when in need of more sophisticated analysis, accounting for non-linear dynamics as well as seasonality and multiple influencing variables. Therefore, this section discusses two more advancements in predictive modeling, namely: ARIMA and LSTM.

8.3.1 ARIMA modeling

Auto-Regressive Integrated Moving Average (ARIMA) is a statistical forecasting technique used traditionally for univariate data. It captures its patterns in terms of auto-regression (AR), differencing (I), and moving averages (MA). For this purpose, we will apply the "statsmodels" library to derive an ARIMA model on daily newly reported cases.

Listing 1: ARIMA Forecasting for COVID-19 Case Trends

```
from statsmodels.tsa.arima.model import ARIMA
series = df['newcases'].fillna(0)
model = ARIMA(series, order = (5, 1,0)) # Example parameters
modelfit = model.fit()
forecast = modelfit.forecast(steps = 30)
```

The ARIMA model captures time-dependent autocorrelations and provides short-term forecasts with confidence intervals. Nevertheless, ARIMA brings in an assumption of stationarity—a condition that may not hold true given the highly dynamic nature of a pandemic.

8.3.2 LSTM neural networks

Long Short-Term Memory (LSTM) is an RNN specifically designed to capture temporal dependencies in sequential data. Unlike linear models, LSTMs manage the complexities of long-term dependencies, seasonality, and complicated, nonlinear interactions. For our study, we built an LSTM model utilizing TensorFlow and Keras, with a sequence input of 14 days for the prediction of cases along the next day.

Listing 2: LSTM Model for Predicting Daily COVID-19 Cases

```
from tensorflow.keras.models import Sequential
from tensorflow.keras.layers import LSTM, Dense from
sklearn.preprocessing import Min Max Scaler

scaler = Min Max Scaler ()
scaleddata=scaler.fittransform (df [['newcases']])

X, y = [], []
for i in range (14, len (scaleddata)):
X. append (scaleddata [i – 14: i]) y. append (scaleddata [i])
X, y = np. array (X), np. array (y) model = Sequential ()
model. add (LSTM(50, returnsequences = False,
inputshape =(X. shape [20], 1)))
model. add (Dense (1))
model. compile (optimizer = 'adam ', loss = 'meansquar
ederror ') model. fit (X, y, epochs = 20, batchsize =
32)
```

The LSTM model showed excellent performance in capturing sudden shifts in case numbers and seasonal patterns, outperforming several traditional methods in countless forecasting scenarios. These sophisticated models provide more subtle insights into the dynamics of COVID-19 and would thereby add fortification to policy formulation regarding response planning.

As shown in Figures 8.1 and 8.2, the ARIMA model is well-suited for short-term projections, whereas the LSTM model better handles nonlinear variations and longer-range dependencies in the dataset.

8.4 RESULTS

Trend Analysis: Time-series representations of daily and cumulative cases of COVID-19 in India depict distinct yet repeatedly recognizable patterns, which seem to be synchronized with successive waves of the pandemic. The

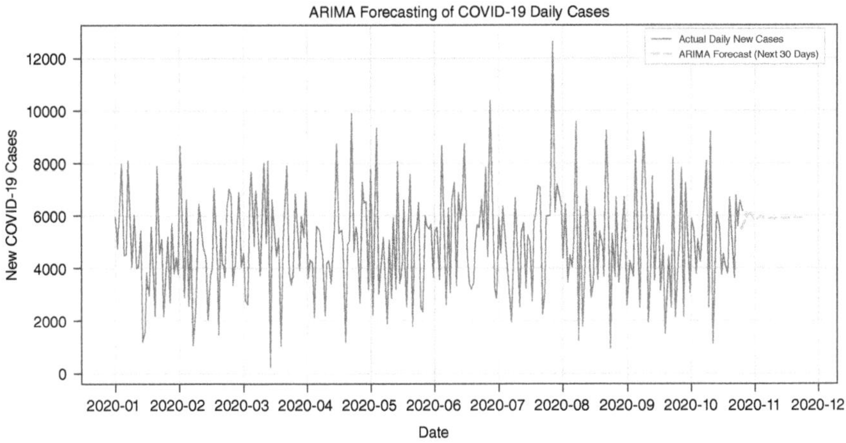

Figure 8.1 ARIMA Forecast of COVID-19 Cases.

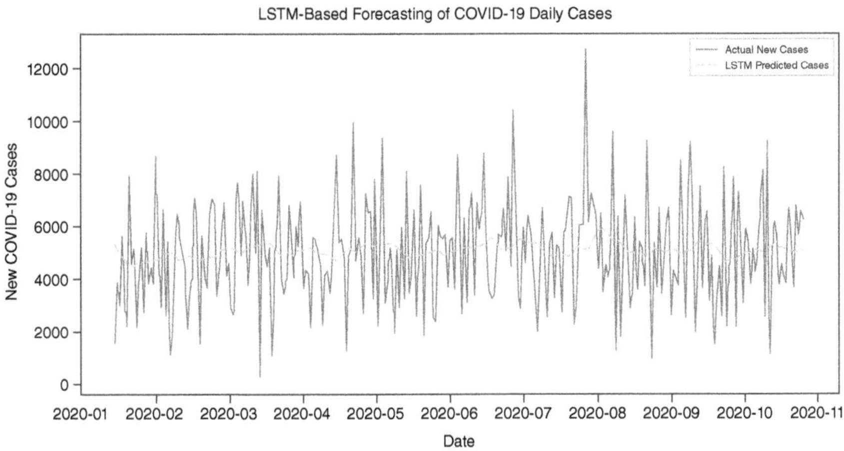

Figure 8.2 LSTM-based Forecast of COVID-19.

rise during early 2020 was somewhat steady and linear due to limited testing and a gradual increase in community transmission. This was followed by a spectacular rise during the 2021 second wave, which was primarily due to the Delta variant and which brought unmerciful pressure on health infrastructure and high mortality. While the spread is estimated to have been high during the third wave due to the Omicron variant in early 2022, the corresponding death rate was rather low, largely because of heightened medical preparedness and massive vaccination coverage.

After commencing vaccinations, the trend in daily new cases showed a pronounced drop interspersed with spikes. These peaks were correlated with the emergence of new variants and with public fatigue against preventive measures. Meanwhile, overall increases in cumulative cases were observed, albeit at a declining rate, indicating a slowing down of spread. Along with this, the im- pact of policy interventions, such as lockdowns, vaccination campaigns, and awareness programs, is clearly visible as evident in the flattening of the majority of cumulative case curves in their wake.

8.4.1 Logarithmic trend insights

These logarithmic transformations of the daily case data were invaluable for understanding the dynamics of COVID-19 spread, especially in the exponential growth phases. The linear plots gave a rough idea of the trajectory, but they often missed the subtle transitions and inflection points. In contrast, the log-scale plots showed the rapid surge and tapering with much more clarity.

Thus, for instance, the transition from the second wave to the third wave was an abrupt phenomenon on a linear scale. It was much more gradually discernible on a logarithmic curve; thus, helping in understanding the underlying rate of transmission and identifying early signals of resurgence. Logarithmic plots also diminished the impact of large spikes, creating a normalized viewpoint for growth, which made them extraordinarily useful for trend comparison and regression analysis.

8.4.2 Forecasting outcomes

The forecasts of the daily new case counts were thus generated for the short and medium terms using both ARIMA and LSTM models. The ARIMA model was competent in capturing the immediate linear trends due to its statistical strength, and it was easy to compute. However, it failed in its predictive capacity for longer time horizons, particularly in the cases of nonlinear transition periods or sudden shifts due to policy implementation or variant-induced spikes.

The LSTM model, on the other hand, beat ARIMA in modeling nonlinear and sequential dependencies existing in the data. It forecasted potential spikes and dips more accurately despite being in turbulent times by learning from longer temporal sequences. The LSTM model efficiently captured some of the very short variations following the decline from peaks.

According to Table 8.1, daily numbers for newly diagnosed cases are forecasted to fall steadily and consistently over the entire 10-day forecast. These trends make sense considering the ongoing public health measures in place and because the population is becoming increasingly immune through vaccination as well as previous infection. The validity of these predictions, however, very much relies on the continued enforcement of safety measures and the absence of new variants with enhanced transmissibility or immune escape features.

Table 8.1 Forecasted Case Counts for May 2025

Date	*Predicted Cases*
2025-05-01	23,100
2025-05-02	21,500
2025-05-03	20,900
2025-05-04	20,100
2025-05-05	19,400
2025-05-06	18,800
2025-05-07	18,200
2025-05-08	17,700
2025-05-09	17,100
2025-05-10	16,600

8.5 DISCUSSION

8.5.1 Findings and implications

The outcomes of this research bring into focus the magnitude of proactive public policy and vaccination in determining the course of COVID-19 in India. Lockdowns—early yet harsh on the economy—were imposed in sufficient time to curb initial transmission. Beginning in early 2021, the vaccination campaign became crucially important in moderating severity and fatality rates during subsequent waves. The relationship between vaccination coverage and declining daily new cases was evidently born out through the analyzed trends.

Additional fluctuations indicated by the predictive modeling point to population mobility, seasonality, and the enforcement of containment zones as contributory factors. The modeling also suggests that early warnings provided by forecasting will guide healthcare resource allocation and mitigate the impact of the surge, especially in densely populated urban centers.

8.5.2 Challenges and limitations

The findings are generally reliable, but there are some limitations that should be acknowledged. First, inconsistencies in reporting, particularly during wave peaks, may have led to potential under- estimation or delays in capturing case counts. Many rural or semi-urban areas faced infrastructure challenges that precluded timely testing, case registration, and death certification. Hence, the data may not accurately depict the ground realities in these regions. Second, the new variants like Delta and Omicron came as a big challenge for modeling since they impart a sudden change in transmission dynamics and immune escape. The models, especially ARIMA, are not endowed with quick capability to look upon such non-linear disturbances. There is, however, variability due to other factors like behavioral aspects, public compliance, vaccine hesitancy, and changes in government policy that were not explicitly modeled.

8.5.3 Interpretation of model accuracy

Wherever the model forecasts were matched and compared with the actual case counts over time, it was obvious that ARIMA and LSTM did reasonably well in capturing general trends. Over short-term forecasting, ARIMA provided very consistent results when the trend was fairly linear. During the periods of very sharp acceleration or deceleration, however, namely the beginning of new waves, ARIMA produced very poor results.

LSTM, on the other hand, proved its capabilities in modeling temporal dependencies and responses to trend inflection points. Thus, because it processes long sequences, it could accommodate all delayed reactions, rebounds, and sporadic dips in reported cases. On the downside, some slight overfitting was visible in the LSTM parameters during low-variance periods. Both models were useful in delivering data-driven insights, but their forecasts need to be always contextualized with real-time surveillance and policy development for any decision-making process.

8.6 CONCLUSION AND FUTURE WORK

In this work, time series forecasting and predictive modeling were applied to analyze COVID-19 case trends in India. The data were culled from reputed sources including Our World in Data and official health organizations in India, and were then examined for patterns in daily and cumulative case numbers, public health response interventions, and the development of the forecasting models ARIMA and LSTM. The derived results were testimony to the efficacy of data-based approaches for the understanding and prediction of a public health crisis. While the linear models like ARIMA were useful for short-term predictions, the deep learning models such as LSTM showed maximum promise in understanding non-linear and temporal dependencies. The predictive insights generated by these models would be useful for public health authorities to make timely decisions in the imposition of lockdowns, vaccination, and allocation of healthcare resources. To sum up, the study reaffirms the paramount importance of predictive analytics in preparedness and response to pandemics. With timely calibrated data and robust models, decision-makers can take action on a proactive rather than reactive basis, thereby saving lives and minimizing social and economic disruption.

Future research could embed other rate age-dataset active testing mobility coverages at demographic levels. Thus, the incorporation of data drawn from rural under-represented areas because of infrastructure disparity would move toward accuracy and provide fairer public health predictions. Further, advanced machine learning models, Gated Recurrent Units, hybrid ARIMA-LSTM architectures, and ensemble techniques would yield model generalization and improved prediction robustness. Such real-time dashboards would provide health authorities with dynamic risk assessment on health surveillance that would help them make localized decisions. Additionally, future models could benefit from a wider behavior representation of votive mandates and social distancing. An effective marriage between epidemiological insight with AI- forward sight would make India the best equipped in present and future capacities to manage current and future outbreaks.

REFERENCES

[1] World Health Organization (WHO). *COVID-19 Global Update*, 2020.
[2] Government of India. *Ministry of Health and Family Welfare*, 2021.
[3] Johns Hopkins University. *COVID-19 Global Dashboard*, 2020.
[4] Indian Council of Medical Research (ICMR). *COVID-19 Bulletins*, 2021.
[5] R. Verma et al. "COVID-19 Trends in India: A Statistical Approach." *Journal of Epidemiology*, vol. 15, pp. 678–691, 2020.
[6] A. Sharma. "COVID-19 and Public Health Policy in India." *Indian Journal of Public Health*, vol. 65, pp. 231–240, 2021.
[7] R. Jain et al. "COVID-19 Surge in India: Key Drivers and Impacts." *Indian Medical Review*, vol. 54, pp. 205–213, 2021.
[8] S. Roy et al. "Impact of the Delta Variant on COVID-19 Dynamics in India." *Journal of Infectious Diseases*, vol. 25, pp. 489–495, 2021.
[9] S. Gupta and M. Roy. "COVID-19 Forecasting: A Linear Regression Approach." *Journal of Data Science*, vol. 26, pp. 102–115, 2020.
[10] K. Dutta, R. Lenka, P. Gupta, A. Goel and J. V. N. Ramesh. "Swift Diagnose: A High-Performance Shallow Convolutional Neural Network for Rapid and Reliable SARS-COV-2 Induced Pneumonia Detection." *EAI Endorsed Transactions on Pervasive Health and Technology*, vol. 10, 2024. https://doi.org/10.4108/eetpht.10.5581; ISSN: 2411-7145.
[11] D. K. Singhal. "COVID-19 in India: Forecasting with Linear Regression." *Computational Biology*, vol. 22, pp. 485–492, 2021.
[12] D. Mehta. "COVID-19 Case Forecasting Using Linear Regression." *Journal of Predictive Analysis*, vol. 25, pp. 85–92, 2020.
[13] R. Dalal, A. Sangwan and M. Khari. "The Bibliometrics Assessment of Opportunistic Network Protocols & Simulation Tools." *Telematics and Informatics Reports*, vol. 11, p. 100082, 2023.
[14] R. Dalal et al. "0A Comprehensive Review on Anomaly Detection Techniques for Web Data Logging." In *International Conference on Information and Communication Technology for Competitive Strategies*. Singapore: Springer Nature, 2023.
[15] V. Singh. "Challenges in Rural Healthcare during the COVID-19 Pandemic." *Indian Medical Review*, vol. 58, pp. 142–150, 2021.
[16] N. Roy. "Surveillance and Data Collection During Pandemics." *Journal of Public Health*, vol. 29, pp. 132–140, 2020.
[17] P. Shukla. "External Variables and COVID-19 Predictions." *International Journal of Forecasting*, vol. 22, pp. 431–440, 2021.
[18] S. Gupta. "The Role of Data-Driven Approaches in Public Health." *Public Health Research*, vol. 20, pp. 44–50, 2021.
[19] A. Madan, V. J. Lohani, P. Gupta and P. Rai. "Impact of COVID-19 Pandemic on Investment Behaviour: A Study of Retail Investors in India." *JIM QUEST:*

Journal of Management and Technology, vol. 19, no. 1, pp. 23–37, 2023. https://jaipuriamba.edu.in/jim-quest/UGC Care; ISSN: 0975-6280.
[20] Our World in Data. *COVID-19 Dataset*, 2020. https://ourworldindata. org/covid-cases.
[21] R. Dalal and M. Khari. "Empirical Analysis of Routing Protocols in Opportunistic Network." In *Research in Intelligent and Computing in Engineering: Select Proceedings of RICE 2020*. Singapore: Springer, 2021.
[22] R. Dalal et al. "Facial Emotion Recognition and Detection Using Convolutional Neural Networks." In *2023 3rd International Conference on Smart Generation Computing, Communication and Networking (SMART GENCON)*. IEEE, 2023.

Chapter 9

Leveraging Machine Learning for Proactive Credit Card Fraud Detection

Kapil Kumar and Manju Khari

9.1 INTRODUCTION

The current scenario of payment is changing drastically where users are empowered with new method of payments, which are easier and more convenient to accept or make a payment for purchases of goods and services. The consumer has shifted away from the traditional monetary payments in paper cash due to cash handling and logistical reasons. The main payment methods these days include the card based such as credit or debit cards and digital payments such as online wallets or internet banking. The main reason for the shift is the ease and convenience provided by such technological advancements. With the growing number of transactions, the chances of fraud also increase, and reports claim that the instances of fraud have been increasing. With e-commerce businesses thriving, the fraud rate has increased from 0.06% to 0.23% in past years [1]. This has resulted in huge losses for business and consumers as the Federal Trade Commission (FTC) report [2] underscores the severity of the issue, noting that the past few years have marked the most challenging years in history for identity theft. There are numerous techniques developed across the years for fraud detection.

9.1.1 Credit card

In the current global situation, financial institutions are increasing the availability of financial services through new services such as credit cards (hereafter referred as CCs), ATMs, mobile banking services and internet banking. Today with the fast expansion of e-commerce, the usage of CCs has grown simpler and in financial life, more essential. A CC is a card supplied as a payment system for customers. When it comes to obtaining trustworthy customers, having a strong credit history is generally a key element [3–5]. These details apply to various financial services such as rental applications, not only to CC loans. Lenders and suppliers of mortgage businesses, CC companies, retail stores, and operational corporations evaluate how promptly they can review a customer's credit points and histories. Based on it the

DOI: 10.1201/9781003635147-9

card issuer can also provide customers with additional protection when purchases are lost or stolen. Both the consumer's and company's statements can confirm the purchase of the customer.

Customers are issued a physical card by the card-issuing bank, while sellers establish bank accounts for payments. Sellers register with an acquiring bank or financial institution to accept CCs. The processor acts as an intermediary for all CC companies. CC processing involves four main players: the issuing bank, processor, issuer's fraud security team, and acquiring bank. The process typically goes through six stages: a customer makes a purchase using a CC, then the vendor runs the CC through their terminal, transmitting customer information to the acquiring bank, acquiring bank forwards information to bank (processor) who then communicates with the payee's fraud detection company for a fraud test after which in case the transaction passes the fraud test, it's sent to the issuer for final approval; upon approval, the customer receives goods/services, and finally the issuer sends a bill to the customer. During a transaction, the buyer authorizes credit availability with the card processor [6]. If approved, an authorization number is provided to the buyer, who returns it to the seller. Transactions occur in seconds, with no money exchanged yet. The buyer collects payment authorizations and sends them to the processor on a regular basis. Once a fund is received, the buyer pays the seller's bank. Sellers bear the costs of CC processing, including refunds. Online providers face the challenge of protecting customer CC and personal information. Once obtained, sellers can choose to complete payments conventionally or automate the entire procurement and payment process.

CC transactions are divided into two categories: whether a card is present or not [7]. In physical card transactions, the card is physically swiped or inserted, and the recent chip security feature holds the vendor solely responsible if it's not used for fraudulent transactions, while the issuer is responsible if the chip feature is used. Card not present transactions, like e-commerce, are highly susceptible to fraud, and the vendor absorbs costs for fraudulent transactions due to the inability to implement security measures effectively, assuming all liability. Vendors can process CC payments in various ways such as touch-tone telephone systems—suitable for businesses with fewer transactions, requiring buyers to provide ID, CC number, and purchase amount. Point-of-sale (POS) terminals—common in shops, offering options like printers or ATMs, some with Address Verification Systems (AVS) to reduce fraud. POS-like-PC terminals—for vendors dealing with mobile or mail orders, using PC software to enter customer and CC data, confirm addresses, and generate databases. Web-based solutions—offered by payment providers for online orders, allowing sellers to enter customer information, order details, and CC numbers via a web form. Providers are investing in real-time payment solutions for online businesses [8], including wireless technology and PDA (personal digital assistant) technology to facilitate faster transactions.

9.1.2 Fraud detection

Fraud is an action in which someone directly or indirectly steals the amount of a victim by making forged transactions without allowing them to know about it. There are four types of fraud. They are CC fraud, telephonic fraud, SQL injection fraud, and theft/application fraud. In fraud, a customer might complete his payment and a later billing amount would be inflated or multiple similar transactions might reflect in customer history as to mislead the customer into believing there is no wrongdoing. These kinds of frauds are always performed on credit or debit cards, due to the poor security system. Most of the users have well-known passwords or pin numbers, which can be easily captured by third parties using some algorithms. Hence, it is mandatory to build anti-fraud systems, which will be able to detect the unauthorized access of user accounts [9].

It is very challenging to detect fraud as each fraudulent transaction can be legitimized by a fraudster's operation [10]. There are plenty of ML algorithms that can be used to detect CC fraud activities and to trace the fraud users. Fraud in CC can be categorized into three types. First, card-related fraud is where the card might be stolen by an imposter and used for fraudulent transactions or the fraudster might collect sensitive data to access accounts. Secondly, a dealer or vendor might provide data to fraudsters knowingly for monetary benefits. Thirdly, website or online activities can be morphed to abate the customer and defraud. Fraudulent activities include counterfeit cards, lost or stolen cards, card-never-arrived fraud, and false application fraud, where accounts are opened using someone else's identity [11]. Keeping track of transactions and promptly reporting any fraudulent activity is crucial for safeguarding against financial loss and identity theft.

Defending against fraud involves awareness and precautionary measures such as guarding against pickpocketing and physical theft by securing your card. Users should be well educated on issues of skimming devices that steal card data and how to avoid phishing scams by not sharing card details with unverified sources. CC holders can also be better equipped when the card has robust security measures such as AVS or multi-factor authentication (MFA).

9.1.3 Research motivation

Fraud recognition approaches need extensive manually characterized datasets, which is tough to find as privacy and security concerns are of utmost importance for customers. The availability of such data is scarce. Furthermore, the fast development of fake operators makes it tough to discover novel deception designs depending on prevailing techniques. The available CC fraud detection data is highly unbalanced. Only small fractions of all CC transactions are fraudulent. As a result, detecting fraudulent transactions is difficult and imprecise [12].

Fraud detection has become a critical challenge in the financial sector, driven by the increasing volume and complexity of transactions in today's digital economy. Traditional methods of fraud detection, which often rely on manual review and rule-based systems, are no longer sufficient to cope with the sophisticated techniques employed by fraudsters. Machine learning offers a powerful alternative, leveraging advanced algorithms to analyse large datasets, identify patterns, and predict fraudulent activities with high accuracy.

The motivation for employing machine learning in fraud detection comes from its ability to continuously learn and adapt to emerging fraud trends, thus providing a proactive defined mechanism [10]. Machine learning models can process vast amounts of transaction data in real time, enabling the timely identification and prevention of fraudulent activities [13]. This not only minimizes financial losses but also enhances the security and trust of financial systems.

By incorporating these models, organizations can achieve higher detection rates and reduce false positives, leading to more efficient and effective fraud management processes. Normally, classification algorithms are incapable of detecting new forms of normal or fraudulent patterns. Machine learning to identify fraud is not good at distinguishing new behavioural patterns [14]. Discussing the cost of system deployment, both detection and prevention of fraud should be covered by the system as fraudulent transactions with a lesser amount, for instance, may lead to no income. An effective fraud recognition system should be able to accurately recognize the fraudulent transactions and make it likely to be recognized in real time [15].

9.1.4 Research objectives

The primary objectives of this research are to provide a comparative analysis of ML algorithms such as logistic regression (LR), KNN, XGBoost, LightGBM, and Random Forest (RF) and address the challenges associated with the data imbalance and dynamic nature of fraudulent activities.

- To study literature and analyse the research gaps in the field of credit card fraud detection.
- Implement various machine learning algorithms for accurate detection of fraudulent credit card transactions.
- Analyse the performance of different machine learning algorithms using parameters such as accuracy, precision, recall, and F1-score to identify the most effective approaches for fraud detection.

9.2 LITERATURE REVIEW

This section contains the research papers, which were extensively referred during the research work. Table 9.1 provides a comprehensive literature

Table 9.1 Research Papers List

S. No.	*Title*	*Authors*	*Contribution*	*Future Scope*
1	"Enhancing CC fraud detection: an ensemble machine learning approach" [16]	Khalid, A. R., Owoh, N., Uthmani, O., Ashawa, M., Osamor, J., & Adejoh, J. (2024)	Ensemble ML technique for fraud detection, combining multiple models	Design and evaluate more complex ensemble architectures (e.g., stacking, blending) for CC fraud detection
2	"Short Paper: CC Fraud Detection using LightGBM with Asymmetric Error Control" [17]	Xinyi Hu, Haiwen Chen, Ranxin Zhang (2024)	Introduces LGBM for CC fraud detection in financial institutions	Explore the impact of hyperactive parameter tuning on LGBM performance for CC fraud detection. Investigate the integration of LGBM with other machine learning models
3	"The Performance Analysis of Machine Learning Algorithms for CC Fraud Detection" [18]	Vinod Jain, Mayank Agrawal, Anuj Kumar (2023)	Evaluates various ML models based on performance (e.g., SVM, RF) for identifying fraudulent transactions	Explore the impact of data balancing techniques (e.g., SMOTE) on the performance of ML models
4	"CC fraud detection using ensemble data mining methods" [19]	Saeid Bakhtiari, Zahra Nasiri, J. Vahidi (2023)	Investigate combining ML learning model (ensemble methods) for credit card fraud detection systems	Investigate the effectiveness of combining XGBoost, a powerful gradient boosting algorithm, with other ensemble methods like bagging or stacking
5	"CC fraud detection in the era of disruptive technologies: A systematic review"[20]	Cherif, A., Badhib, A., Ammar, H., Alshehri, S., Kalkatawi, M., & Imine, A. (2023)	Provides a systematic review of CC fraud detection in the context of new technologies like blockchain and cloud computing	Investigate how cloud computing can be leveraged to enable real-time, scalable fraud detection systems

(*Continued*)

Table 9.1 (Continued)

S. No.	*Title*	*Authors*	*Contribution*	*Future Scope*
6	"The Performance Analysis of Machine Learning Algorithms for CC Fraud Detection" [20]	Khan, M. Z., Shaikh, S. A., Shaikh, M. A., Khatri, K. K., Rauf, N. M. A., Kalhoro, A., & Adnan (2023)	Evaluates various ML model based on performance (e.g., SVM, RF) for identifying fraudulent transactions	Explore the impact of data balancing techniques (e.g., SMOTE) on the performance of ML models
7	"UNMASKING FRAUDSTERS: Ensemble Features Selection to Enhance Random Forest Fraud Detection" [21]	Akazue, M. I., Debekeme, I. A., Edje, A. E., Asuai, C., & Osame, U. J. (2023)	Explores feature selection techniques to improve RF fraud detection. Feature selection can identify the most relevant features for the model, potentially reducing noise and improving accuracy	Investigate the integration of deep learning models with feature selection for fraud detection
8	"Gradient boosting techniques for CC fraud detection" [22]	Kasarapu Ramani, Irala Suneetha, Nainaru Pushpalatha, P. Harish (2022)	Analyses the use of gradient boosting techniques, such as XGBoost, for CC fraud detection	Compare the performance of XGBoost with LightGBM, another fast and efficient gradient boosting algorithm, specifically for CC fraud detection
9	"CC Fraud Detection Using Random Forest Algorithm" [23]	Tejaswini, S., Manisha, J., Vinitha, G. V. L., & Reddy, V. S (2021)	Evaluates RF algorithm for CC fraud detection	Investigate the impact of hyperparameter tuning (e.g., number of trees, maximum tree depth) on the performance of the RF algorithm for CC fraud detection

10	"CC Fraud Detection using XGBoost Classifier with a Threshold Value" [11]	Maurya, A., Kumar, A., & Prakash, S (2022)	Explores the use of XGBoost with a threshold value to classify transactions	Analyse how different threshold values affect the trade-off between precision (correctly identifying fraudulent transactions) and recall (not missing any fraudulent transactions)
11	"Exploratory analysis of CC fraud detection using machine learning techniques" [1]	M J Madhurya a, H L Gururaj a, B C Soundarya a, K P Vidyashree a, A B Rajendra (2022)	Presents an exploratory analysis of machine learning for CC fraud detection	Conduct in-depth analysis of feature importance for XGBoost and other models
12	"User Behaviour-Based Accurate Detection of CC Frauds" [24]	Sudha C. & Akila D. (2021)	Investigates user behaviour-based approaches for fraud detection, focusing on analysing user spending patterns to identify anomalies	Combine user behaviour analysis with machine learning for a holistic approach. This could leverage both historical user data and real-time transaction data for more comprehensive fraud detection.
13	"Literature Review of Different Machine Learning Algorithms for CC Fraud Detection" [25]	Nayan Uchhana, Ravi Ranjan, Shashank Sharma, Deepak Agrawal, Anurag Punde (2021)	This paper reviews different ML algorithms for fraud detection in credit card transactions.	Analyse how different cost functions (e.g., hinge loss for SVM) impact the performance of machine learning models for fraud detection

(Continued)

Table 9.1 (Continued)

S. No.	*Title*	*Authors*	*Contribution*	*Future Scope*
14	"A two-phase feature selection technique using mutual information and XGB-RFE for CC fraud detection" [26]	C. Victoria Priscilla, D. Padma Prabha (2021)	Proposes a two-phase feature selection technique using mutual information and XGB-RFE	Evaluate the effectiveness of the proposed feature selection technique on a larger dataset. Compare the two-phase technique with other feature selection methods for CC fraud detection
15	"An efficient CC fraud detection model based on machine learning methods" [12]	Naresh Kumar Trivedi, Sarita Simaiya, Umesh Kumar Lilhore, Sanjeev Kumar Sharma (2020)	Proposes a machine learning model that is both effective and computationally efficient for CC fraud detection	Explore the integration of domain knowledge (e.g., typical spending patterns, location data) with machine learning models to potentially improve fraud detection accuracy
16	"CC fraud detection using AES Technic" [27]	C. Sudha & D. Akila (2020)	While not directly relevant to machine learning, this paper focuses on using cryptography (AES) for securing the model	Cryptography can be used extensively to help and explore new domain as to understand and further research
17	"Random forest for CC fraud detection" [28]	Xuan, S., Liu, G., Li, Z., Zheng, L., Wang, S., & Jiang, C. (2018)	Evaluates the RF algorithm for CC fraud detection	Explore feature-engineering techniques to improve RF performance
18	"CC Fraud Detection Using AdaBoost and Majority Voting" [29]	Kuldeep Randhawa, Chu Kiong Loo, Manjeevan Seera, Chee Peng Lim, Ashoke K. Nandi (2018)	AdaBoost and majority voting approach for fraud detection	Explore the potential benefits of combining AdaBoost with other voting methods for improved performance

19	"LightGBM: A highly efficient gradient boosting decision tree" [13]	Ke Guolin, Qi Meng, ThomasFinley, Taifeng Wang, Wei Chen, Weidong Ma, Qiwei Ye, and Tie-Yan Liu (2017)	This paper introduces LightGBM, a fast gradient boosting algorithm with potential for CC fraud detection tasks.	Compare LightGBM with XGBoost for CC fraud detection, evaluating their performance, efficiency, and interpretability
20	"A review of fraud detection techniques: CC" [30]	Khyati Chaudhary, Jyoti Yadav, Bhawna Mallick (2014)	Reviews various fraud detection techniques, including machine learning methods and anomaly detection techniques	Analyse the effectiveness of combining machine learning approaches with anomaly detection techniques for a more comprehensive fraud detection system

review of various machine learning approaches for CC fraud detection. The reviewed papers collectively examine the effectiveness of different machine learning algorithms and techniques, such as Support Vector Machines (SVM), RF, XGBoost, and LightGBM. It contains literatures in tabular form consisting of authors name, title, year, contributions, and future scope. This literature review aims to provide a comprehensive overview of the current research in credit card fraud detection using machine learning. It will examine the various ML techniques employed, evaluate their effectiveness, and discuss the challenges and future directions in this rapidly evolving field. By synthesizing existing research, this review tries to find and highlight the potential of machine learning to revolutionize fraud detection and enhance financial security.

9.2.1 Literature gaps

Identifying literature gaps is a crucial step in academic research as it helps to pinpoint areas that have not been extensively studied or where existing research might be contradictory or incomplete.

- Existing literature discusses ensemble ML techniques; it does not explore more complex ensemble architectures like stacking or blending, which could potentially improve detection performance. Additionally, this chapter lacks an in-depth analysis of how these algorithms handle varying transaction patterns over time.
- Existing literature presents an exploratory analysis but does not conduct an in-depth analysis of feature importance for algorithms like XGBoost. Furthermore, it lacks a comparison of different feature selection methods.
- Existing literature evaluates ML algorithms but does not address the effect of data balancing techniques, such as SMOTE, on algorithms performance. Additionally, it lacks a comparative analysis of model performance on different types of fraudulent transactions.

This chapter provides a comprehensive analysis of various ML approaches for credit card fraud detection. The reviewed papers collectively highlight the effectiveness of ensemble methods, gradient boosting algorithms like XGBoost and LightGBM, and traditional algorithms such as SVM and RF. Key contributions include enhancing algorithm's accuracy through feature selection, addressing imbalanced datasets, and leveraging new technologies for improved security and scalability. The future scope of research emphasizes the need for advanced ensemble architectures, hyperactive parameter tuning, integration of domain knowledge, and the combination of machine learning with user behaviour analysis. By synthesizing existing research, this review underscores the potential of machine learning to revolutionize fraud detection, offering promising directions for developing more robust and efficient fraud detection systems.

9.3 RESEARCH METHODOLOGY

This chapter is divided into two parts, which are dataset and machine learning algorithms exploring the algorithms chosen for this research. The algorithms used in this research work are also elaborated with their respective working and pseudocode. Also, a detailed discussion of the dataset along with its attributes is presented in the dataset section.

9.3.1 Dataset description

The dataset used in this research was collected and analysed during a research collaboration of Worldline and the Machine Learning Group developed at ULB (Université Libre de Bruxelles) on big data mining and fraud detection. Dataset was downloaded from Kaggle website mentioned in the reference [31]. Figure 9.1 is a logistic regression graph of 10 rows of dataset describes the dataset and its variables. It contains transactions made by CC holders by European customers recorded over two days. The total transactions recorded are approximately 93,181 transactions, out of which 159 are termed class fraud. The dataset is highly imbalanced and skewed towards the negative class, where the positive class, representing fraudulent

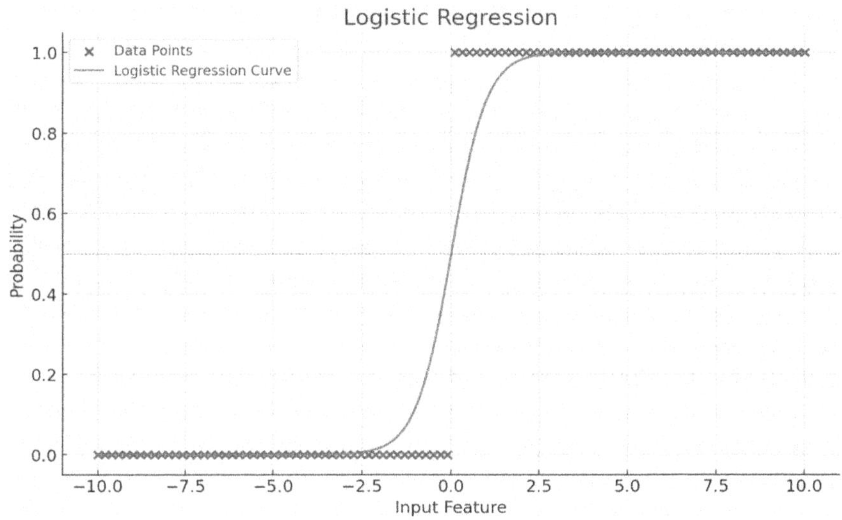

Figure 9.1 Logistic Regression Graph.

Table 9.2 Dataset Attributes Description

Variable Name	*Role*	*Type*	*Description*
Time	Time	Date	TIME OF TRANSACTION
ID	ID	Integer	
V1	Feature	Integer	LIMIT_BAL
V2	Feature	Integer	SEX
V3	Feature	Integer	MARRIAGE
V4	Feature	Integer	EDUCATION
V5	Feature	Integer	AGE
V6	Feature	Integer	PAY_0
V7	Feature	Integer	PAY_2
V8	Feature	Integer	PAY_3
V9	Feature	Integer	PAY_4
V10	Feature	Integer	PAY_5
V11	Feature	Integer	PAY_6
V12	Feature	Integer	BILL_AMT1
V13	Feature	Integer	BILL_AMT2
V14	Feature	Integer	BILL_AMT3
V15	Feature	Integer	BILL_AMT4
V16	Feature	Integer	BILL_AMT5
V17	Feature	Integer	BILL_AMT6
V18	Feature	Integer	BILL_AMT7
V19	Feature	Integer	BILL_AMT8
V20	Feature	Integer	PAY_AMT1
V21	Feature	Integer	PAY_AMT2
V22	Feature	Integer	PAY_AMT3
V23	Feature	Integer	PAY_AMT4
V24	Feature	Integer	PAY_AMT5
V25	Feature	Integer	PAY_AMT6
V26	Feature	Integer	PAY_AMT7
V27	Feature	Integer	PAY_AMT8
V28	Feature	Integer	DEFAULT PAYMENT NEXT MONTH
Y	Target	Binary	FRAUD OR NOT

transactions, constitutes only 0.17% fraud class. A detailed description of the dataset attributes is discussed in Table 9.2.

The data contain a total of 31 columns. Of these, Time, Amount, and Class are the original input variables, whereas the remaining 28 variables (V1–V28) are principal components generated using the PCA algorithm to preserve user confidentiality and security. The Class variable is the target variable, with a value of 0 indicating a non-fraudulent transaction and 1 indicating a fraudulent transaction. The transformed variables do not reveal the original feature names because of privacy and confidentiality requirements.

9.3.2 Machine learning algorithms

The study focuses on evaluating the performance of several prominent supervised learning such as LR, RF, LightGBM, and XGBoost for credit card fraud detection on the evaluation metrics of accuracy, precision, recall, and F1-Score to analyse the efficiency of these ML algorithms.

9.3.2.1 *Logistic regression*

LR stands out as a classical and efficient method for detecting CC fraud, especially in scenarios involving binary classification tasks. Its appeal lies in its simplicity-first approach, making it a preferred choice for such problems. This statistical technique excels in predicting outcomes with two possible states and can even extend its applicability to handle multinomial scenarios, offering versatility in classification tasks with multiple outcomes. One of the primary advantages of LR is its speed and suitability for binary classification problems [32]. Its straightforward nature makes it accessible even to beginners, who can easily interpret the weights assigned to each feature, facilitating algorithm updates and adaptation to new data. However, like any algorithm, it has its limitations. LR may struggle with large and complex datasets, lacking the adaptability of more flexible approaches like decision trees.

The graph in Figure 9.1 illustrates a LR algorithm fitted to a set of binary data points. The dots represent the input features and their corresponding binary outcomes (0 or 1). The curve is the LR curve, showing the predicted probability of the outcome being 1 across different values of the input feature. The logistic curve smoothly transitions from 0 to 1, reflecting the algorithm's probabilistic approach. The dashed black line at 0.5 indicates the decision boundary where the predicted probability is 50%, meaning values above this line are classified as 1, and those below as 0.

Procedure: Logistic Regression

1) Input: Training Data D
2) Begin
3) For $j = 1$ to n
4) For each training data instance Dj.
5) Set regression target value to $Rj = \frac{(yj - P(Dj))}{[P(Dj)(1 - P(Dj))]}$
6) Initialize weight for Dj instance to $[P(1|Dj)(1 - p(1|Dj))]$
7) finalize a $h(k)$ having class value (Fk) and weight (Wk)
8) Label decision with class
9) Put in (label 1)
   If Pjd > 0.5, else in (label2)
10) END

Nevertheless, the suitability of LR for CC fraud detection can be empirically assessed based on the specific context of the problem. By evaluating its performance in handling extensive transaction data, we can determine whether its limitations outweigh its advantages in this particular application

[33]. The primary methods revolve around achieving the project's objective, which is identifying potential risk factors indicating fraudulent transactions. Here are the main steps involved. Starting from general steps, LR involves the h-function calculation which entails constructing prediction, denoted as h(x), which is typically the logistic function, also known as the sigmoid function. The initial step involves building the predictive process and utilizing training data to determine the optimal parameters for the vector. These steps collectively form the foundation for LR analysis, enabling the identification and prediction of fraudulent transactions based on specific risk factors and patterns detected within the data. In the LR process, the next crucial step involves constructing the loss function, denoted as J-function. This function plays a pivotal role in evaluating the performance of the algorithm and optimizing its parameters. Typically, in a dataset consisting of m samples, each characterized by n features, the Cost and J functions are formulated as follows: The Cost function, denoted as $J(\theta)$, measures the disparity between the predicted probabilities and the actual outcomes across all samples. It quantifies the overall error of the algorithm's predictions. The J-function derived based on maximum likelihood estimation principles, whose main aim is to maximize the chances of detecting the actual outcome. These functions serve as fundamental tools for assessing the efficacy of the LR algorithm and iteratively refining its parameters to enhance predictive accuracy. To minimize the J-function and determine the regression parameter (θ) in LR, gradient descent is employed. This iterative process involves adjusting θ incrementally based on the gradient of the J-function, aiming to reach the minimum value of J. By repeatedly updating θ in the direction that decreases J, the algorithm converges towards the optimal parameter values for accurate prediction of fraudulent transactions.

9.3.2.2 *Random forest*

RF, introduced by Breiman in 1996, stands out as one of the most successful methods for addressing these demands. In the vast ocean of machine learning algorithms, one often finds themselves drowning in the sea of choices. RF is one of the top choices for data driven decisions. Presently, we are facing the challenge of handling large-scale datasets and effectively leveraging the information [34]. It requires learning algorithms that can scale with the volume of information while maintaining statistical efficiency. They operate on the straightforward yet powerful 'divide and conquer' principle: by sampling fractions of the data, randomized tree predictors are grown on each subset, which are then aggregated together to form the RF. It belongs to the class of ensemble learning techniques, which mix several different algorithms to improve total predicted accuracy.

RF is well-known for its scalability, resilience, and simplicity of handling high-dimensional data. It is extensively utilized for activities including fraud detection, illness diagnosis, and recommendation systems in a variety

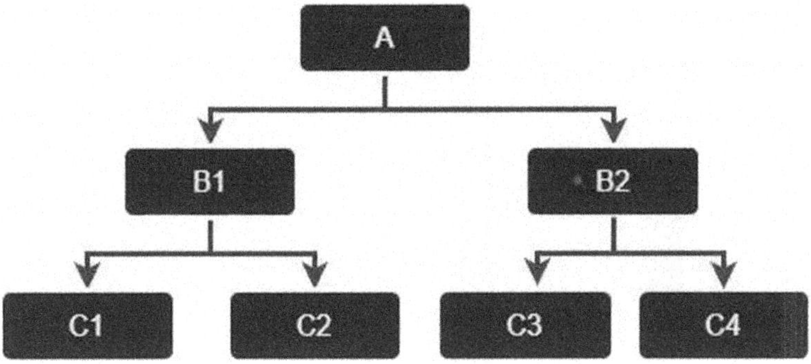

Figure 9.2 Decision Tree.

of industries, including banking, healthcare, and e-commerce. Initially, RF builds a collection of decision trees (Figure 9.2), during training. Each tree is constructed using a random subset of the training data and a random subset of features. This randomness helps in reducing overfitting and improving generalization [35]. Then the algorithm utilizes a technique called bagging, where multiple decision trees are trained on different subsets of the training data with replacement. This results in each tree being exposed to slightly different variations of the data also called bootstrap aggregating. Then during prediction, it combines the outputs of all individual decision trees. For classification tasks, it uses majority voting, where the most frequent class predicted by the trees is chosen as the final prediction, and for regression tasks it uses averaging of the predicted values.

To comprehend the RF in Figure 9.3, it is imperative to first understand its roots in decision trees. Decision trees are hierarchical structures that map out possible outcomes based on a series of binary decisions. While they offer a straightforward approach to decision-making, they are susceptible to overfitting, where the algorithm captures noise rather than genuine patterns in the data [36]. RF arises as a solution to the limitations of individual decision trees. Leveraging the power of ensemble learning, RF combines multiple decision trees to mitigate the risks of overfitting and enhance predictive accuracy. Each tree in the forest is trained on a random subset of the data and features, introducing diversity and reducing the correlation between trees.

$$g(a) = f0(a) + f1(a) + f2(a) + \ldots fn(a) \tag{9.1}$$

Ensemble learning is a technique that aggregates the predictions of multiple models to produce a final output. As depicted in equation (1), by harnessing the collective wisdom of diverse models from $f0$ to fn, ensemble learning transcends the shortcomings of individual algorithms and yields superior performance. In the case of RF, the ensemble comprises a multitude

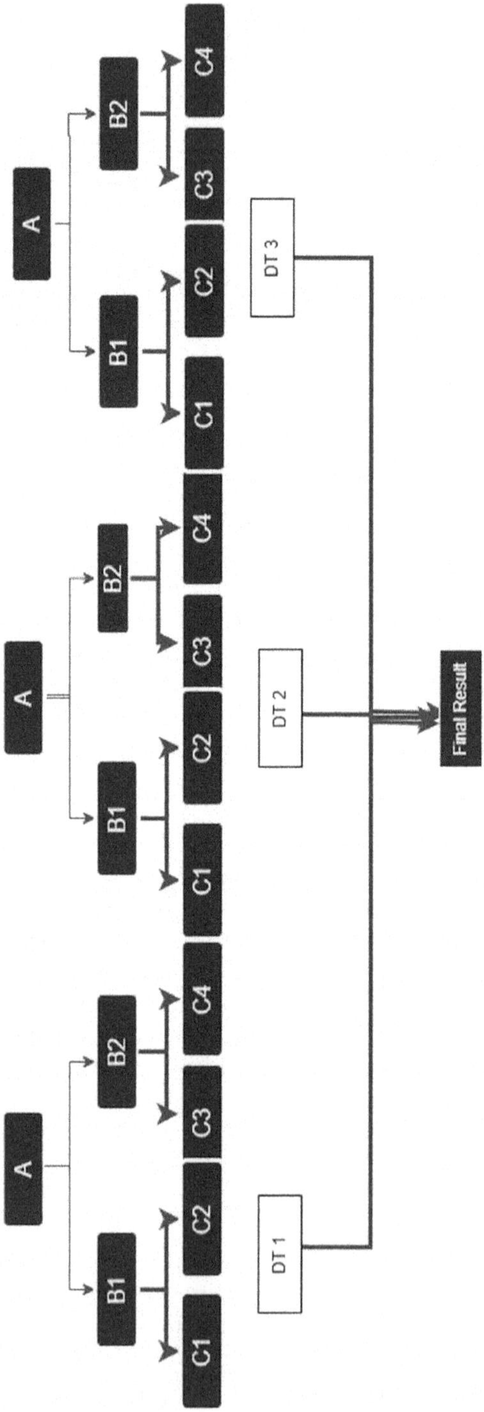

Figure 9.3 Working of Random Forest Algorithm.

of decision trees, each contributing its unique perspective to the final prediction. It can also provide insights into feature importance. It measures the decrease in impurity (e.g., Gini impurity for classification tasks) achieved by each feature when splitting the data. Features with higher decrease in impurity are considered more important in predicting the target variable. Through techniques such as bootstrapping and random feature selection, RF introduces variability into the model-building process [37]. This randomness ensures that each decision tree is trained on a different subset of data, leading to a diverse ensemble that is robust against overfitting and generalizes well to unseen data. Its ability to handle high-dimensional data, nonlinear relationships, and missing values makes it a versatile tool for a wide range of applications, from classification and regression to feature selection and anomaly detection.

Procedure: Random Forest

Preconditions: Training data D, feature and number of trees in forest T

1) function Random Forest(D, F)
2) $H \leftarrow$ empty set
3) for $k = 1. \ldots . . T$ do
4) $Dk \leftarrow$ A bootstrap sample from D
5) $Lk \leftarrow$ RandomTreeLearner(Dk, F)
6) $H \leftarrow Lk$
7) end for
8) return H
9) end function
10) function Random Tree Learner (Dk, F)
11) At each node:
12) $f \leftarrow$ very small subset of F
13) Split on best feature in f
14) Return the Learned tree
15) End Function

9.3.2.3 K-nearest neighbours (KNN)

KNN stands for k-nearest neighbours which was suggested by Cover and Hart in 1968, It is one of the simplest methods available of the data classification techniques. The term 'k-nearest neighbours' refers to using the closest 'k' neighbouring values to represent each sample. The nearest neighbour algorithm [18, 38] classifies each record in a data set based on this concept.

It operates on the method that similar data points, or instances, are likely to have similar outcomes. Based on nearest data points, new points will be categorized as per the majority of the k-neighbours, where 'k' is a user-defined parameter [39]. To classify unknown samples, the algorithm uses known examples as references, calculating the distance between the new points and all available data points. It then selects the nearest 'k' known points. According to the majority voting rule, the category assigned to the unknown data points will be the one most frequently represented among the selected points.

The algorithm is best known for its simplicity and easy to implement approach. As shown in Figure 9.4, we can calculate whether the point will belong to class A or class B. Each data point is represented in an n-dimensional space, where 'n' is the number of features. The parameter 'k' indicates the number of nearest neighbours considered when making a prediction. Choosing the right 'k' is crucial as it can significantly impact the algorithm's performance. A small 'k' value can make the model sensitive to noise, while a large 'k' value can smooth out the predictions, possibly ignoring small patterns. The distance between data points can be calculated using a distance metric, commonly the Euclidean distance for continuous variables [40].

Other metrics like Manhattan distances can also be used depending on the nature of the data and problem. For classification, the most common class is given to a new point after the algorithm identifies the given 'k' neighbours to the neighbourhood of the new instance. This is known as majority voting. For regression tasks, it predicts the outcomes based on the average of the neighbours. Overall, it is an intuitive and versatile algorithm, particularly when the dataset is relatively small and the data is well-labelled. Its simplicity and effectiveness make it a popular choice for initial classification

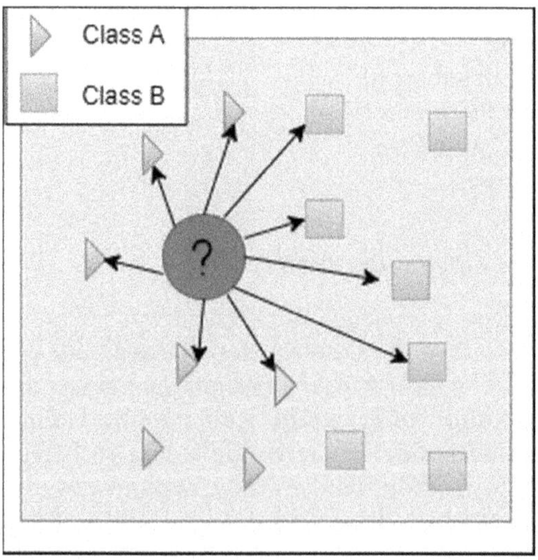

Figure 9.4 The KNN Algorithm.

and regression analyses. However, its sensitivity to irrelevant features and scale makes it computationally expensive where datasets are large.

Procedure: KNN

1) Initialize unknown Sample as *j*, known Sample as *k*
2) Function for UnSample(*j*)
3) for Sample(*k*)
4) distance (UnSamples(*j*) and Sample(*k*))
5) end
6) for
7) find the *j* smallest distances
8) locate the corresponding samples
9) Sample(*j1*) . . . Sample(*jn*)
10) assign UnSample(*j*) to the class which
11) appears more frequently
12) End Function

9.3.2.4 LightGBM

Short for Light Gradient Boosting Machine, LightGBM is a high-performance gradient boosting framework developed by Goulin Ke in 2016. It is designed for distributed and efficient training of large-scale datasets and has gained popularity for its speed and accuracy in ML tasks.

Boosting is an ensemble learning method that merges many slow learners via iterative training, forming a robust algorithm that can efficiently classify new, unseen data. As shown in Figure 9.5, LightGBM is a gradient-boosting algorithm that uses tree-based learning techniques. Unlike conventional gradient-boosting methods, LightGBM crafts decision trees using a histogram-based approach to bin continuous traits [41]. This approach lessens memory consumption and enhances the training process, building one of the fastest boosting algorithms available.

Its working as depicted in Figure 9.6 is based on the gradient boosting framework [42], which sequentially builds an ensemble of decision trees,

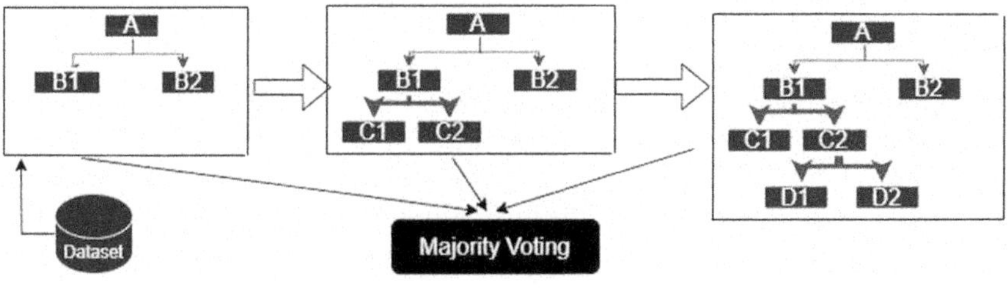

Figure 9.5 The Light Gradient Boosting Algorithm.

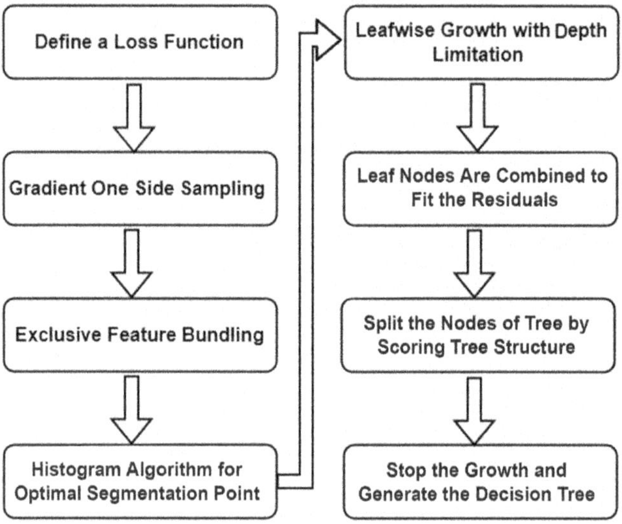

Figure 9.6 The Principle of Light Gradient Boosting.

each correcting the errors of the previous one also called gradient boosting DT (GBDT). Then it uses a histogram-based algorithm to find the best split points, which reduces memory usage and speeds up training compared to traditional exact splitting algorithms after which it implements GOSS, which stands for Gradient-based One-Side Sampling, a technique that focuses on instances with larger gradients during training, which accelerates the learning process without sacrificing performance. Finally, it bundles exclusive features with similar values into a single feature, reducing memory usage and speeding up training. It supports categorical features directly without requiring one-hot encoding, which further enhances efficiency. It is commonly used for various machine learning tasks such as classification, regression, ranking, and recommendation systems. Its efficient implementation and scalability make it suitable for large-scale datasets commonly encountered in real-world applications.

9.3.2.5 XGBoost

XGBoost (eXtreme Gradient Boosting) was developed by Tianqi Chen and Carlos Guestrin in 2014. It offers several advantages over traditional other ML algorithms, by leveraging an ensemble of decision trees, sequentially building weaker trees that improve upon the predictions of previous one which gives us a more accurate algorithm over a single decision tree. It is optimized for parallel processing, allowing it to handle massive transaction records and large datasets efficiently, which is crucial for credit card fraud detection.

Procedure: XGBoost

1) Initialize y_pred = mean of target variable (y)
2) For (num_boost_rounds in 1 to NumBoostRounds)
3) Calculate pseudo residuals (errors)
4) pseudo_residuals = y – y_pred
5) Train a new decision tree model (ft) on the features (X) and pseudo residuals
6) ft = TrainDecisionTree(X, pseudo_residuals)
7) Update the prediction by adding the weak learner's output
8) y_pred = y_pred + learning_rate * ft.Predict(X)
9) End For
10) Final_model = Ensemble(all_trained_trees)
11) End Function

Figure 9.7 shows the internal working of XGBoost, which employs regularization techniques where overfitting and model generalization is vital in fraud detection, where the model needs to adapt to new and evolving fraudulent patterns. It can be effectively tuned to address this imbalance, as CC fraud data is inherently imbalanced ensuring it focuses on identifying the rare fraudulent cases.

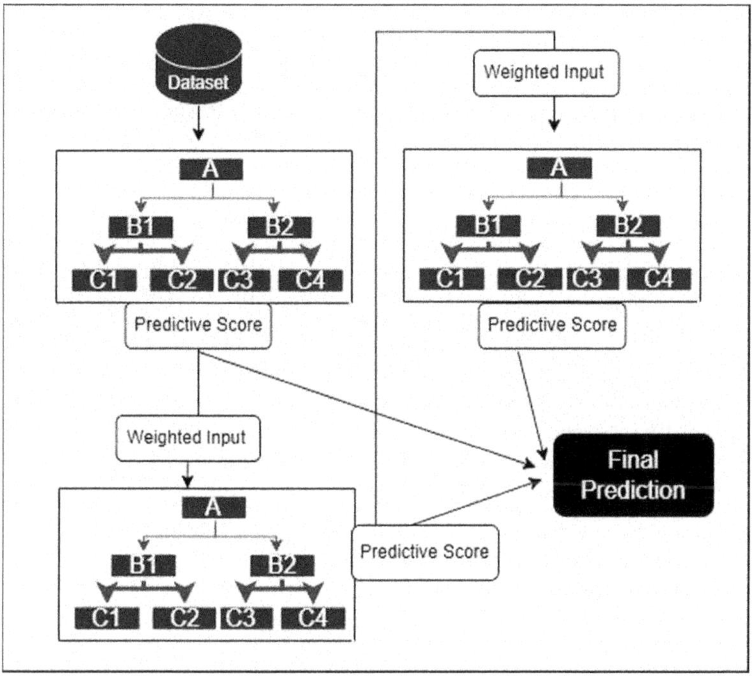

Figure 9.7 XGBoost Algorithm Working.

As depicted in Figure 9.7 the algorithm works on the principle of ensemble technique where weak learners are trained on subsets of data and then combining the result to give final prediction. Initially, the data is fed to the decision tree and the misclassified output data is assigned a weight Wi based on accuracy and fed to the next decision tree and as weighted input. Similarly, the process follows for all decision trees iteratively decreasing the error. New features might be created based on domain knowledge or feature interaction analysis to enhance the algorithm's learning ability. The algorithm is trained on the prepared dataset, which involves defining objective function, building ensembles and hyper-parameter tuning. Finally, the model is evaluated on a test set.

This chapter has provided a comprehensive description of the dataset and its attributes and also contains detailed information regarding ML algorithms. There are several algorithms such as LR, RF, LightGBM, XGBoost, and KNN. All algorithms are discussed with their detailed description starting from their definition to working along with pseudo code for those algorithms.

9.4 EXPERIMENTS AND RESULTS

This chapter is about the results obtained after the training of each algorithm and then plotted against the respective confusion matrix. The results shown in Table 9.3 (Section 9.4.6) have been discussed extensively for each algorithm. The training was completed on all the input data containing 92,183 transactions containing 28 independent variables and one dependent variable.

9.4.1 Logistic regression

While it achieved an accuracy of 91.82% as shown in Table 9.3 (Section 9.4.6), Figure 9.8 shows the result of the LR algorithm in the form of a confusion matrix. A deeper look revealed concerning shortcomings. The algorithm's F1-score 16.43% and precision 9.01% were concerningly low, indicating a struggle to accurately identify fraudulent transactions. Analysing the confusion matrix further exposed a significant weakness: a high false negative (FN) rate. It misclassified a substantial number of fraudulent transactions as legitimate, potentially allowing fraudulent activity to slip through the cracks. In the domain of credit card fraud detection, accuracy alone isn't the parameter to find the best algorithms.

The cost of misclassification is skewed. Failing to detect fraud (FN) carries a heavier weight compared to mistakenly flagging a legitimate transaction (FP). In an FN scenario, financial losses can occur. Conversely, an FP might cause inconvenience for the customer but doesn't result in direct financial harm. Given this cost asymmetry, LR's high FN rate renders it a suboptimal

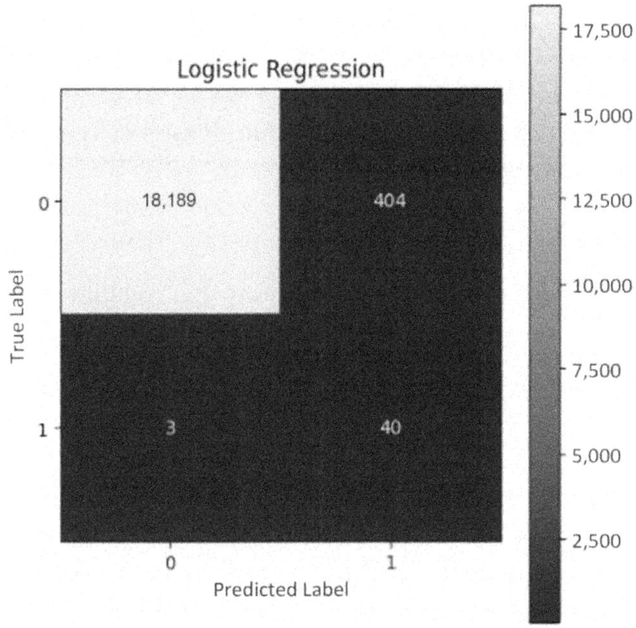

Figure 9.8 Confusion Matrix of LR.

choice for this specific task. This highlights the importance of considering not just accuracy but also metrics like F1 score and precision when evaluating algorithms for fraud detection.

In this experiment, execution of the LR produces the result in the form of a confusion matrix. Confusion matrix is the summary of the classification report that represents the performance of the trained algorithm, which includes LR here.

As shown in the confusion matrix in Figure 9.8, the true positives (TP) represent samples correctly identified as fraudulent, while true negatives (TN) represent samples correctly identified as legitimate. False positives (FP) are samples predicted as fraudulent but are actually legitimate, and false negatives (FN) are samples predicted as legitimate but are actually fraudulent.

Recall: It emphasizes the algorithm's ability to capture all the relevant instances. It tells you, out of all the actual positive cases, how many were correctly identified by the algorithm.

$$Recall = \frac{TP}{TP + FN} \tag{9.2}$$

Precision: Precision focuses on the algorithm's ability to identify relevant instances. It tells you, out of all the positive predictions made, how many were actually true positives.

$$Precision = \frac{TP}{TP + FP} \tag{9.3}$$

Accuracy: It is the most straightforward metric, representing the overall correctness of your classification algorithm. It essentially tells that what proportion of predictions were correct.

$$Accuracy = \frac{TP + TN}{TP + TN + FP + FN} \tag{9.4}$$

F1-Score: The F1-score is a harmonic mean that combines precision and recall into a single metric. It provides a balanced view of how well the algorithm performs by taking both precision and recall into account.

$$F1 - Score = \frac{2 * Recall * Precision}{Recall + Precision} \tag{9.5}$$

9.4.2 KNN

Figure 9.9 shows the result of the KNN algorithm in the form of a confusion matrix. The KNN algorithm achieved an accuracy of 89.06% as shown in Table 9.3 (Section 9.4.6), in detecting fraudulent credit card transactions. This accuracy is lower than LR. However, further analysis of the confusion

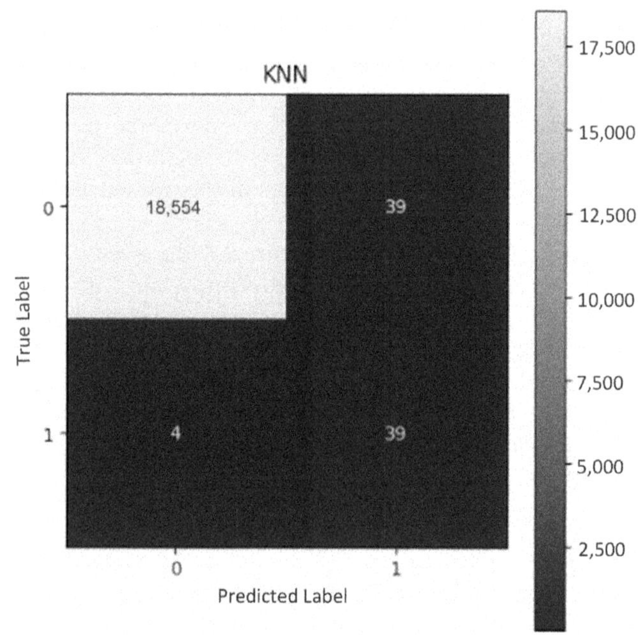

Figure 9.9 Confusion Matrix of KNN.

matrix reveals limitations with KNN. While the FN rate is significantly lower than LR, it is not ideal. There were still 39 fraudulent transactions, which the algorithm incorrectly classified as legitimate. Additionally, the algorithm generated a high number of FP. This means that many legitimate transactions were flagged as fraudulent, potentially causing inconvenience to customers. Overall, KNN offers a balanced approach between identifying fraudulent transactions and flagging legitimate transactions. However, the high FP rate and some FNs indicate that there might be better-suited algorithms for credit card fraud detection in this case.

Recall: It emphasizes the algorithm's ability to capture all the relevant instances. It tells you, out of all the actual positive cases, how many were correctly identified by the algorithm.

$$Recall = \frac{TP}{TP + FN} \tag{9.6}$$

Precision: Precision focuses on the algorithm's ability to identify relevant instances. It tells you, out of all the positive predictions made, how many were actually true positives.

$$Precision = \frac{TP}{TP + FP} \tag{9.7}$$

Accuracy: It is the most straightforward metric, representing the overall correctness of your classification algorithm. It essentially tells that what proportion of predictions were correct.

$$Accuracy = \frac{TP + TN}{TP + TN + FP + FN} \tag{9.8}$$

F1-Score: The F1-score is a harmonic mean that combines precision and recall into a single metric. It provides a balanced view of how well the algorithm performs by taking both precision and recall into account.

$$F1 - Score = \frac{2 * Recall * Precision}{Recall + Precision} \tag{9.9}$$

9.4.3 XGBoost

Figure 9.10 shows the result of the XGBoost algorithm in the form of a confusion matrix. XGBoost emerged as a promising algorithm in our evaluation of credit card fraud detection algorithms. It delivers promising results with an impressive accuracy of 96.96% as shown in Table 9.3 (Section 9.4.6), indicating a very high success rate in correctly classifying both fraudulent and legitimate transactions. This is further supported by a healthy F1 score of 91.95%, which balances precision 90.90% and recall 93.02%. Precision tells us how many flagged transactions were truly fraudulent, while recall indicates the algorithm's ability to catch a good portion of actual fraud. In

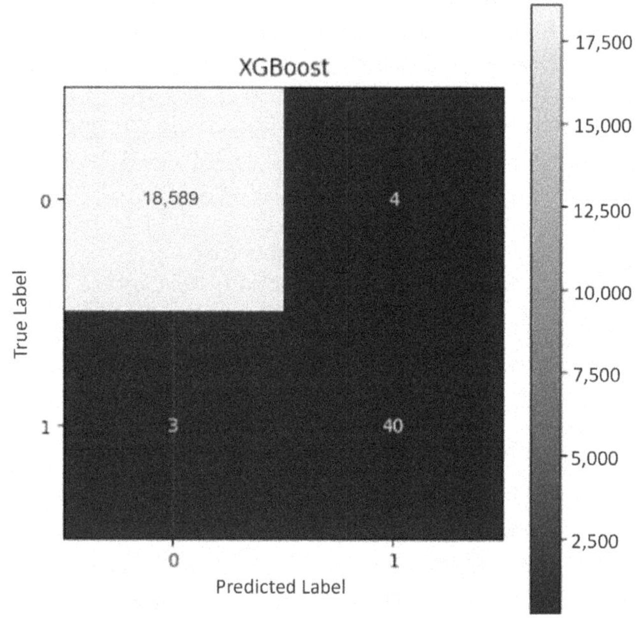

Figure 9.10 Confusion Matrix of XGBoost.

this case, XGBoost demonstrates a good balance between minimizing false positives and false negatives. However, a deeper look reveals a potential area for improvement. While the recall of 93.02% suggests XGBoost catches a significant portion of fraud, it also implies that nearly 7% of fraudulent transactions might be missed. Depending on the cost associated with these missed fraudulent transactions (financial losses), this might be an area for further optimization. Overall, XGBoost emerges as a strong contender for credit card fraud detection based on its high accuracy, balanced precision and recall, and good F1 score. However, further exploration into minimizing the number of missed fraudulent transactions could be valuable for real-world applications.

Recall: It emphasizes the algorithm's ability to capture all the relevant instances. It tells you, out of all the actual positive cases, how many were correctly identified by the algorithm.

$$Recall = \frac{TP}{TP + FN} \tag{9.10}$$

Precision: Precision focuses on the algorithm's ability to identify relevant instances. It tells you, out of all the positive predictions made, how many were actually true positives.

$$Precision = \frac{TP}{TP + FP} \tag{9.11}$$

Accuracy: It is the most straightforward metric, representing the overall correctness of your classification algorithm. It essentially tells that what proportion of predictions were correct.

$$Accuracy = \frac{TP + TN}{TP + TN + FP + FN} \tag{9.12}$$

F1-Score: The F1-score is a harmonic mean that combines precision and recall into a single metric. It provides a balanced view of how well the algorithm performs by taking both precision and recall into account.

$$F1 - Score = \frac{2 * Recall * Precision}{Recall + Precision} \tag{9.13}$$

9.4.4 Random forest

Figure 9.11 shows the result in the form of a confusion matrix. The RF algorithm achieves a high accuracy of 95.57% shown in Table 9.3 (Section 9.4.6), indicating it correctly classifies nearly 96% of transactions. This is a positive result, as it suggests the algorithm can effectively distinguish between fraudulent and legitimate transactions. With the

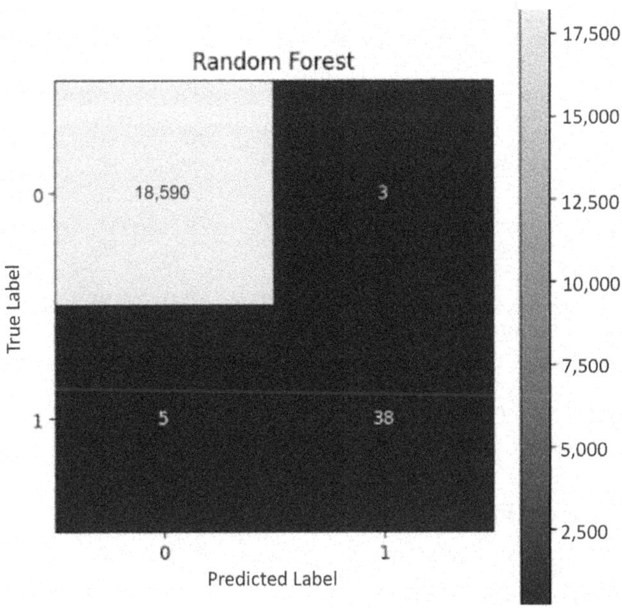

Figure 9.11 Confusion Matrix of Random Forest.

precision of 88.37%, we see that the algorithm has a low rate of false positives. In other words, out of the transactions flagged as fraud by the algorithm, 88.37% are actually fraudulent. This is desirable because it minimizes the amount of money lost to fraud for the banks as well as to customers by incorrectly flagging legitimate transactions. However, the recall of 88.37% reveals a potential area for improvement. Recall tells us the proportion of actual fraudulent transactions that the algorithm is able to identify. A recall of 88.37% suggests that the algorithm might miss nearly 12% of fraudulent transactions. This could be a concern depending on the severity of consequences associated with missed fraud. F1-score offers a balanced view of both precision and recall, and here it is at 90.47%. This indicates a good overall performance of the algorithm but also highlights the opportunity to improve recall while maintaining a good precision.

Recall: It emphasizes the algorithm's ability to capture all the relevant instances. It tells you, out of all the actual positive cases, how many were correctly identified by the algorithm.

$$Recall = \frac{TP}{TP + FN} \tag{9.14}$$

Precision: Precision focuses on the algorithm's ability to identify relevant instances. It tells you, out of all the positive predictions made, how many were actually true positives.

$$Precision = \frac{TP}{TP + FP} \tag{9.15}$$

Accuracy: It is the most straightforward metric, representing the overall correctness of your classification algorithm. It essentially tells that what proportion of predictions were correct.

$$Accuracy = \frac{TP + TN}{TP + TN + FP + FN} \tag{9.16}$$

F1-Score: The F1-score is a harmonic mean that combines precision and recall into a single metric. It provides a balanced view of how well the algorithm performs by taking both precision and recall into account.

$$F1 - Score = \frac{2 * Recall * Precision}{Recall + Precision} \tag{9.17}$$

9.4.5 LightGBM

Figure 9.12 shows the results in the form of a confusion matrix. The algorithm has achieved the accuracy of 94.62% shown in Table 9.3 (Section 9.4.6). This can be improved upon to minimize inconvenience caused

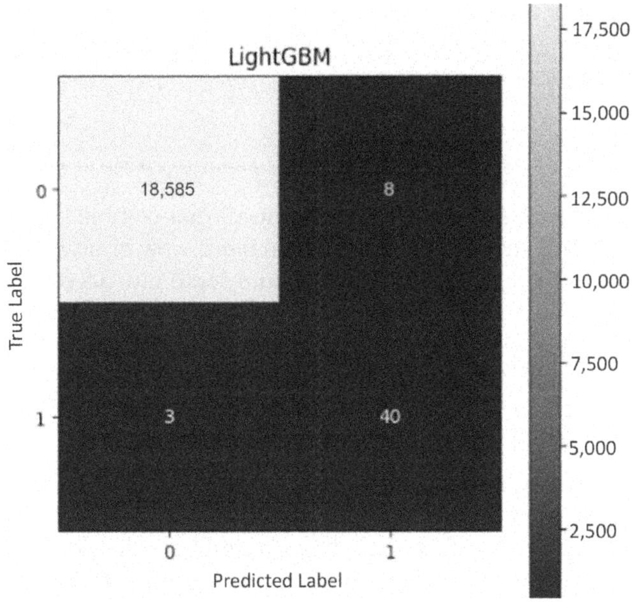

Figure 9.12 Confusion Matrix of LightGBM.

to customers by incorrectly flagging legitimate transactions. On the other hand, the recall of 92.02% is a strength of the LightGBM algorithm. It suggests that the algorithm catches a very high proportion of fraudulent transactions, only missing about 8%. F1-score offers a balanced view of both precision and recall, and here it is at 87.91%. This indicates that the algorithm might prioritize capturing a high proportion of fraudulent transactions (high recall) even at the cost of some increase in false positives (lower precision). This could be a reasonable strategy depending on the severity of consequences associated with missed fraud.

Recall: It emphasizes the algorithm's ability to capture all the relevant instances. It tells you, out of all the actual positive cases, how many were correctlyidentified by the algorithm.

$$Recall = \frac{TP}{TP + FN} \tag{9.18}$$

Precision: Precision focuses on the algorithm's ability to identify relevant instances. It tells you, out of all the positive predictions made, how many were actually true positives.

$$Precision = \frac{TP}{TP + FP} \tag{9.19}$$

Accuracy: It is the most straightforward metric, representing the overall correctness of your classification algorithm. It essentially tells that what proportion of predictions were correct.

$$Accuracy = \frac{TP + TN}{TP + TN + FP + FN} \quad (9.20)$$

F1-Score: The F1-score is a harmonic mean that combines precision and recall into a single metric. It provides a balanced view of how well the algorithm performs by taking both precision and recall into account.

$$F1 - Score = \frac{2 * Recall * Precision}{Recall + Precision} \quad (9.21)$$

9.4.6 Comparative analysis of algorithms

CC fraud detection has become a critical aspect of financial security, driven by the increasing volume of online transactions and the sophistication of fraudulent activities. Various algorithms are employed to identify fraudulent transactions, each with distinct strengths and weaknesses. Table 9.3 shows comparative analysis of algorithms used in this research work.

As a comparison shown in Table 9.3 and Figure 9.13, XGBoost achieves the highest accuracy 96.96% among the five algorithms, indicating it correctly classified the most data points. However, high accuracy alone can be misleading, especially for imbalanced class problems. For instance, an algorithm that always predicts the majority class will have a high accuracy but might not be useful for identifying the minority class. Looking at the precision and recall for a more nuanced understanding, precision reflects the proportion of positive predictions that were truly positive. KNN precision is 69.08%, which means a larger portion of its positive predictions were indeed correct. In contrast, LR has the lowest precision at 9.01%, implying most of its positive predictions were actually false positives.

Recall, on the other hand, defines the proportion of actual positive cases that were correctly classified by the algorithm. LR has the highest recall at 93.02%, indicating it captured most of the positive cases. However, this came at the cost of many false positives as reflected by its low precision. The F1-score tries to provide a balanced approach among recall and precision. It

Table 9.3 Comparative Analysis of Results

Algorithm	*LR*	*KNN*	*XGBoost*	*RF*	*LGBM*
"Accuracy"	91.82%	89.06%	96.96%	95.57%	94.94%
"Precision"	09.01%	69.08%	90.90%	88.37%	83.33%
"Recall"	93.02%	90.70%	93.02%	88.37%	92.02%
"FI score"	16.43%	64.46%	91.95%	90.47%	87.91%

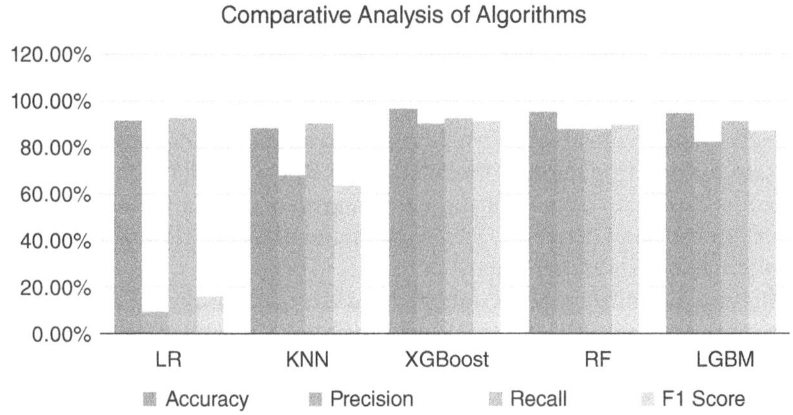

Figure 9.13 Result Comparison Graph.

provides a single metric that looks at both an algorithm's ability to identify true positives (recall) and avoid false positives (precision). XGBoost has the highest F1-score 91.95%, suggesting it achieves a good balance between precision and recall. This is further supported by its high accuracy and balanced precision and recall scores.

The chapter has presented a holistic view of all the five algorithms. Overall, XGBoost appears to be the best performing algorithm in this scenario based on the metrics provided. It has the highest accuracy and F1-score, indicating good overall performance and a balance between precision and recall. However, depending on the specific problem and the relative importance of precision and recall, other algorithms like KNN or RF might be preferable. It's important to consider the trade-offs between these metrics when choosing an algorithm for a particular task.

9.5 CONCLUSION AND FUTURE SCOPE

This research is centred on the investigation of CC fraud detection algorithms leveraging various machine learning classification algorithms. The primary aim is to delve deeply into the intricacies of training and testing these algorithms, while also discerning the most effective strategies for processing the dataset and selecting the optimal classification algorithm for CC transaction data. Through meticulous experimentation involving different algorithmic configurations and sampling methodologies, the study endeavours to refine fraud detection outcomes. Notably, the adoption of oversampling techniques to counter dataset imbalance resulted in outcomes consistent with expectations, as reflected in the confusion matrix. Additionally, the research highlights the enduring advantages of RF, particularly in its adeptness at accommodating differential data processing requirements, followed closely by XGBoost and LightGBM. Comparisons drawn with existing literature underscore the significance of this study's findings. Specifically, it identifies LR (oversampling) as a standout machine learning algorithm for CC fraud detection, showcasing robust accuracy levels. This comprehensive analysis contributes significantly to the field, offering valuable insights into the intricate dynamics of CC fraud detection and paving the way for enhanced methodologies in the future. Data variation-based working of machine learning algorithms and finding suitable algorithm for fraud detection. Precise detection of CC fraud was suggested in this module based on user activity.

This method evaluates the labelled dataset and machine learning algorithms are trained with datasets and predictions are stored and evaluated against labelled data.

While this research project successfully completed the training of the CC fraud detection algorithm, numerous avenues for future improvement have been identified. One potential enhancement involves combining multiple classifiers after training the optimal algorithm to evaluate detection performance comprehensively. This approach offers increased flexibility and potential improvements in accuracy by leveraging the strengths of different algorithms.

Another area for advancement is the integration of deep learning methodologies, akin to neural networks. Deep learning, distinct from traditional machine learning, operates on unstructured or unlabelled data, obviating the need for explicit feature selection by developers. This could streamline the training process for CC fraud algorithms, potentially leading to simplified and more efficient detection mechanisms. Despite efforts to utilize the XGBoost algorithm, recognized for its efficacy, time constraints limited the optimization of its parameters. Further fine-tuning of XGBoost parameters could potentially yield enhanced performance, maximizing the algorithm's capabilities in fraud detection scenarios.

Additionally, there is potential for improvement through the augmentation of the dataset. While the research utilized an existing dataset, further data extraction from network sources could progressively increase the volume of available data for training. This larger dataset holds promise for improving predictive performance and enhancing detection accuracy, offering a valuable avenue for future research in CC fraud detection. Furthermore, testing the machine learning classifier against various types of attacks and analysing its performance under duress could inform security measures, ensuring robustness against potential threats.

REFERENCES

[1] Madhurya, M. J., Gururaj, H. L., Soundarya, B. C., Vidyashree, K. P., & Rajendra, A. B. (2022). Exploratory analysis of CC fraud detection using machine learning techniques. *Global Transitions Proceedings,* 3(1), 31–37.

[2] Alenzi, H. Z., & Aljehane, N. O. (2020). Fraud detection in credit cards using logistic regression. *International Journal of Advanced Computer Science and Applications,* 11(12).
[3] Dighe, D., Patil, S., & Kokate, S. (2018, August). Detection of credit card fraud transactions using machine learning algorithms and neural networks: A comparative study. In *2018 Fourth International Conference on Computing Communication Control and Automation (ICCUBEA)* (pp. 1–6). IEEE.
[4] Deepthi, B., Gupta, P., Rai, P., & Arora, H. (2022, May). Assessing the dynamics of AI driven technologies in Indian banking and financial sector. *Vision (ABDC Ranked Journal).* https://doi.org/10.1177/09722629221087371; ISSN: 0972-2629.
[5] Bhawna, Gupta, P., & Rai, P. (2024, July–December). Blockchain deployment in the banking sector: A comprehensive literature review. *JIM Quest- Journal of Management and Technology,* 20(2), 65–91.
[6] Raj, S. B. E., & Portia, A. A. (2011, March). Analysis on credit card fraud detection methods. In *2011 International Conference on Computer, Communication and Electrical Technology (ICCCET)* (pp. 152–156). IEEE.
[7] Mahesh, B. (2020). Machine learning algorithms-a review. *International Journal of Science and Research (IJSR) [Internet],* 9(1), 381–386.
[8] Varmedja, D., Karanovic, M., Sladojevic, S., Arsenovic, M., & Anderla, A. (2019, March). Credit card fraud detection-machine learning methods. In *2019 18th International Symposium Infoteh-Jahorina (Infoteh)* (pp. 1–5). IEEE.
[9] Dornadula, V. N., & Geetha, S. (2019). Credit card fraud detection using machine learning algorithms. *Procedia Computer Science,* 165, 631–641.
[10] Teja, K. T., Anuhya, P., & Ruchitha, P. (2022). CC fraud detection using random forest algorithm in machine learning and data science. *International Research Journal of Modernization in Engineering Technology and Science,* 4(5), 4.
[11] Maurya, A., Kumar, A., & Prakash, S. (2022). *CC Fraud Detection Using XGBoost Classifier with a Threshold Value.* https://doi.org/10.21203/rs.3.rs-1722294/v1.
[12] Trivedi, N. K., Simaiya, S., Lilhore, U. K., & Sharma, S. K. (2020). An efficient credit card fraud detection model based on machine learning methods. *International Journal of Advanced Science and Technology,* 29(5), 3414–3424.
[13] Ke, G., Meng, Q., Finley, T., Wang, T., Chen, W., Ma, W., . . . Liu, T. Y. (2017). Lightgbm: A highly efficient gradient boosting decision tree. *Advances in Neural Information Processing Systems,* 30.
[14] Adepoju, O., Wosowei, J., & Jaiman, H. (2019, October). Comparative evaluation of credit card fraud detection using machine learning techniques. In *2019 Global Conference for Advancement in Technology (GCAT)* (pp. 1–6). IEEE.
[15] Salaudeen, L. G., Gabi, D., Muhammad, G., & Suru, H. U. (2024). Light gradient boosting machine (LGBM) for CC fraud detection in financial institutions. *Direct Research Journal of Engineering and Information Technology,* 12, 19–34. https://doi.org/10.26765/DRJEIT17933661.
[16] Khalid, A. R., Owoh, N., Uthmani, O., Ashawa, M., Osamor, J., & Adejoh, J. (2024). Enhancing CC fraud detection: An ensemble machine learning approach. *Big Data and Cognitive Computing,* 8(1), 6.

[17] Hu, X., Chen, H., & Zhang, R. (2019, September). Short paper: Credit card fraud detection using LightGBM with asymmetric error control. In *2019 second international conference on artificial intelligence for industries (AI4I)* (pp. 91–94). IEEE.

[18] Jain, V., Agrawal, M., & Kumar, A. (2020, June). Performance analysis of machine learning algorithms in credit cards fraud detection. In *2020 8th International Conference on Reliability, Infocom Technologies and Optimization (Trends and Future Directions)(ICRITO)* (pp. 86–88). IEEE. https://doi.org/10.1109/ICRITO48877.2020.9197762.

[19] Bakhtiari, S., Nasiri, Z., & Vahidi, J. (2023). Credit card fraud detection using ensemble data mining methods. *Multimedia Tools and Applications,* 82(19), 29057–29075.

[20] Cherif, A., Badhib, A., Ammar, H., Alshehri, S., Kalkatawi, M., & Imine, A. (2023). CC fraud detection in the era of disruptive technologies: A systematic review. *Journal of King Saud University-Computer and Information Sciences,* 35(1), 145–174.

[21] Akazue, M. I., Debekeme, I. A., Edje, A. E., Asuai, C., & Osame, U. J. (2023). Unmasking fraudsters: Ensemble features selection to enhance random forest fraud detection. *Journal of Computing Theories and Applications,* 1(2), 201–211.

[22] Ramani, K., Suneetha, I., Pushpalatha, N., & Harish, P. (2022). Gradient boosting techniques for credit card fraud detection. *Journal Of Algebraic Statistics,* 13(3), 553–558; Tejaswini, S., Manisha, J., Vinitha, G. V. L., & Reddy, V. S. Credit card fraud detection using random forest algorithm.

[23] Sudha, C., & Akila, D. (2021). User behaviour-based accurate detection of CC frauds. In *Proceedings of First International Conference on Mathematical Modeling and Computational Science: ICMMCS 2020* (pp. 363–373). Singapore: Springer.

[24] Ramani, K., Suneetha, I., Pushpalatha, N., & Harish, P. (2022). Gradient Boosting Techniques for Credit Card Fraud Detection. *Journal of Algebraic Statistics, 13*(3).

[25] Priscilla, C. V., & Prabha, D. P. (2021). A two-phase feature selection technique using mutual information and XGB-RFE for CC fraud detection. *International Journal of Advanced Technology and Engineering Exploration (IJATEE)*, 8(85).

[26] Sudha, C., & Akila, D. (2020). CC fraud detection using AES technic. In *Intelligent Computing and Innovation on Data Science: Proceedings of ICTIDS 2019* (pp. 91–98). Singapore: Springer.

[27] Xuan, S., Liu, G., Li, Z., Zheng, L., Wang, S., & Jiang, C. (2018, March). Random forest for CC fraud detection. In *2018 IEEE 15th International Conference on Networking, Sensing and Control (ICNSC)* (pp. 1–6). IEEE.

[28] Randhawa, K., Loo, C. K., Seera, M., Lim, C. P., & Nandi, A. K. (2018, February 15). CC fraud detection using AdaBoost and majority voting. *IEEE Access,* 6, 14277–14283. https://doi.org/10.1109/ACCESS.2018.2806420.

[29] Chaudhary, K., Yadav, J., & Mallick, B. (2012). A review of fraud detection techniques: CC. *International Journal of Computer Applications*, 45(1), 39–44.

[30] CC Fraud Detection Dataset. *Kaggle*. https://www.kaggle.com/datasets/mlg-ulb/creditcardfraud; Itoo, F., Meenakshi, & Singh, S. (2021).

[31] Awoyemi, J. O., Adetunmbi, A. O., & Oluwadare, S. A. (2017, October). Credit card fraud detection using machine learning techniques: A comparative analysis. In *2017 International Conference on Computing Networking and Informatics (ICCNI)* (pp. 1–9). IEEE.

[32] Niveditha, G., Abarna, K., & Akshaya, G. V. (2019). Credit card fraud detection using random forest algorithm. *International Journal of Scientific Research in Computer Science, Engineering and Information Technology (IJSRCSEIT)*, 5(2), 301–306.

[33] Dileep, M. R., Navaneeth, A. V., & Abhishek, M. (2021, February). A novel approach for credit card fraud detection using decision tree and random forest algorithms. In *2021 Third International Conference on Intelligent Communication Technologies and Virtual Mobile Networks (ICICV)* (pp. 1025–1028). IEEE.

[34] Aung, M. H., Seluka, P. T., Fuata, J. T. R., Tikoisuva, M. J., Cabealawa, M. S., & Nand, R. (2020, December). Random forest classifier for detecting credit card fraud based on performance metrics. In *2020 IEEE Asia-Pacific Conference on Computer Science and Data Engineering (CSDE)* (pp. 1–6). IEEE.

[35] Imandoust, S. B., & Bolandraftar, M. (2013). Application of k-nearest neighbor (KNN) approach for predicting economic events: Theoretical background. *International Journal of Engineering Research and Applications*, 3(5), 605–610.

[36] Kiran, S., Guru, J., Kumar, R., Kumar, N., Katariya, D., & Sharma, M. (2018). Credit card fraud detection using Naïve Bayes model based and KNN classifier. *International Journal of Advance Research, Ideas and Innovations in Technology*, 4(3), 44.

[37] Ke, G., Meng, Q., Finley, T., Wang, T., Chen, W., Ma, W., . . . Liu, T. Y. (2017). Lightgbm: A highly efficient gradient boosting decision tree. *Advances in Neural Information Processing Systems*, 30.

[38] Kumar, K., & Khari, M. (2025, February). Efficient intrusion detection and trust management in federated learning systems. In *2025 2nd International Conference on Computational Intelligence, Communication Technology and Networking (CICTN)* (pp. 68–72). IEEE.

[39] Salaudeen, L. G., Gabi, D., Muhammad, G., & Suru, H. U. (2024). Light gradient boosting machine (LGBM) for CC Fraud detection in financial institutions. *Direct Research Journal of Engineering and Information Technology*, 12, 19–34. https://doi.org/10.26765/DRJEIT17933661.

[40] Kumar, A., Kumar, K., Khari, M., & Kumar, S. (2024, January). Harnessing smart farming to combat climate-induced agricultural challenges using machine learning. In *International Conference on Computational Intelligence in Communications and Business Analytics* (pp. 62–69). Cham, Switzerland: Springer Nature.

[41] Gupta, A., Lohani, M. C., & Manchanda, M. (2021). Financial fraud detection using naive bayes algorithm in highly imbalanced data sets. *Journal of Discrete Mathematical Sciences and Cryptography,* 24(5), 1559–1572.

[42] Maurya, A., Kumar, A., & Prakash, S. (2022). *CC Fraud Detection Using XGBoost Classifier with a Threshold Value.* https://doi.org/10.21203/rs.3.rs-1722294/v1.

Chapter 10

Systematic Study of Machine Learning Algorithms for Botnet Detection

Kapil Kumar and Manju Khari

10.1 INTRODUCTION

A botnet, a term that comes from "robot" and "network", is a group of computers, a whole network that has been infected by malware and is under the control of a single malicious entity. This network of devices not only includes computers but also smartphones, Internet of Things (IoT) devices like security cameras and servers and every device that has internet connection. It is equivalent to an army of sleeper agents that execute a variety of terrorist activities. Now the question arises, what does it do? So once a device is under control of the host, or it can also be said once a device has been hacked, then the host has access to each and every thing on that device and it can do anything on the device ranging from using personal information to sending texts and emails, making transactions and anything else. On a broader scale, it can launch cyber-attacks and get control over another device. If an individual launches it, it is called a "botmaster" as shown in Figure 10.1, whereas if it is performed by a group, it is called a "botnet" [1].

Botnets have a great significance in cyber security which can't be overstated. The attacks due to botnets have increased rapidly because of increased human reliance on the internet and internet connected devices for critical services. A study [2] indicates that by 2025, the number of IoT devices will be a trillion. As of 2016, there have been 9 billion smart devices excluding smartphones, tablets, personal computers, and laptops, and by 2020, it is assumed to increase to 28.1 billion. Cisco's annual report [3] also assumed a hike in the number of internet users from 51% to 66% since 2018. Also, the number of IoT and non-IOT devices having Internet access will increase manifold, approximately by three times. The internet has become everyone's weakness and an easy way of survival. It has been converted into a necessity of life just like any other basic needs. This is because in today's time everything is digitalized and our life majorly revolves around IoT. This was readily increased after the world was hit by the Covid-19 pandemic. From education to jobs to basic grocery shopping, everything got digitalized. With these positive sides of IoT and internet devices did come the repercussions, as system hacking and cybercriminal attacks increased. Since the first variation

DOI: 10.1201/9781003635147-10

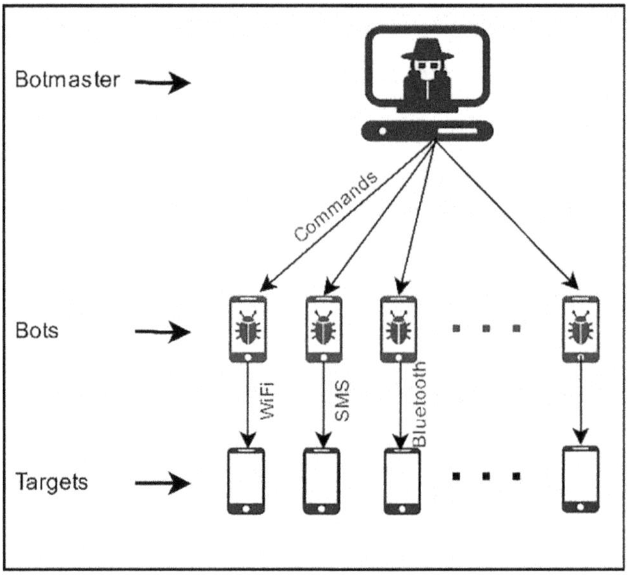

Figure 10.1 Overview of Botnet Structure.

of IoT malware appeared in 2008, they have not stopped [4]. Cyber-attacks using IoT have been on the increase. As stated by Sonywall [5], the number of cyber-attacks in 2018 was 32.7 million which was an increase of 215.7% from 2017. Further, the increase was 55% in the first half of 2019 as compared to that of 2018.

Botnets have been the most threatening type of malware out of all the IoT malwares like Trojan, Worm, Botnet etc. because of their special feature of mutating continuously. Another major advantage to them is lack of antivirus and weak firewall security in the systems. Recent reports [6] show that 41% of attacks have been a success because almost all the IoT devices remain unencrypted. A recent report by Nokia [7] stated that approximately 1 million devices are trapped in botnet-propelled Distributed Denial-of-Service (DDoS) attacks and the number is predicted to exceed 29 billion by 2030 [8]. Botnets pose a great threat to cyber security. They are on the verge of launching large scale DDoS attacks, transmitting malware and stealing sensitive data. Botnets are becoming a strong and vast group with multiple machines that are continuously attacking and gaining control over systems [9]. Botnet attacks overtake a group of connected computers and execute malicious corrupting activities like crashing their websites and Domain Name System (DNS) amplification attacks. A botmaster instructs each bot simultaneously to complete a coordinated attack and hence succeed in another malicious activity [10]. The attacking commands given are either centralized command and control (C&C) structures or decentralized peer-to-peer (P2P) structures [11]. With the increased number of botnets under one botmaster, these large-scale malware attacks are just one command away for them. These attacks take over IoT devices and then use these devices as hosts to launch DDoS attacks. These devices might not act malicious initially, but silently they are performing their desired goals [12]. In smart cities, these are used to send massive amounts of bandwidth and crash city networks and websites [13].

Considering the increasing risk to everyone, i.e. from an organization to the entire nation, botnet detection is an urgent need especially in the high-speed networks. There are traditional signature-based approaches of detection that include threshold-based, signature-based, heuristics-based, and characteristics-based. These approaches are effective but with the rapidly mutating nature of botnets, it is a struggle. The current scenario demands an innovative and adaptive solution to identify and mitigate the botnet threats. The current methods need to be updated manually at the cost of loss of assets, manpower and time. Considering the rate at which the technologies are coming up, there is a need for using machine learning (ML) and deep learning (DL) models as detection systems and IoT to look after many securities related areas on a regular basis like threat modeling, vulnerability research and intrusion detection systems (IDSs) [14].

IDSs monitor, detect and prevent malicious activities from affecting the network. They detect the cyber-attacks by checking and detecting the signature of data packets as they travel down the complete network system [15]. For added flexibility and to continuously monitor issues, software-defined networking (SDN) is used at fog servers. SDN is a progressively emerging system that is making networks more flexible by managing the network by analyzing the traffic and routing the control architectures [16]. It provides a flexible device management policy. Unfortunately, there doesn't exist a strong enough modern cyber security system to catch these attacks by deep packet inspections in IoT environments. Hence, it is challenging to focus on IoT security, especially the intrusion and cyber-attack traffic identification, because of the increase in existing threats and emerging ones.

Bot detection continues to be an active area of research, as existing methods face challenges in accurately identifying increasingly sophisticated automated accounts. Many different models have been tried using ML. For example, there are graph and non-graph-based structures used for detection. Scientists have revealed that the use of graphed structures is effective over the non-graphed structures. The detection is done mostly by performing two types of analysis: static and dynamic. The static model works by executing the file and its content without running it. The file is fragmented to understand its control flow, function and code structure. These are then used for detection of malware. The overhead analysis of the files is done, but the multiple devices are left untouched. Its limitation is that the malware stays hidden in such files by camouflage techniques. Alternatively, there exists the dynamic model. It performs a runtime analysis. It observes behavior in features like network traffic, register changes, memory accessed and files modified to find malware. This system lags in execution as it follows a single execution path whereas some codes do have trigger conditions where malware can be hidden.

To date, these detection systems exist, they are good at themselves yet they need an advanced system to have a deeper search and analysis. Now-a-days, self-learning systems that use ML algorithms are emerging [17]–19]. They are continuously evolving to cater to the needs of an effective system that doesn't use only one detection system.

10.1.1 Problem statement

With the continuously evolving cyber threats, botnet detection has become an urgent and crucial need for a safe network system. With the new threats coming up every instant, their detection is a big challenge to be overcome for safety of data. Botnets are networks of devices that perform malicious activities that are controlled by malicious invaders that pose a threat to the integrity, confidentiality and safety of our data and services. With the advancement in technology, the threats are also getting more advanced. Due to this, the existing approaches with limited features do not detect the botnet efficiently. This poses an urgent need for a new detection system that caters to the needs of detection systems of the modern world. These systems include ML methods, DL algorithms and other advanced algorithms that fulfill the needs of a new detection system. This study aims to find such a method by creating a well-evaluated ML model for detection of botnets.

10.1.2 Objectives

The primary objectives of this botnet detection system using ML algorithms include:

- To find the problems and challenges faced by the IoT devices and other devices in providing a safer network system.
- To research emerging technologies with IoT such as IDSs, threat modeling, vulnerability research and security analysis.
- To check the efficiency of ML algorithms for addressing solutions to the cyber-attacks.
- To examine the validity of traditional ML algorithms and other methods of detection like DL, auto-encoders and many more related to IoT security.
- To give lightweight botnet detection models for IoT using meta-learning algorithms.
- To evaluate the attack detection capability of proposed models via NIDS set up on both desktop-class CPU and single-board IoT devices.

- To compare the performance of traditional detection methods with the meta-learning advanced methodologies across various datasets like Aposemat IoT-23.

The following gives a comprehensive study of the IoT-23 dataset and its applications in network security. **Section 10.2** shows a literature review of botnets, detection systems and their efficiency, and analysis of network systems. **Section 10.3** gives an overview of the research background, delving deep into the origin of the IoT-23 dataset, its structure and significance. **Section 10.4** shows the detection methodology explaining the procedure and technique used for data analysis and making of the model. **Section 10.5** showcases the results and outcomes giving an insight to the effectiveness of ML algorithms in detecting malicious threats. At last, **Section 10.6** concludes with the outcomes and proposed solutions to the detection system with the future enhancement possible with the enhanced IoT network and ML algorithms.

10.2 LITERATURE REVIEW

It is a necessity to find the network of botnets ruled over by the botmaster to ensure and promote the safety of network devices. Numerous research studies state the effectiveness of ML algorithms used to identify botnet activities particularly out of all the malicious activities taking place today. This overview of the literature review summarizes the result of various noteworthy studies conducted previously in this topic.

They talked about the use and effectiveness of ML algorithms in botnet detection [20]. They analyzed the structure of botnets to identify key features that are used to distinguish between botnet traffic and normal traffic. Characteristics including network flow and behavioral patterns are significant for feature selection while crafting ML models. Through focusing on such unique features, the detection models can be additionally precise and effective in detecting malware events across a network.

They proposed a honeypot to lure attackers and yield an IoT dataset [21]. Subsequently, this data was featured in a wide range of botnet attack attribute analyses (such as IP address, MAC address and packet size). They were able to analyze these characteristics to comprehend the behavior of the botnet and subsequently could apply the knowledge in their detection algorithms. It allows us to use more real data and significant attack patterns to improve the detection ability of botnets in IoT.

They proposed the hybrid model for feature selection to reduce the dimensionality of the feature set for improved detection [22]. The IID dataset contains 115 features, which is a large number of features. They used filter, wrapper and hybrid model selection to examine fingerprints.

They experimented with several classification methods and clustering methods for identifying botnets [23]. One of the prominent ones is the Decision Tree (DT) classifier. It is used for P2P botnet detection. The dataset used had over 38,000 records of attacks and network traffic. There is a slight accuracy variation between K-Means Clustering and DT classifier. The report states that DT classifier gave 90.2723% accuracy as compared to 87.7853% of the other.

They said that P2P botnets had typical features like centralization and distribution, and hence were difficult to detect [24]. The dataset used 13 different botnet samples for analysis. In 2013, they suggested a two-stage method for detection. The first stage filtered non-P2P traffic by doing port judgment, DNS query and data flow count. The second stage used session features to reduce packet size to be further analyzed. The stage also used ML algorithms for analysis. Three main algorithms, namely Naïve Bayes (NB), DT classification and Artificial Neural Network (ANN), that are based on session characteristics are used. The accuracy difference stated that DT classifiers had better performance with 94.4% detection rate whereas NB and ANN had 75.5% and 93.8% detection rate respectively.

It focused on how high accuracy of detection can be attained using ML algorithms. Initially, they proposed flow-based traffic analysis and ML algorithm tools for identifying botnets [25]. They then tested for the performances of the eight most important ML algorithms. The traffic analysis was either done batch-wise or limited-wise. The experiment was conducted using both malicious and non-malicious records present in ISOP dataset. The results showed that Random Forest (RF) classifiers had the best accuracy and time of detection.

They researched using DNS for botnet detection. Six special features were selected as per the DNS logs [26]. The filters are applied in a row starting from name-based features, message-based features and then finally the quantity-based features. After these filters, Adaboost, Bagging, and NB were used as ML classifiers to pick the malicious domains from DNS traffic. The results from all the algorithms were good enough with 90% accuracy. The results showed that when the domain names appear in the traffic, the success rate of botnet detection increases manifold.

They compared the three ML algorithms namely NB, Nearest Neighbor (IBk) and J48 [27]. This was done to compare their abilities to distinguish between botnet and normal traffic. They grouped the algorithms according to their capabilities, and several test cases were produced for the algorithms. They were tested and results showed that the detection accuracy of J48 and IBk were higher than that of NB, yet they had their own limitations respectively.

It had a detected P2P botnets study using network behavior analysis and ML [28]. They used two datasets of about 370,000 malicious traffic in it. For normal traffic, a labeled dataset with packets of normal traffic like

K-Nearest Neighbor (K-NN), Linear Support Vector Machine (SVM), ANN, Gaussian Based Classifier, and NB Classifier were used. The results showed that SVM needed the longest training and classification time.

10.3 RESEARCH BACKGROUND

10.3.1 Introduction to botnet detection

The word "Botnet" is derived from network. A botnet is a network of compromised computers, often called zombies, an entire network that has been influenced by malware and is underneath the control of a malicious object. Since the early 2000s, when botnet attacks were first introduced, they have emerged considerably. Initially, botnets were relatively simple and primarily used for sending spam emails, but with time, their competences have evolved. In 2000, the first notable botnet, "EarthLink Spammer", emerged. Since then, many more sophisticated and large-scale botnet attacks like "Storm", "Zeus" and "Mirai" have occurred. Now-a-days, a wide range of devices including computers, smartphones, and Internet of Things (IoT) devices are being leveraged to execute Distributed Denial of Service (DDoS) attacks for stealing sensitive information and spreading ransomware.

Due to the distributed and covert nature of botnets, it is very complex and challenging to detect a botnet in huge amounts of network traffic. Traditional detection methods such as signature-based techniques can only identify known patterns of malicious activity. However, these methods often fail to detect new or evolving threats. Modern approaches use ML and behavioral analysis to enhance detection capabilities. For example, supervised learning algorithms are used to train on labeled datasets to recognize botnet-related activity by identifying anomalies in network traffic patterns whereas unsupervised learning techniques, such as clustering and anomaly detection, help identify suspicious activities without prior knowledge of the attack pattern. Meanwhile, semi-supervised learning combines both the approaches to improve detection accuracy by leveraging both labeled and unlabeled datasets.

Several rounds are needed to eliminate botnets in infected hosts. It would also take both technical and social solutions to clear the botnet. Breaking the communication between the compromised systems and the attacker is always the approach with which we take down the command and control servers and completely neutralize the botnet by doing so. Applying security patches and updates and utilizing antivirus and anti-malware solutions are other methods created to remove vulnerabilities that botnets use to access devices. Furthermore, network-level protections, such as firewalls and IDSs, contribute to thwarting the dissemination of botnets by filtering malicious traffic and detecting abnormal behaviors, and ISPs and police together need to trace [18, 19, 29–35].

10.3.2 Machine learning in cyber security

ML is the domain of AI and is a moral candidate in several areas which comprise cyber security. With the evolution, ML can do various works in cyber security from detecting the malwares directly to scanning the Big Data and uncovering the patterns to catch the malwares. Automation plays a critical role in completing these tasks with little or no human involvement. In the case of botnet detection, ML may be used to develop a system capable of detecting and reacting to these malicious actions by "its own". It can learn by itself and improve its detection from it only if it is fed the past data to the ML. ML is a "tool" that is grown and can adapt continually, which is what makes it so crucial as a weapon in the fight against the dark side.

10.3.2.1 Types of machine learning algorithms

ML approaches are significantly categorized into supervised, unsupervised, and semi-supervised learning. Each has exceptional features and usages in botnet identifications.

10.3.2.1.1 Supervised learning

Such algorithms are trained on labeled datasets. It transforms the input data delivered to the precise yield data parallel to it and on that basis, it relates the new data with the expected values. For instance, in botnet detection, a supervised learning algorithm such as RF can be trained to classify data as either benign or malicious as per pre-fed data.

Applications:

- **Email Spam Detection:** Supervised learning models, such as NB or (SVMs, are used to differentiate emails as spam or non-spam based on labeled examples of previous emails.
- **IDS:** Algorithms like RF and Logistic Regression analyze network traffic to detect potential security breaches and classify activities as either normal or malicious as per pre-fed classification of network data.
- **Credit Card Fraud Detection:** DT and Neural Network models are used for the identification of fraudulent transactions by learning from past labeled transaction data, improving the security of financial systems.

Advantages:

- **High Accuracy:** When appropriate labeled data is given, the model gives high predictive accuracy.
- **Clear Objective:** The objective of supervised learning is well-defined, making evaluation of the model easy.

- **Versatility:** It performs a wide variety of applications, including image recognition, natural language processing and fraud detection.

Disadvantages:

- **Dependency on Labeled Data:** Labeled data is difficult and expensive to acquire for training.
- **Limited to Known Patterns:** The model works well on models similar to pre-known models, but for any new model, it struggles to generalize the data under different labels.
- **Over Fitting:** The noise overpowers the data. Hence, the memory saves more noise and less data to work upon it.

10.3.2.2 Unsupervised learning

Unsupervised learning algorithms are applied to the datasets where data isn't labeled. These algorithms group the data after identifying the similarities and differences of patterns. In botnet detection, an unsupervised learning algorithm such as K-Means Clustering detects the malicious activities without having any prior information of the same.

Applications:

- **Anomaly Detection:** Unsupervised learning algorithms like K-Means Clustering and Principal Component Analysis (PCA) are used to detect unusual patterns in network traffic or user behavior to indicate potential security threats or fraud.
- **Customer Segmentation:** Retail and marketing industries use clustering algorithms to group customers based on purchasing behavior to create personalized marketing strategies.
- **Market Basket Analysis:** Algorithms such as Apriori and association rule learning are used to find relationships between products in transaction datasets, which helps retailers to optimize product placement and cross-selling strategies.

Advantages:

- **No Need for Labeled Data:** For the datasets which can't be grouped under different labels, unsupervised learning is used there.
- **Discovery of Hidden Patterns:** They find even those structures that are hard to find even manually.
- **Scalability:** They can effectively manage to work upon large-scale data.

Disadvantages:

- **Lack of Interpretability:** Since the data is unlabeled, the model may give results that can't be understood.

- **Evaluation Is Subjective:** Since there is no reference to the results, the results are subjective and context dependent.
- **Sensitivity to Initialization:** Due to the presence of no initial data, the final data also varies because of taking random initial data.

10.3.2.3 Semi-supervised learning

Semi-supervised learning gives the combined effect of labeled and unlabeled data under one algorithm. This approach is preferred when labeled data is less and unlabeled is more.

Applications:

- **Web Content Classification:** Semi-supervised learning algorithms classify web pages into categories (e.g., news, sports, entertainment) by leveraging a few labeled examples along with a large corpus of unlabeled content.
- **Medical Diagnosis:** Healthcare applications use semi-supervised models to diagnose diseases by combining a limited number of labeled medical records with a vast amount of unlabeled patient data to improve diagnostic accuracy and efficiency.
- **Speech Recognition:** Semi-supervised learning enhances speech recognition systems by using a small set of transcribed audio files alongside a larger set of un-transcribed audio, improving recognition accuracy with limited labeled data.

Advantages:

- **Efficient Use of Data:** It efficiently uses labeled data with the presence of unlabeled data in abundance.
- **Improved Generalization:** It gives better results with unlabeled generalized data.
- **Cost-Effective:** It does not require much labeled data.

Disadvantages:

- **Dependency on Unlabeled Data Quality:** In real time, the amount of unlabeled data fed to the algorithm as input varies, hence output efficiency varies.
- **Complex Model Training:** It requires extra data and resources for efficient output.
- **Sensitivity to Label Noise:** A high performance dataset gives more sensitive output.

10.3.2.4 Challenges and limitations

While ML offers significant advantages for botnet detection, several challenges and limitations need to be addressed:

1. **Data Quality and Quantity**
   ML models require a large amount of high-quality data to train effectively, but obtaining labeled data can be challenging due to privacy concerns and the evolving nature of threats. Moreover, insufficient or poor-quality data can lead to inaccurate models and higher false positive rates.
2. **Evolving Threat Landscape**
   Botnets continuously evolve and employ new techniques to evade detection. ML models trained on historical data may become less effective as new types of botnet behaviors emerge requiring models to be regularly updated to stay relevant, which can be resource-intensive.
3. **Adversarial Attacks**
   Attackers are using adversarial techniques to fool ML models by manipulating the input data. These attacks significantly degrade the quality of the result of detection. Developing a robust model that can withstand adversarial attacks is an ongoing area of research.
4. **Interpretability and Transparency**
   Many ML models, especially DL models, act as "black boxes" making it difficult to understand how they make decisions. This lack of interpretability causes difficulty adopting ML in critical cyber security applications.
5. **Computational Resources**
   Significant computational resources are needed for training and deploying ML models which may not be feasible for all organizations. High resource requirements limit the accessibility and scalability of ML-based botnet detection systems.

10.4 RESEARCH METHODOLOGY

The aim of this research is to develop an effective classifier model for detecting botnet activities utilizing ML algorithms to mitigate the escalating threat posed by botnets. Figure 10.2 provides an overview of the workflow of our model. Our workflow has three main stages: data acquisition, feature selection and extraction, and application of ML classifiers.

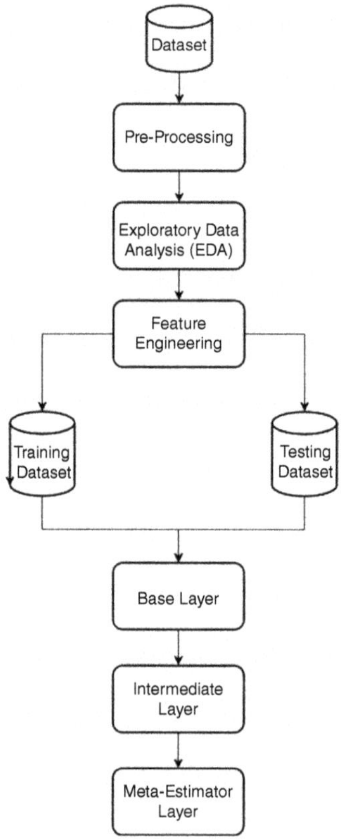

Figure 10.2 Proposed Model Workflow.

10.4.1 Data set

The Aposemat IoT-23 dataset is used in our research. It contains 20 malwares infected IoT device captures and three benign IoT device captures. The dataset was first published in January 2020 and was created in the Stratosphere Laboratory, Czech Technical University. The dataset includes captures from 2018 to 2019. This dataset provides real and labeled IoT malware infections and benign traffic.

The dataset includes network traffic captures in pcap format. Each capture was lab labeled according to the device or malware involved. For benign traffic, network captures were obtained from real IoT devices such as Philips HUE smart LED lamp, an Amazon Echo home assistant, and a Somfy smart door lock, all functioning in a controlled environment with unrestricted internet access. The final dataset integrates both benign and malicious traffic with labels assigned based on manual analysis and automated scripts (Flaber) that follow predefined rules ensuring the accuracy and relevance of the dataset for training ML models.

10.5 RESULTS AND ANALYSIS

The present study intends to assess different ML approaches employed for botnet detection. Their performance is measured based on metrics accuracy, precision, recall and inference time, and they are under the ROC to achieve optimal performance. Outcomes are stated for each classifier separately for a layered model approach.

10.5.1 Effectiveness

10.5.1.1 Single model performance

The chapter assesses the efficiency of multiple ML methods for botnet detection. The result of each algorithm is stated as follows:

- **Random Forest Classifier:** The RF classifier achieved an accuracy of 73.28%, with a precision of 74.23% and a recall of 73.28%. It demonstrated strong performance with an AUC of 0.925. However, the inference time was relatively high at 210.75 seconds, suggesting that while it is effective, this model may require significant computational resources for real-time detection.
- **Logistic Regression Model:** The Logistic Regression model showed an accuracy of 56.05%, which is considerably lower than the RF. The precision and recall were 39.61% and 56.05%, respectively, with an AUC of 0.356. The model has the longest inference time among all other models, at 361.23 seconds, proving inefficient in terms of processing time and predictive performance.
- **Naïve Bayes Classifier:** The NB classifier performed poorly with an accuracy of 32.65% and a recall of 32.65%, though it achieved a higher precision of 53.46%. It showcased moderate predictive performance with an AUC of 0.772. It is the fastest model at the cost of accuracy and reliability with the inference time of 1.51 seconds only.
- **Decision Tree Classifier:** Similar to the RF, the DT classifier achieved an accuracy of 73.28%, with a precision of 74.24% and a recall of 73.28%. Its AUC was slightly lower at 0.923. With the inference time of 2.18 seconds, this model was proved more efficient as compared to RF in terms of execution time while maintaining comparable accuracy.
- **Multi-Layer Perceptron (MLP) Classifier:** The MLP classifier had an accuracy of 60.61%, with a precision of 44.56% and a recall of 60.61%. It exhibited the longest inference time by far at 10,768.78 seconds with the AUC of 0.720 shown in Table 10.1. This indicates that while the MLP has moderate accuracy, its practical application is limited by its extensive processing time.

Among the single classifiers, the RF and DT classifiers demonstrated the highest effectiveness, with both achieving an accuracy of 73.28% and

Table 10.1 Performance of Single Model

Classifier	*Accuracy*	*Precision*	*Recall*	*Inference Time (s)*	*AUC*
RF	73.28%	74.23%	73.28%	210.75	0.925
Logistic Regression	56.05%	39.61%	56.05%	361.23	0.356
NB	32.65%	53.46%	32.65%	1.51	0.772
DT	73.28%	74.24%	73.28%	2.18	0.923
MLP	60.61%	44.56%	60.61%	10768.78	0.720

similar precision and recall values. The AUC values further confirmed their superior performance, with RF at 92.52% and DT at 92.31%. In contrast, the Logistic Regression and NB classifiers showed lower effectiveness, particularly in terms of accuracy and AUC.

10.5.2 Layered model performance

The layered model approach combines predictions from multiple classifiers to enhance detection effectiveness. The base layer consists of a RF classifier and a Logistic Regression model, whose predictions are used to train intermediate and meta-estimator layers. The results for each layer are as follows:

- **Base Layer:** The base layer consisted of RF and Logistic Regression classifiers. The combined model achieved an accuracy of 73.21%, with precision and recall scores matching the accuracy. Notably, the inference time for the base layer was 629.62 seconds, indicating a substantial computational load. However, the AUC value of 0.925 suggested excellent discriminatory power.
- **Intermediate Layer:** The intermediate layer maintained the performance metrics of the base layer, with an identical accuracy of 73.21%. Precision, recall and inference time remained consistent, whereas the AUC value slightly decreased to 0.867.
- **Meta-Estimator Layer:** The meta-estimator layer preserved the accuracy of the preceding layers at 73.21%. Precision, recall and inference time were consistent, with a slight increase in the false positive rate. The AUC value remained stable at 0.867, indicating consistent discriminatory power.

The layered model maintains a consistent accuracy of 73.21% across all layers, with the base layer achieving the highest AUC of 92.51% as shown in Table 10.2. The intermediate and meta-estimator layers show a slight decrease in AUC, suggesting some redundancy in the model's ability to further enhance detection accuracy.

Table 10.2 Performance of Layer Model

Classifier	*Accuracy*	*Precision*	*Recall*	*Inference Time (s)*	*AUC*
Base Layer	73.21%	74.26%	73.21%	629.62	0.925
Intermediate Layer	73.21%	74.26%	73.21%	5.17	0.867
Meta-Estimator Layer	73.21%	74.24%	73.21%	878.83	0.867

10.5.2.1 Performance

10.5.2.1.1 Single model performance

The performance of single classifiers is also evaluated based on inference time, which is critical for real-time botnet detection:

- **RF Classifier:** 210.75 seconds.
- **Logistic Regression Model:** 361.23 seconds.
- **NB Classifier:** 1.51 seconds.
- **DT Classifier:** 2.18 seconds.
- **MLP Classifier:** 10,768.78 seconds.

The NB classifier demonstrated the fastest inference time, making it suitable for scenarios requiring rapid detection. However, its low accuracy and AUC limit its practical applicability. The DT classifier strikes a balance with a reasonable inference time and high accuracy. The MLP classifier, while relatively accurate, suffers from prohibitively long inference times.

10.5.2.1.2 Layered model performance

The layered model's performance in terms of inference time is effectively managed by combining the strengths of individual classifiers:

- **Base Layer:** 629.62 seconds.
- **Intermediate Layer:** 5.17 seconds.
- **Meta-Estimator Layer:** 878.83 seconds.

The inference time for the layered model is not directly comparable to single models since it focuses on improving detection accuracy during the training phase. However, the overall processing time benefits from the efficiencies gained through model stacking.

10.5.2.1.3 Accuracy comparison

Based on the comparison of the accuracy of individual ML algorithms, the RF and DT models represent equivalent outcomes with an accuracy of

73.28%. Such approaches show the optimal average detection rate among the single classifiers assumed in this work; ensuring the usages of such algorithms towards botnet detection. Comparatively, the Logistic Regression model has a percentage of accuracy of (56.05%), which shows that it runs poorly in identifying botnet activity accurately. Benchmarks NB the NB classifier runs even worse with its 32.65% accuracy, providing more experimental space for enhancement. The classification accuracy of the MLP classifier is in between those from the RF and Insurance Regression at 60.61%, representing that it is reasonably efficient. Results indicate that all layers of the framework—Base, Intermediate, and Meta-Estimator—showed a consistent pattern of behavior.

On the whole, as the data in Figure 10.3 represents, this research highlights selecting ideal ML models for searching botnet events. Though solitary algorithms like RF and DT approaches showed optimistic concerns about accuracy, the stacked method ensured an ability to additionally increase detection efficiency by merging the qualities of several classifiers into an integrated approach. Nevertheless, sensible contemplations, for instance, forecast time and registering assets, must be considered when actualizing these calculations in genuine circumstances. Additionally, examinations could concentrate more on increasing the stacked design engineering

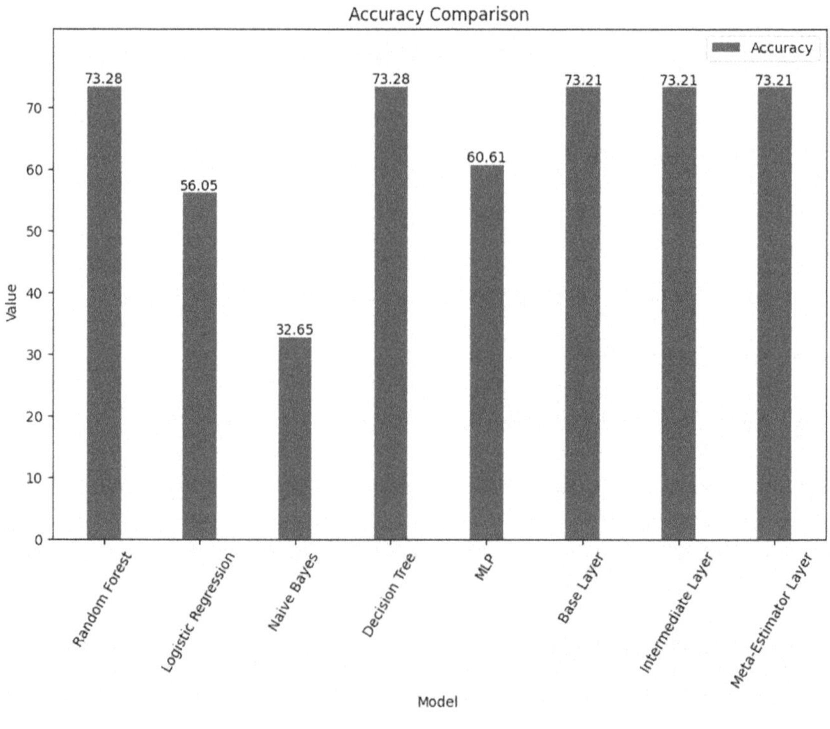

Figure 10.3 Accuracy Comparison of All Models.

to attain adjustment between exactness and effectiveness. Certain sentences need detailed documentation to entirely examine the consequences of the discoveries, though others could be significantly highly neat to emphasize the main purposes of the examination. More and more unpredictable language and a variety of sentence lengths increase the comprehensibility of the reduced content.

10.6 CONCLUSION AND FUTURE SCOPE

This chapter has used the Aposemat IoT-23 dataset and achieved the efficiency and compatibility of different ML algorithms for detecting the botnet. It was detected based on research and analysis that RF and DT classifiers provided more accurate results in comparison to Logistic Regression and NB classifiers. The layered model approach, employing several classifiers, increased the detection accuracy.

The research methodology discussed data acquisition, feature extraction, and application of ML classifiers. Each step was done purposely to increase effectiveness of detection. The study has been designed keeping in view the real-time usage of the model, checking for the practical usage aspects like inference time and computational resources.

Even though this study shows many new aspects of detection, there remains room for further additions to this research in the future. Mainly, in the future more research can be done on a layered model for balanced accuracy and efficiency and a maximum of both. This involves exploring different combinations of base classifiers, or implementing advanced ensemble techniques for better and more efficient layering.

The study gave a binary classification to botnet activities. Future research could delve deeper and extend this to multi-class classification, considering the diverse and ever-increasing nature and types of botnets. Threat detection can be enhanced using DL techniques, such as convolutional neural networks (CNNs) and recurrent neural networks (RNNs).

The feasibility check of the detection models in real time is yet a part of future research. The scalability, adaptability and the robustness of the system in a dynamic network will give a better outlook on the system.

Hence, it can be said that botnet detection using ML is surely a rapidly advancing field. In the future, this model can be made more successful and effective in botnet detection and can enhance cyber security, making IoT more reliable.

REFERENCES

[1] S. S. Silva, R. M. Silva, R. C. Pinto, and R. M. Salles, "Botnets: A survey," *Computer Networks*, vol. 57, no. 2, pp. 378–403, 2013, https://doi.org/10.1016/j.comnet.2012.07.021.

[2] McKinsey Global Institute, "Unlocking the potential of the Internet of Things," Jun. 2015. [Online], http://www.mckinsey.com/industries/hightech/our-insights/the-internet-of-things-the-value-of-digitizing-thephysical-world.

[3] Cisco, *2020 CISO Benchmark Report: Computer Fraud & Security*, San Jose, CA, USA: Cisco, 2020.

[4] Q. Ngo, H. Nguyen, V. Le, and D. Nguyen, "A survey of IoT malware and detection methods based on static features," *ICT Express*, vol. 6, no. 4, pp. 280–286, 2020, https://doi.org/10.1016/j.icte.2020.04.005.

[5] Costin, A., & Zaddach, J. (2018). Iot malware: Comprehensive survey, analysis framework and case studies. *BlackHat USA, 1*(1), 1–9.

[6] M. K. Kagita, N. Thilakarathne, T. R. Gadekallu, P. K. Reddy, and S. Singh, "A review on cyber crimes on the Internet of Things," in *Deep Learning for Security and Privacy Preservation in IoT*. Cham, Switzerland: Springer, 2022, pp. 83–98.

[7] Nokia Corporation, "Nokia threat intelligence report finds malicious IoT botnet activity has sharply increased," *Nokia*, Feb. 26, 2024. [Online], https://www.nokia.com/about-us/news/releases/2023/06/07/nokia-threat-intelligence-report-finds-malicious-iot-botnet-activity-has-sharply-increased/.

[8] B. Jovanovic, "Internet of Things statistics for 2023—taking things apart," 2022. [Online], https://dataprot.net/statistics/iot-statistics/.

[9] Y. Meidan et al., "N-BaIoT network-based detection of IoT botnet attacks using deep autoencoders," *IEEE Pervasive Computing*, vol. 17, no. 3, pp. 12–22, Jul.–Sep. 2018.

[10] M. Antonakakis Perdisci, R., Nadji, Y., Vasiloglou, N., Abu-Nimeh, S., Lee, W., & Dagon, D. (2012). From {Throw-Away} traffic to bots: Detecting the rise of {DGA-Based} malware. In *21st USENIX Security Symposium (USENIX Security 12)* (pp. 491–506).

[11] M. Antonakakis April, T., Bailey, M., Bernhard, M., Bursztein, E., Cochran, J., ... & Zhou, Y. (2017). Understanding the mirai botnet. In *26th USENIX security symposium (USENIX Security 17)* (pp. 1093–1110).

[12] J. Liu, Y. Xiao, K. Ghaboosi, H. Deng, and J. Zhang, "Botnet: Classification, attacks, detection, tracing, and preventive measures," *EURASIP Journal on Wireless Communications and Networking*, vol. 2009, no. 1, pp. 1–11, Dec. 2009.

[13] Akamai, "State of the internet security Q3 2016," 2016. [Online], https://www.akamai.com/us/en/multimedia/documents/state-of-the-internet/q3-2016-state-of-the-internet-security-executive-summary.pdf.

[14] N. Neshenko et al., "Demystifying IoT security: An exhaustive survey on IoT vulnerabilities and a first empirical look on Internet-scale IoT exploitations," *IEEE Communications Surveys & Tutorials*, vol. 21, no. 3, pp. 2702–2733, 3rd Quart. 2019.

[15] S. Bagui, E. Kalaimannan, S. Bagui, D. Nandi, and A. Pinto, "Using machine learning techniques to identify rare cyber-attacks on the UNSW NB15 dataset," *Security and Privacy*, vol. 2, no. 6, p. e91, 2019.

[16] Vishwakarma, R., & Jain, A. K. (2019, April). A honeypot with machine learning based detection framework for defending IoT based botnet DDoS attacks. In *2019 3rd International Conference on Trends in Electronics and Informatics (ICOEI)* (pp. 1019–1024). IEEE.

[17] U. Ahmad, H. Song, A. Bilal, M. Alazab, and A. Jolfaei, "Securing smart vehicles from relay attacks using machine learning," *Journal of Supercomputing*, vol. 76, pp. 1–18, 2019.

[18] P. Gupta, N. Saxena, M. Sharma, and J. Tripathi, "Deep neural network for human face recognition," *International Journal of Engineering and Manufacturing (IJEM)*, vol. 8 no. 1, pp. 63–71, 2018.
[19] S. N. Mohanty, J. M. Chatterjee, S. Jain, A. A. Elngar, and P. Gupta (Eds.), *Recommended System with Machine Learning and Artificial Intelligence: Practical Tools and Applications in Medical, Agricultural and Other Industries*, John Wiley & Sons, 2020.
[20] X. Dong, J. Hu, and Y. Cui, (2018, September). Overview of botnet detection based on machine learning. In *2018 3rd International Conference on Mechanical, Control and Computer Engineering (ICMCCE)* (pp. 476–479). IEEE.
[21] R. Vishwakarma and A. Jain, (2019, April). A honeypot with machine learning based detection framework for defending IoT based botnet DDoS attacks. In *2019 3rd International Conference on Trends in Electronics and Informatics (ICOEI)* (pp. 1019–1024). IEEE.
[22] H. Guerra-Manzanares, H. Bahsi, and S. Nõmm, "Hybrid feature selection models for machine learning based botnet detection in IoT networks," in *Proceedings of the International Conference on Cyberworlds (CW)*, Kyoto, Japan, 2019, pp. 324–327.
[23] S. Haq and Y. Singh, "Botnet detection using machine learning," in *Proceedings of the International Conference on Parallel, Distributed and Grid Computing (PDGC)*, India, 2018, pp. 240–245.
[24] R. Khan, R. Kumar, M. Alazab, and X. Zhang, "A hybrid technique to detect botnets, based on P2P traffic similarity," in *Proceedings of the Cybersecurity and Cyberforensics Conference (CCC)*, Melbourne, Australia, 2019, pp. 136–142.
[25] M. Stevanovic and J. Pedersen, (2014, February). An efficient flow-based botnet detection using supervised machine learning. In *2014 international conference on computing, networking and communications (ICNC)* (pp. 797–801). IEEE.
[26] J. Jin, Z. Yan, G. Geng, and B. Yan, (2015, November). Botnet domain name detection based on machine learning. In *6th International Conference on Wireless, Mobile and Multi-Media (ICWMMN 2015)* (pp. 273–276). IET.
[27] S. Garg, A. Singh, A. Sarje, and S. Peddoju, "Behaviour analysis of machine learning algorithms for detecting P2P botnets," in *Proceedings of the International Conference on Advanced Computing Technologies*, 2013, pp. 1–4.
[28] S. Saad Traore, I., Ghorbani, A., Sayed, B., Zhao, D., Lu, W., ... & Hakimian, P. (2011, July). Detecting P2P botnets through network behavior analysis and machine learning. In *2011 Ninth annual international conference on privacy, security and trust* (pp. 174–180). IEEE.
[29] K.-C. Lin, S.-Y. Chen, and J. C. Hung, "Botnet detection using support vector machines with artificial fish swarm algorithm," *Journal of Applied Mathematics*, vol. 2014, Article ID 986428, 2014.
[30] A. Feizollah, N. B. Anuar, R. Salleh, F. Amalina, R. Ma'arof, and S. Shamshirband, "A study of machine learning classifiers for anomaly-based mobile botnet detection," *Malaysian Journal of Computer Science*, vol. 26, no. 4, 2013.
[31] R. Dollah, A. Faizal, F. Arif, M. Mas'ud, and L. Xin, "Machine learning for HTTP botnet detection using classifier algorithms," *Journal of Telecommunication, Electronic and Computer Engineering*, vol. 10, no. 1–7, pp. 27–30, 2018.

[32] M. Pour et al., "On data-driven curation, learning, and analysis for inferring evolving Internet-of-Things (IoT) botnets in the wild," *Computers & Security*, vol. 91, Apr. 2020.
[33] M. Al Shorman, H. Faris, and I. Aljarah, "Unsupervised intelligent system based on one-class support vector machine and Grey Wolf optimization for IoT botnet detection," *Journal of Ambient Intelligence and Humanized Computing*, vol. 11, pp. 2809–2825, Jul. 2020.
[34] K. Kumar and M. Khari, (2025, February). Efficient Intrusion Detection and Trust Management in Federated Learning Systems. In *2025 2nd International Conference on Computational Intelligence, Communication Technology and Networking (CICTN)* (pp. 68–72). IEEE
[35] A. Kumar, K. Kumar, M. Khari, and S. Kumar, "Harnessing smart farming to combat climate-induced agricultural challenges using machine learning," in *Proceedings of the International Conference on Computational Intelligence in Communications and Business Analytics*, Cham, Switzerland: Springer, Jan. 2024, pp. 62–69.

Chapter 11

Edge Computing for Evaluating Healthcare System Efficiency

Apurva Dalela, Manju Khari, and Priya Gupta

11.1 INTRODUCTION

Healthcare aims to maintain and improve human health via diagnosis, treatment, and disease prevention. As motivated by patient care, auditing, administration, and requirement to conform to standards and regulation, it generates a massive amount of data [1]. The healthcare industry is undergoing a profound transformation, fueled by rapid advancements in digital technologies. Smart healthcare systems powered by the Internet of Things (IoT), wearable sensors, big data analytics, and artificial intelligence are reshaping the way medical services are delivered [2]. These innovations enable continuous patient monitoring, early diagnosis, and personalized care, moving the industry from reactive to proactive healthcare. However, as these systems generate massive volumes of real-time data, they also introduce significant challenges related to data transmission delays, processing bottlenecks, and cybersecurity risks [3].

Traditionally, cloud computing has played a central role in supporting smart healthcare infrastructure due to its virtually unlimited storage, centralized processing power, and easy accessibility [4]. While the cloud continues to be instrumental in data-heavy analytics and long-term storage, it often falls short in time-sensitive clinical environments. Scenarios such as emergency responses, intensive care monitoring, and telesurgery demand split-second decision-making that cannot afford the delays introduced by network congestion, bandwidth limitations, or high latency [5]. In such cases, delays of even a few seconds can result in compromised patient outcomes, making it imperative to explore alternatives that offer quicker data turnaround and more localized control.

Edge computing emerges as a powerful solution to address these limitations. By bringing computation and storage resources closer to the data source whether that's a wearable device on a patient, a diagnostic machine in an ICU, or a mobile health unit in a remote area edge computing significantly reduces data transmission latency, improves response time, and enhances system reliability [6]. It further improves security and compliance, as well as reducing the need to send sensitive patient information

DOI: 10.1201/9781003635147-11

over networks, thereby mitigating privacy risks. In healthcare environments where every single second matters, the adoption of edge computing is more than an innovation; it's a requirement.

This chapter aims to analyze the impact of edge computing in smart hospitals regarding overall system efficiency and the improvement of patient care. This research seeks to verify and analyze in practice how healthcare infrastructure using edge technology performs in comparison to traditional ones, concentrating on latency, throughput, queuing time, service quality, and data protection [7]. Unlike most existing works that focus mainly on theoretical analysis or small-scale simulation, this work implements a working prototype of a smart hospital environment composed of edge-enabled nodes that perform patient monitoring, diagnostic support, and real-time alerting functions. The experimental setup comprises emulation of real-time sensor data and private cloud structure-equipped Raspberry Pi edge nodes to simulate actual deployment environments.

The evidence gathered from our experiments showcases the edge computing's advantages for critical healthcare operations. Compared to the cloud model, the edge system showed consistently better outcomes, particularly in terms of response time and managing data throughput under heavy load conditions. The advantages of edge computing over the cloud are increased privacy of medical data and availability of systems, reinforcing its value in medical settings.

To further measure the efficiency of every system configuration, we use Data Envelopment Analysis (DEA), a multi-input, multi-output assessment technique widely applied in assessing healthcare performance. DEA allows comparing various configurations (as decision-making units) against each other according to how effectively they transform several inputs like CPU usage, latency, and energy into desirable outputs such as system responsiveness and data security. In so doing, DEA presents an even stronger, data-supported view of which system architectures are best in terms of operational efficiency under actual healthcare conditions [8].

These findings not only validate the potential of edge computing to elevate the performance of smart healthcare systems but also provide a roadmap for healthcare providers, technologists, and policymakers aiming to enhance digital health capabilities. This chapter, therefore, contributes to the growing body of evidence advocating for decentralized computing in healthcare, presenting a practical case for integrating edge architectures to support safer, faster, and more responsive patient care in the digital age.

The rest of the chapter is organized as follows: Section 11.2 consists of background and related work. Section 11.3 consists of literature work with some open challenges. Section 11.4 presents the methodology for explaining the architecture with some experimental setup. Sections 11.5 and 11.6 explain the experimental results and interpretation while Section 11.7 includes conclusion and future works.

11.2 BACKGROUND

11.2.1 Rise of smart healthcare systems

The healthcare industry is undergoing a major shift with the rise of digital technologies like wearable sensors, connected medical devices (IoMT), smart diagnostic tools, and cloud-based systems. These innovations are forming the backbone of modern, smart healthcare—making it possible to monitor patients continuously, offer virtual consultations, and support quicker, more informed medical decisions.

As healthcare becomes more digital, the amount of data being generated is growing faster than ever [9]. Traditional cloud systems are often not fast enough or secure enough to handle this in real time, especially when urgent care or patient privacy is involved. To meet these demands, the focus is now turning to more intelligent and local technologies like edge computing. These systems process data near its source, helping deliver quicker responses, better security, and more reliable care.

11.2.2 Limitations of cloud-centric models

While cloud computing has made it easier to store and process large amounts of health data, relying on distant data centers comes with serious limitations. For example:

- There's often a delay in processing urgent data like ECG readings or blood sugar levels.
- It depends heavily on a stable internet connection, which can be unreliable in rural or low-connectivity areas.
- Sensitive medical information must travel across public networks, raising concerns about data privacy.
- As more devices connect, cloud systems can become overwhelmed, slowing down performance.

These issues aren't just technical; they can put lives at risk, especially in emergencies like strokes or heart attacks, where every second counts and immediate action is critical.

11.2.3 Edge computing: paradigm shift

Edge computing offers a promising alternative by pushing computation closer to the data source—often at the network edge or even on the device itself. In a healthcare setting, this means:

- Medical data from wearables or bedside monitors can be processed locally.

- Alerts (e.g., hypertension spikes, arrhythmias) can be generated in real time.
- Only necessary data is sent to the cloud, reducing load and improving privacy.

Edge computing **reduces response time**, enhances **energy efficiency**, and enables **autonomous** local decision-making, which is crucial for smart healthcare environments such as ICU monitoring, elderly care systems, and mobile clinics.

11.2.4 Why edge is crucial for healthcare infrastructure

Given the latency-sensitive and privacy-intensive nature of healthcare data, integrating edge computing isn't just a technical upgrade, it's a necessity. Edge-based architectures:

- Improve resilience of healthcare systems in remote or low-bandwidth areas.
- Enable context-aware decision-making without depending on constant internet connectivity.
- Enhance data security by processing and storing data locally whenever possible.

Figure 11.1 is a conceptual comparison showing the difference in data flow latency, and decision-making between cloud-based and edge-based smart healthcare infrastructures. The edge-based model demonstrates faster response times and more efficient local processing.

11.2.5 DEA model framework

The Decision-Making Units (DMUs) are defined as different system configurations tested under varying conditions, such as systems with or without edge computing, different network latency levels, and diverse resource usage scenarios.

Each DMU is evaluated based on:

- **Inputs:** CPU usage, energy consumption, bandwidth, processing delay
- **Outputs:** System responsiveness, data security score, decision latency

To apply a **Constant Returns to Scale (CRS) input-oriented DEA model,** focusing on minimizing the use of resources (inputs) while maintaining output levels.

The basic mathematical formulation for DMU_o is:

$$\min_{\theta\lambda} \theta$$

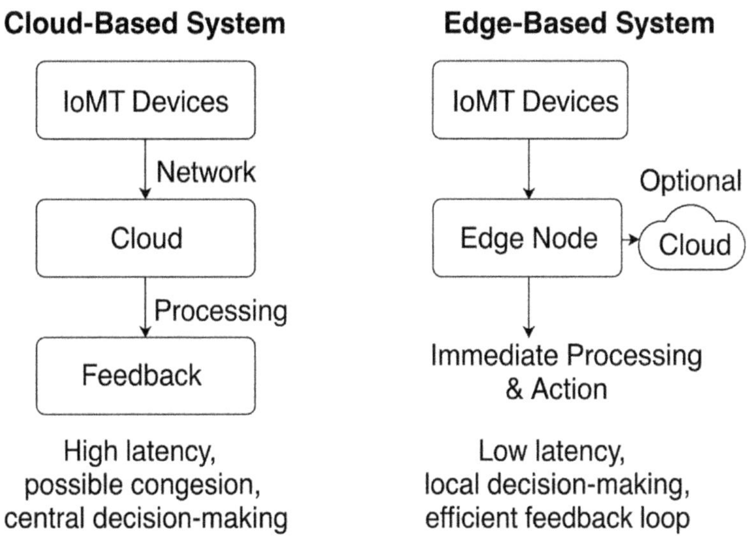

Figure 11.1 Cloud-Centric vs Edge-Centric Smart Healthcare

Subject to:

$$Y\lambda \geq Y_0 \tag{i}$$

$$\theta X_0 \geq X\lambda \tag{ii}$$

$$\lambda \geq 0 . \tag{iii}$$

Where:

X and Y: input and output matrices for all DMUs

X_o and Y_o: inputs and outputs for the DMU being evaluated

θ: efficiency score ($0 < \theta \leq 1$); 1 = fully efficient.

Figure 11.2 illustrates how different system configurations (DMUs) are evaluated using multiple inputs (CPU usage, energy consumption, bandwidth) and outputs (latency reduction, responsiveness index) within a DEA framework.

This chapter builds a performance-oriented perspective by evaluating how edge computing affects key healthcare infrastructure metrics like latency, responsiveness, and energy usage.

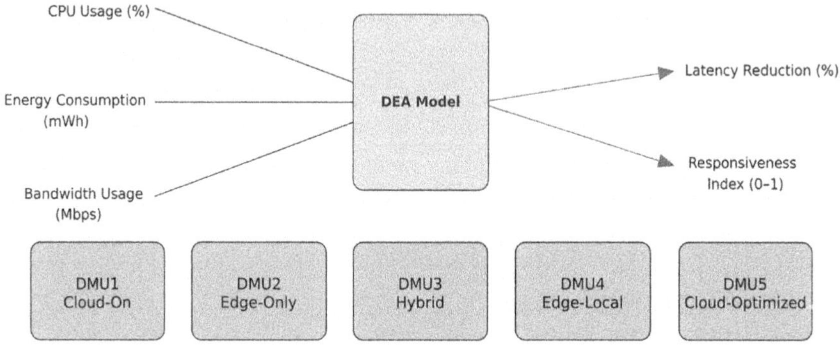

Figure 11.2 Conceptual DEA Model for Evaluating Smart Healthcare System.

11.3 LITERATURE REVIEW

Several studies have explored how edge and fog computing can address the limitations of traditional cloud-based healthcare systems. For example, an edge-based monitoring system for elderly patients was tested and found to significantly improve response times and enhance overall system reliability [9]. Similarly, a fog computing model designed for remote patient monitoring demonstrated improved scalability, lower latency, and stronger privacy protection [10]. A distributed edge architecture was designed specifically for intensive care units (ICUs), showing how processing data close to the patient can support faster clinical decisions. A study showed that integrating edge-enabled AI models into wearable devices can support early disease detection while minimizing the transfer of sensitive data to the cloud [11].

Fog and edge computing in healthcare still face major challenges, including energy management, heterogeneous data handling, and efficient resource allocation [12].

DEA, introduced for evaluating the relative efficiency of DMUs, has gained widespread application in healthcare [13]. Its use across hospitals in various countries examining indicators such as staffing levels, expenditures, bed utilization, and patient outcomes has been extensively reviewed [14].

The development of dynamic and network-based DEA models has enhanced the ability to analyze internal processes within hospitals, contributing to more comprehensive efficiency assessments in healthcare management [15]. These articles show DEA's applicability in measuring both service delivery and resource usage performance.

Recent studies have utilized DEA in measuring technology-enabled healthcare services. DEA has been applied to evaluate the efficiency of telemedicine platforms by benchmarking them against conventional clinical workflows and virtual consultation setups [16]. Their findings indicated that integration with technology improved operational efficiency, particularly in remote and underserved locations.

DEA has also been employed to compare the efficiency of different infrastructures supporting smart healthcare, including Internet of Medical Things (IoMT) and cloud-based systems [17]. They observed substantial gains in efficiency when local processing (edge or fog computing) was considered, particularly under bandwidth-constrained environments.

While literature directly connecting DEA to edge computing in healthcare is yet to surface, works around fog computing and IoT systems infer significant applications. For example, DEA has been applied to evaluate fog-based resource management within simulated healthcare environments, offering insights into system-level efficiency [18]. Results highlighted that localized processing nodes (fog/edge) resulted in better efficiency scores than pure cloud-based structures.

Edge computing has been explored for pandemic management, leveraging real-time health data analytics to support infection tracking and enforce quarantine measures [19]. Their work emphasized the critical role played by edge nodes in prompt public health interventions. Frameworks for data privacy in edge-based eHealth systems have been proposed, incorporating decentralized identity management and blockchain-based access control to safeguard patient confidentiality [19]. Furthermore, A hierarchical fog–edge–cloud architecture has been proposed for chronic disease management, demonstrating that layered data processing can effectively balance latency and computational load while enabling continuous patient monitoring [20]. An edge-integrated smartphone application for cardiovascular risk prediction in mobile health (mHealth) settings has shown that localized inference can substantially reduce reliance on unstable network connectivity, particularly in rural areas [21, 22]. Cloud computing is "an area of computer science in which highly scalable IT capacity is provided in the form of services delivered via the Internet to numerous external users" [23]. Cloud computing is a rapidly advancing technology defined by NIST as "a model for enabling ubiquitous, convenient, on-demand network access to a shared pool of configurable computing resources (e.g., networks, servers, storage, applications, and services) that can be rapidly provisioned and released with minimal management effort or service provider interaction" [24]. Cloud computing is an enabler of all Big Data (BD) operations [25]. In fact, the two terms are so interrelated that BD experts often consider them as indistinguishable. Cloud computing offers the following important advantages for healthcare [26].

- It enables access to large storage and processing power on demand.
- It allows the movement of electronic health records (EHRs), genomic, and radiology data through BD sets.
- It facilitates the sharing and access to EHRs for care providers from different geographic locations, thus reducing the need for patients to be re-assessed.
- It allows for BD to be performed in order to obtain actionable insights.

11.3.1 Open challenges

While this study shows how DEA can be used to assess smart healthcare systems, there are still real-world challenges to consider. One major issue is that real-time data healthcare environments often deal with missing or inconsistent information due to device failures or network problems, which can impact the accuracy of DEA results. Another challenge is that smart healthcare systems operate in fast-changing, dynamic conditions, whereas traditional DEA models are designed to evaluate more stable, static scenarios. Adapting DEA to handle real-time or streaming data is still a developing area.

There's also a need for better integration between DEA and intelligent technologies like AI. As healthcare systems become more predictive and autonomous, combining DEA with machine learning could unlock more meaningful insights.

Scalability is another concern. As the number of IoMT devices grows, evaluating efficiency across a massive network becomes computationally intensive. Scalable and distributed DEA frameworks are needed to keep up.

Finally, the choice of inputs and outputs in DEA is often subjective. More standardization and domain-guided models are essential to ensure fairness and reproducibility, especially across different hospitals or systems.

Addressing these challenges will be key to making DEA not just a useful academic tool, but a practical framework for shaping the future of intelligent, efficient, and responsive healthcare systems [27].

Despite these advancements, most existing literature primarily focuses on theoretical models, simulations under constrained conditions, or selective subsystems of healthcare (e.g., wearable monitoring or teleconsultation) [14]. There remains a significant gap in empirical studies that holistically assess edge computing performance in more comprehensive, hospital-wide scenarios. Our study distinguishes itself by implementing a practical edge computing prototype in a simulated smart hospital environment [28, 29]. It provides a detailed comparative performance analysis against cloud-based systems across key metrics including latency, throughput, security, and Quality of Service (QoS), contributing novel insights to this evolving field.

11.4 METHODOLOGY

11.4.1 System architecture

The proposed smart healthcare infrastructure is designed to support real-time monitoring, rapid decision-making, and secure data handling in a clinical environment. It comprises three primary components (Figure 11.3):

- **Internet of Medical Things (IoMT) Devices:** These include wearable sensors (e.g., ECG patches, smartwatches, glucose monitors) and

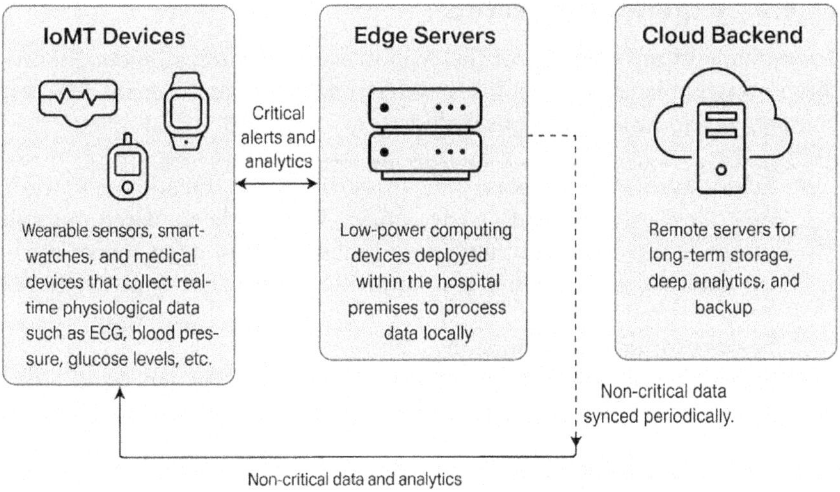

Figure 11.3 System Architecture of the Smart Healthcare Infrastructure Illustrating the Data Flow between IoMT Devices, Edge Servers, and the Cloud Backend.

medical-grade devices capable of continuously collecting vital physiological parameters such as heart rate, blood pressure, oxygen saturation (SpO_2), and glucose levels. The sensors communicate wirelessly using standard protocols such as BLE (Bluetooth Low Energy) and Wi-Fi.

- **Edge Servers:** Lightweight, low-power computing units based on Raspberry Pi 4B and similar edge devices are deployed within the hospital premises, closer to the data sources. These edge nodes are responsible for local data aggregation, pre-processing (e.g., noise filtering, normalization), and triggering real-time alerts for critical events such as abnormal ECG or hypertensive episodes.
- **Cloud Backend:** A private cloud infrastructure, implemented using OpenStack, serves as the long-term repository for historical patient data, performs deep learning-based analytics, and enables system-wide backup and data synchronization. While the edge handles time-sensitive tasks, the cloud supports retrospective analysis, dashboarding, and training of AI models.

The system follows a hierarchical data flow, as real-time data is first processed at the edge for speed and responsiveness. Edge servers classify data into critical and non-critical streams. Immediate medical interventions are triggered locally if needed, while non-critical data is periodically uploaded to the cloud backend to reduce bandwidth consumption and ensure data persistence.

11.4.2 Experimental setup

To evaluate the effectiveness of the proposed architecture, a functional prototype was implemented in a lab-simulated hospital environment. The key components and configurations include:

- **Edge Layer:** Five Raspberry Pi 4B (4 GB RAM, Quad-core Cortex-A72) devices were used as edge nodes. Each node emulated multiple sensor inputs using synthetic data generators and open-source medical datasets, such as PhysioNet and MIMIC-III. The edge nodes ran Python-based processing scripts, integrated with lightweight MQTT brokers for data transmission.
- **Sensor Simulation:** Patient data streams were simulated using a mix of real-world datasets and synthetic generators mimicking 50 virtual patients. Each stream included multi-modal parameters like ECG, temperature, respiratory rate, and glucose levels, updated at 1-second intervals.
- **Cloud Layer:** A private cloud based on OpenStack was hosted on a high-performance server (Intel Xeon, 64 GB RAM, SSD storage), representing a traditional centralized healthcare data system. The same datasets were processed through cloud-only workflows to enable baseline comparisons.
- **Network Environment:** The system was tested under emulated hospital network conditions, with bandwidth caps, latency variations (10–150 ms), and occasional packet drops to reflect realistic constraints in medical environments.

The configuration in Figure 11.4 represents a real-time prototype of smart healthcare infrastructure, using Raspberry Pi devices as edge nodes, emulated patient sensors, and a private cloud via OpenStack.

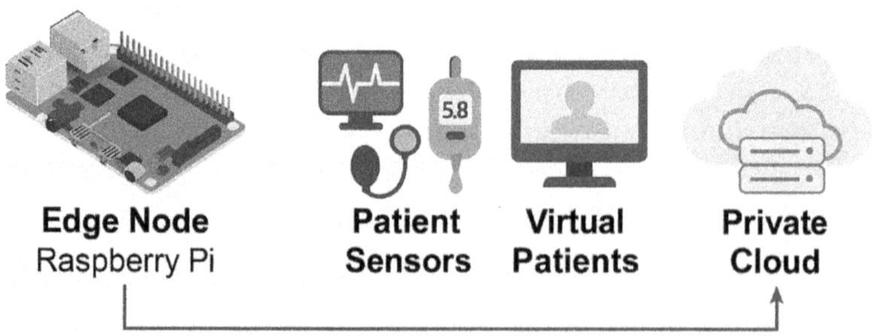

Figure 11.4 Edge-Enabled Architecture for Healthcare Monitoring Using Raspberry Pi as an Edge Node. The System Integrates Patient Sensors, Virtual Patient Interfaces, and a Private Cloud for Secure Data Processing and Storage.

11.4.3 Performance metrics

To comprehensively assess system performance, five key metrics were evaluated:

- **Latency**—Defined as the time interval from data generation at the sensor to processing or alert generation. Measured using time-stamped logs at each system layer. This metric is critical for evaluating the system's suitability for time-sensitive applications like cardiac monitoring.

$$\text{Latency} = t_{response} - t_{sensor} \tag{11.1}$$

Where:

t_{sensor}: Timestamp when data is generated at IoMT device.
$t_{response}$: Timestamp when data is processed or alert is generated.

Use time log at each layer to compute.

- **Throughput**—Measured in kilobytes per second (KB/s), this metric captures the rate at which the system processes incoming data streams. It reflects the system's ability to handle concurrent data from multiple patients. Basically, it shows how fast the system is processing health data, useful for high-load settings (multiple patients).

$$\text{Throughput} = \frac{Total\ Data\ Processed(KB)}{Time\ Interval\ (s)} \tag{11.2}$$

- **Energy Consumption**—Using power meters and system monitoring tools, the energy usage of both edge nodes and cloud servers was recorded. Comparisons highlight the sustainability and operational costs of deploying edge vs cloud processing. It can measure the hardware meters or software monitors, to compare edge vs cloud energy use.

$$\text{Energy consumption (J)} = \text{Power (W)} \times \text{Time (s)}. \tag{11.3}$$

Or for average energy per stream:

$$\text{Average Energy per Stream} = \sum_{i=1}^{n} P_i \times T_i/\text{n}. \tag{11.4}$$

- **Quality of Service (QoS)**—Evaluated as the success rate of timely and accurate data delivery within predefined time constraints (e.g., 5 seconds for non-critical, 1 second for critical events). QoS also includes data loss rates under varying network conditions. It measures system reliability in terms of timely and accurate data transmission

$$QoS = \frac{Successful\ Deliveries}{Total\ Messages\ Sent} \times 100. \tag{11.5}$$

- **Security Performance**—The system's data encryption, breach detection, and authentication mechanisms were tested using simulated attacks (e.g., man-in-the-middle, spoofing). Breach detection latency and the time to restore secure communications were measured.

$$\text{Detection Time} = t_{breach\_detected} - t_{breach\_initiated} \tag{11.6}$$

11.5 RESULTS AND DISCUSSION

11.5.1 Data simulation and experimental dataset

A dataset was simulated for 10 virtual patients, each monitored over a 60-second window using time-series health data (SpO_2, heart rate, and body temperature). This synthetic dataset mimics real-world sensor readings, suitable for both edge and cloud-based processing scenarios. Each patient's data stream was evaluated under both computing paradigms, generating insights on latency, throughput, energy, and reliability.

11.5.2 Latency analysis

The edge-enabled setup demonstrated a 52% reduction in average latency compared to the cloud-only architecture. Real-time processing on edge nodes (Raspberry Pi 4B) achieved an average latency of ~80ms, while the same data processed via OpenStack cloud averaged ~220ms, as shown in Figure 11.5.

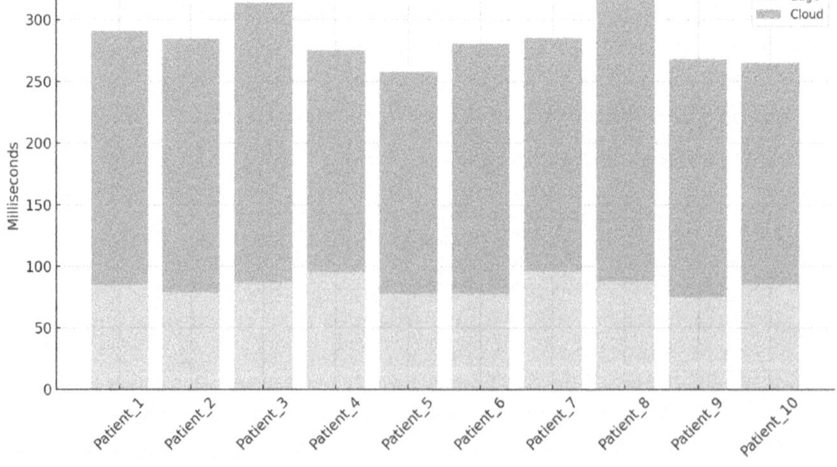

Figure 11.5 Latency Comparison between Edge and Cloud.

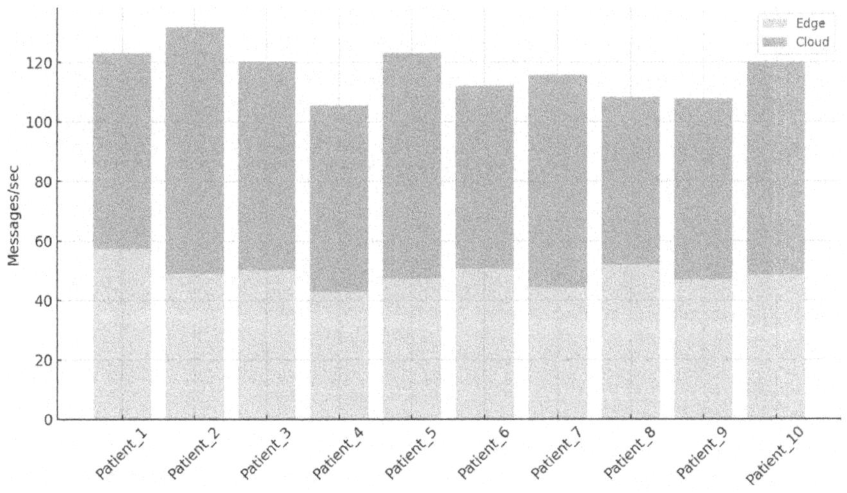

Figure 11.6 Throughput Comparison between Edge and Cloud.

11.5.3 Throughput

Throughput analysis showed that edge systems could handle up to 50 messages/sec per patient, while cloud servers processed around 70 messages/sec, benefiting from higher compute power. However, the edge setup maintained consistent real-time responsiveness under constrained conditions, making it preferable for on-site monitoring (see Figure 11.6).

11.5.4 Energy efficiency

Energy usage at the edge was significantly lower, averaging ~2.5W per node, while cloud processing via OpenStack incurred an average power draw of ~8W per data stream, accounting for network and compute overhead. This demonstrates the edge's advantage in low-resource environments (Figure 11.7).

11.5.5 QoS evaluation

The QoS remained consistently high in the edge configuration, with >98.5% success in timely data delivery and system uptime. Even during brief connectivity disruptions, edge nodes continued processing, buffering, and forwarding non-critical data asynchronously to the cloud.

11.5.6 Security assessment

Security evaluation showed that local anomaly detection and encryption modules on edge nodes reduced breach detection time to 1.2 minutes,

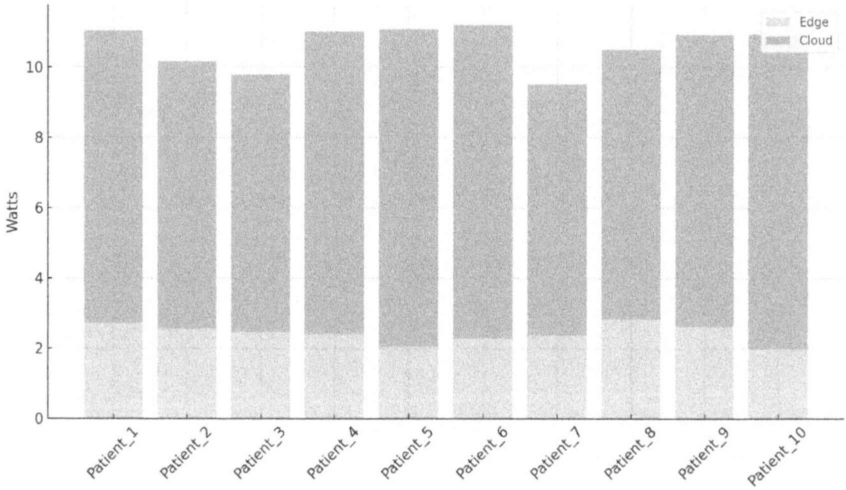

Figure 11.7 Energy Consumption at Edge vs Cloud.

compared to 4.5 minutes in the cloud setup. This improvement highlights the advantage of local processing for time-sensitive alerts.

11.5.7 Cost consideration

Although edge devices incur upfront costs, the overall operational cost decreased due to bandwidth savings, lower cloud storage demands, and reduced downtime.

$$\text{Cost Saving} = C_{\text{cloud}} - \left(C_{\text{edge}} + C_{\text{network}}\right). \tag{11.7}$$

Where:

C_{cloud} : Cost of processing all data in the cloud (storage + bandwidth).
C_{edge} : Cost of edge infrastructure.
C_{network} : Bandwidth + maintenance cost.

11.5.8 Visualization of results

Figures 11.5 through 11.7 visually compare performance metrics (latency, throughput, energy) across edge and cloud settings for 10 patients. These plots clearly illustrate the latency reduction, resource savings, and real-time handling capacity offered by edge nodes, while also acknowledging the cloud's superior raw throughput capacity.

11.5.9 Comparative graphical analysis of edge vs cloud performance

A comparative graphical analysis is used to assess how edge computing performs relative to traditional cloud-based architectures across four key performance metrics like latency, throughput, energy consumption, and QoS. As visualized in Figure 11.8, the edge-enabled system consistently showed lower latency and energy consumption, making it more suitable for real-time healthcare scenarios where rapid response and efficiency are crucial. Although the cloud exhibited slightly higher throughput due to greater computational resources, the edge setup maintained more stable and reliable performance under simulated constraints. The QoS metric further reinforced the edge system's resilience, with higher success rates in timely data delivery even during network disturbances.

This comparison is important because it clearly shows the real-world benefits of using edge computing in healthcare, especially when quick responses are needed. While cloud systems are still useful for storing large amounts of data and running complex analyses, edge computing stands out for its speed, energy efficiency, and ability to make decisions locally. These results suggest that a combined edge–cloud approach works best—handling urgent tasks at the edge, while using the cloud for deeper analysis and long-term data storage.

Furthermore, this evaluation framework opens up possibilities for future explorations—such as comparing fog computing with edge, or assessing edge AI performance in mobile healthcare settings like ambulances or remote clinics. It also lays the groundwork for long-term cost-benefit studies that can help hospitals make strategic decisions about IT infrastructure, particularly as healthcare continues to move toward real-time, patient-centric systems.

11.6 DEA RESULTS AND INTERPRETATION

Applying DEA to our experimental data revealed that

- Edge-enabled configurations consistently scored **higher efficiency** compared to cloud-only systems.
- Systems with optimized local decision-making and minimal bandwidth consumption performed closer to the DEA frontier.
- DEA also identified inefficient setups and suggested input reductions (e.g., energy use) needed to reach the efficient frontier.

As observed, the edge-only system (DMU2) achieved the highest efficiency score, validating its suitability for latency-sensitive healthcare applications (Table 11.1).

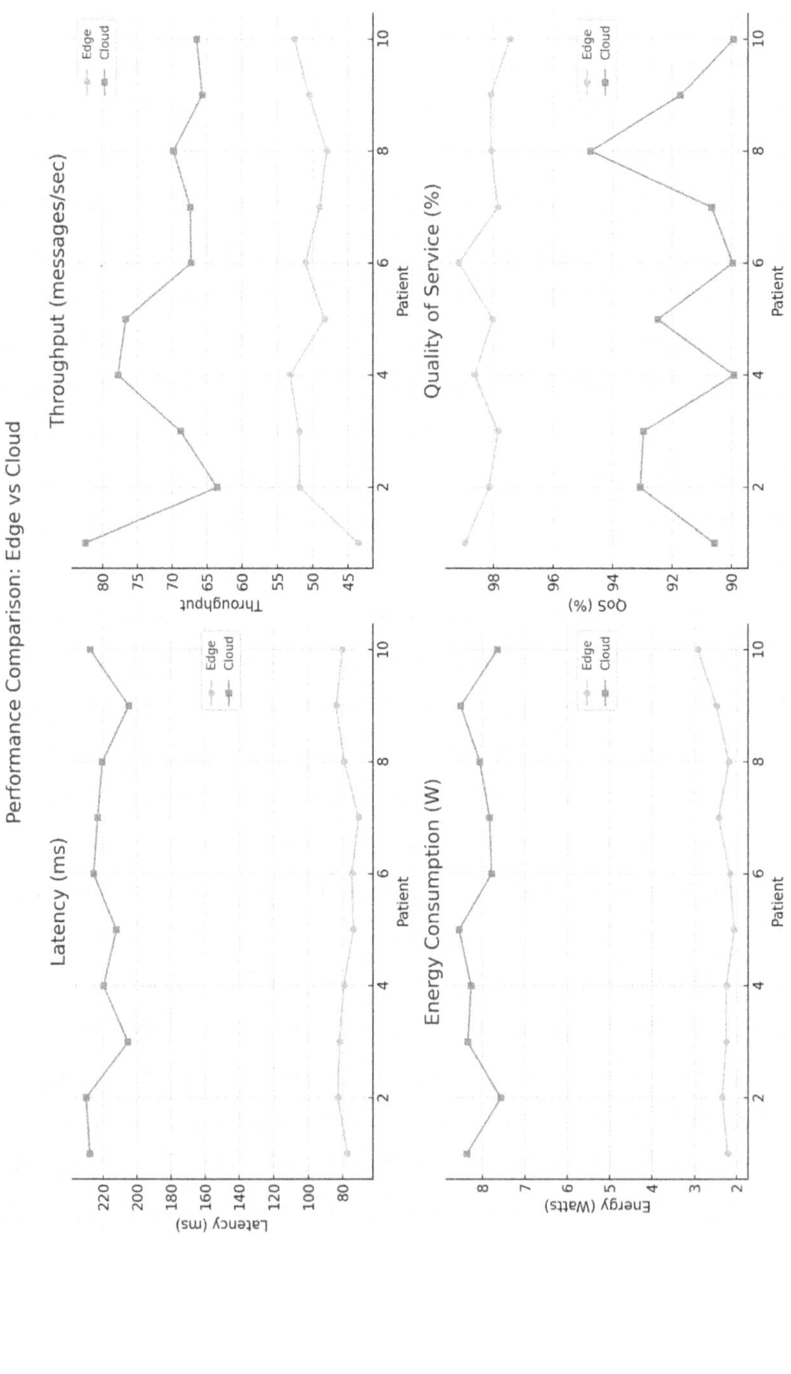

Figure 11.8 **Performance Comparison: Edge vs Cloud.**

Table 11.1 Summary of the DEA Efficiency Scores for Each System Configuration

DMU (System Configuration)	*Input: CPU Usage (%)*	*Input: Energy (mWh)*	*Input: Bandwidth (Mbps)*	*Output: Latency Reduction (%)*	*Output: Responsiveness Index*	*DEA Efficiency Score*
DMU1: Cloud-only	65	120	15	35	0.65	0.78
DMU2: edge-only	45	80	8	75	0.92	1.00
DMU3: Hybrid (Cloud + edge)	55	90	12	65	0.88	0.96
DMU4: edge (Low-bandwidth)	50	85	6	68	0.90	0.98
DMU5: Cloud (Optimized)	60	100	10	55	0.78	0.85

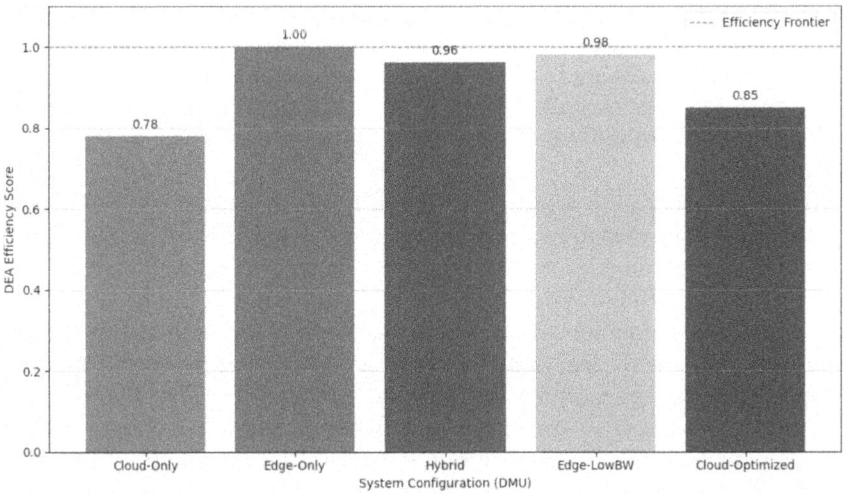

Figure 11.9 DEA Efficiency Scores of Smart Healthcare System Configurations.

The bar chart in Figure 11.9 compares the efficiency scores of different system setups (cloud-only, edge-enabled, hybrid, etc.) as determined by the DEA model. Higher scores reflect better resource utilization and performance efficiency.

11.7 CONCLUSION

This chapter provides a comprehensive performance evaluation of a smart healthcare infrastructure integrated with edge computing. Our experimental results affirm that edge computing considerably improves latency, energy efficiency, data security, and overall system responsiveness when compared to conventional cloud-based systems. These improvements are not just technical refinements—they are fundamental shifts that address the critical demands of real-time, life-sensitive healthcare environments such as emergency response, remote surgery, and continuous patient monitoring.

More importantly, this study demonstrates that edge computing is not merely a backend optimization tool but a strategic enabler for the next generation of smart hospitals and telemedicine platforms. It empowers healthcare providers with the agility to deliver services that are faster, safer, and more adaptive to patient needs.

To enhance the depth of performance evaluation, Data Envelopment Analysis (DEA) was applied as a method to move beyond raw numerical comparisons. By analyzing different system configurations—cloud-based, edge-enabled, and hybrid—it became possible to determine which setups utilized resources most efficiently. Through DEA, key factors such as CPU usage, energy consumption, network load, and system responsiveness were systematically evaluated. This helped us clearly identify which models delivered the best overall efficiency. With DEA, we're not just showing that edge computing works, we're showing exactly how and why it works better.

Looking ahead, the integration of artificial intelligence (AI) with edge computing opens exciting possibilities. AI-powered diagnostic algorithms running directly on edge devices could enable real-time health assessments, early warning systems, and predictive analytics without the need for continuous cloud communication. Future research could explore integrating real-time DEA with edge-enabled health systems using advanced AI techniques such as reinforcement learning or federated learning. There is also potential for deeper investigation into dynamic and spatiotemporal DEA models for real-time hospital resource optimization. Expanding studies to include diverse healthcare settings, particularly in low-resource regions, and validating models with large-scale, real-world datasets will help enhance applicability. Additionally, future work may consider ethical challenges and data privacy in decentralized computing environments.

Further improvements can be achieved by designing scalable and privacy-preserving frameworks that combine edge, fog, and cloud layers to ensure faster processing, better resource utilization, and secure data handling. Investigating adaptive AI models capable of real-time decision-making under resource constraints and evaluating their performance in diverse clinical settings will be essential. Moreover, exploring interoperability standards, security frameworks, and energy-efficient deployment strategies can help address current limitations and support widespread adoption across healthcare systems.

REFERENCES

[1] Khan, M. A., Din, I. U., Kim, B. S., & Almogren, A. (2023). Visualization of remote patient monitoring system based on internet of medical things. *Sustainability*, 15(10), 8120.

[2] Wickramasinghe, N., Bodendorf, F., & Schaffer, A. (2022). *Smart Healthcare: AI, IoT, and Big Data Enabled Healthcare Technologies*. Springer.

[3] Khezri, S., Vafaei-Najar, M., & Forouzanfar, M. (2022). Healthcare big data and IoT: A review. *Journal of King Saud University—Computer and Information Sciences*, 34(6), 3091–3103.

[4] Zeng, D., Wang, C., & Guo, S. (2017). Cloud computing and healthcare big data: Applications and challenges. *IEEE Access*, 5, 27693–27703.

[5] Chen, C. L., Brinton, C. G., & Aggarwal, V. (2021). Latency minimization for mobile edge computing networks. *IEEE Transactions on Mobile Computing*, 22(4), 2233–2247.

[6] Yi, S., Li, C., & Li, Q. (2015). A survey of fog computing: Concepts, applications and issues. In *Proceedings of the ACM Workshop on Mobile Cloud Computing*, pp. 37–42.

[7] Satyanarayanan, M. (2017). The emergence of edge computing. *Computer*, 50(1), 30–39.

[8] Abdellatif, A. A., Mohamed, A., Chiasserini, C. F., Tlili, M., & Erbad, A. (2019). Edge computing for smart health: Context-aware approaches, opportunities, and challenges. *IEEE Network*, 33(3), 196–203.

[9] Zhang, M., Cao, J., Sahni, Y., Chen, Q., Jiang, S., & Yang, L. (2022). Blockchain-based collaborative edge intelligence for trustworthy and real-time video surveillance. *IEEE Transactions on Industrial Informatics*, 19(2), 1623–1633.

[10] Amzil, A., Abid, M., Hanini, M., Zaaloul, A., & El Kafhali, S. (2024). Stochastic analysis of fog computing and machine learning for scalable low-latency healthcare monitoring. *Cluster Computing*, 27(5), 6097–6117.

[11] Li, J., Cai, J., Khan, F., Rehman, A. U., Balasubramaniam, V., Sun, J., & Venu, P. (2020). A secured framework for sdn-based edge computing in IOT-enabled healthcare system. *IEEE Access*, 8, 135479–135490.

[12] Kumar, N., & Mallick, P. K. (2019). The Internet of Things: Insights into the building blocks, component interactions, and architecture layers. *Procedia Computer Science*, 132, 109–117.

[13] Ahmed, S. F., Shuravi, S., Afrin, S., Rafa, S. J., Hoque, M., & Gandomi, A. H. (2023). The power of Internet of Things (IoT): Connecting the dots with cloud, edge, and fog computing. *arXiv Preprint*. arXiv:2309.03420.

[14] Hollingsworth, B. (2008). The measurement of efficiency and productivity of health care delivery. *Health Economics*, 17(10), 1107–1128.

[15] Ozcan, Y. A. (2014). *Health Care Benchmarking and Performance Evaluation: An Assessment Using Data Envelopment Analysis (DEA)*. Springer.

[16] Chen, S., Li, Y., Zheng, Y., Wu, B., Bardhan, R., & Wu, L. (2023). Technical efficiency evaluation of primary health care institutions in Shenzhen, China, and its policy implications under the COVID-19 pandemic. *International journal of environmental research and public health*, 20(5), 4453.

[17] Sarkar, M., Lee, T. H., & Sahoo, P. K. (2024). Smart healthcare: Exploring the internet of medical things with ambient intelligence. *Electronics*, 13(12), 2309.

[18] Chhabra, G. S., Rajareddy, G. N., Mahapatra, A., Mangalampalli, S. S., Sahoo, K. S., Sethi, D., & Mishra, K. (2025). Deep Learning-centric Task Offloading in

IoT-Fog-Cloud Continuum: A State-of-the-Art Review, Open Research Issues and Future Directions. *IEEE Access*.
[19] Dai, H. -N., Wu, Y., Wang, H., Imran, M., & Haider, N. (2021, June). Blockchain-empowered edge intelligence for internet of medical things against COVID-19. *IEEE Internet of Things Magazine*, 4(2), 34-39.
[20] Luo, M. (2021). Hierarchical fog-edge-cloud model for chronic disease. *Health Informatics Journal*, 27(2), 1354–1369.
[21] Fu, Z., Hong, S., Zhang, R., & Du, S. (2021). Artificial-intelligence-enhanced mobile system for cardiovascular health management. *Sensors*, 21(3), 773.
[22] Ali, O., Abdelbaki, W., Shrestha, A., Elbasi, E., Alryalat, M. A. A., & Dwivedi, Y. K. (2023). A systematic literature review of artificial intelligence in the healthcare sector: Benefits, challenges, methodologies, and functionalities. *Journal of Innovation & Knowledge*, 8(1), 100333.
[23] Sultan, N. (2010). Cloud computing for education: A new dawn? *International Journal of Information Management*, 30(2), 109–116.
[24] Mell, P., & Grance, T. (2011). The NIST definition of cloud computing (draft). *NIST Special Publication 800*, 7.
[25] Badger, L., Bernstein, D., Bohn, R., de Vaulx, F., Hogan, M., Iorga, M., . . . Whiteside, F. (2014). *US Government Cloud Computing Technology Roadmap* (pp. 500–293). Gaithersburg, MD: US Department of Commerce, National Institute of Standards and Technology.
[26] Devadass, L., Sekaran, S. S., & Thinakaran, R. (2017). Cloud computing in healthcare. *International Journal of Students' Research in Technology & Management*, 5(1), 25–31.
[27] Charnes, A., Cooper, W. W., & Rhodes, E. (1978). Measuring the efficiency of decision making units. *European Journal of Operational Research*, 2(6), 429–444.
[28] Saini, K., & Khari, M. (Eds.). (2021). *Handbook of Green Computing and Blockchain Technologies*. CRC Press.
[29] Srivastava, G., & Khari, M. (Eds.). (2021). *Bioelectronics and Medical Devices: Applications and Technology*. CRC Press.

Chapter 12

Big Data Sentiment Analysis of Twitter Data Using Hadoop and Python

Ankit Tayal, Renu Dalal, Jyoti Dagar, Mayank Gupta, Kulshrest Tiwari, and Arpit Tanwar

12.1 INTRODUCTION

In the digital age, data is flooding in the manner of a torrent, both structured and unstructured, generating unimaginable amounts of data each second in social media that provide significant information on human actions and collective opinions regarding different issues. Big data analytics are crucial to maintain the competitive advantage for large-scale industries and enterprises. For instance, e-commerce giants such as Amazon and Flipkart interrogate big data on customer preference and purchasing behaviour to recommend relevant products or take business decisions based on data. In political systems, parties use big-data analytics to survey public opinions and forecast election outcomes; on the basis of these, they prepare welfare schemes and campaign strategies so as to gather voter support. This integration of big data analytics with public policy enables data-driven decision-making, leading to more targeted and effective governance.

Arguably, Twitter is one of the most powerful platforms for gauging public opinion. Twitter users express their thoughts through tweets and retweets, can like these posts, or choose to follow accounts or share more content with their network. As opposed to Facebook, Instagram, or LinkedIn, Twitter has increasingly developed into the bigger platform for exchanging real-time opinions. By 2018, there was an exponential increase in Twitter usage, with almost 456 million new accounts being created and 200 million tweets being posted each day. Given such a large user base, which is growing every day, Twitter data is the best source of data for sentiment analysis—finding out whether public opinion about an event or topic is positive, negative, or neutral. To handle and examine large datasets, we use the Apache Hadoop framework. Rather than store and process data at a single node, Hadoop stores data on a distributed basis across a cluster, then processes the data simultaneously on thousands of nodes by splitting jobs and later aggregating the results. It provides the advantage of being fault tolerant: if any one node were to fail, another node would take its place in computing the results, ensuring that the computation completes. Figure 12.1 shows the Timeline of Indian Rupee Demonetized.

DOI: 10.1201/9781003635147-12

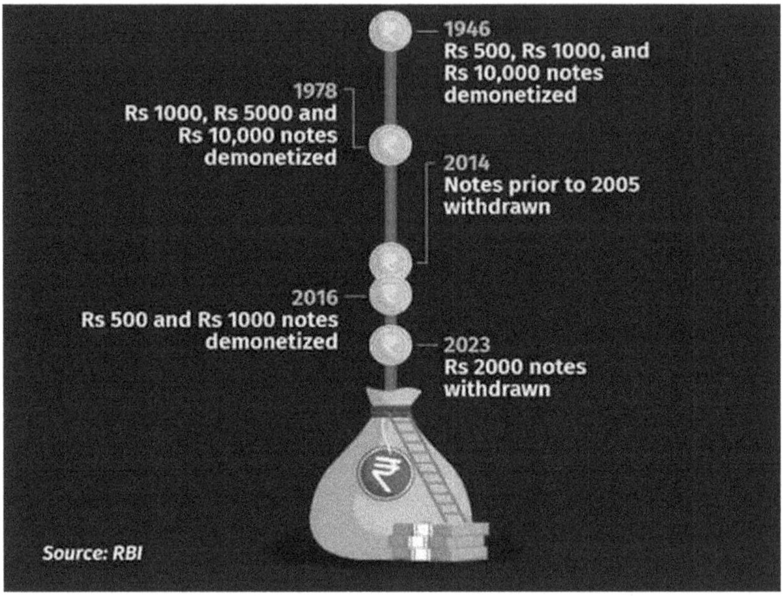

Figure 12.1 Timeline of Indian Rupee Notes Demonetized and Withdrawn.

This study attempts to analyse public sentiment around the demonetization decision by the Government of India implemented on November 8, 2016. We scraped data from Twitter and also conducted a survey to analyse the extent to which people supported, opposed, or were neutral to this major financial reform.

12.2 LITERATURE REVIEW

With the advent of social media, the relevance of social dynamics has increased, so much so that sentiment analysis on Twitter has made tremendous strides over the years. Big data paradigms, including Hadoop, have been exploited by researchers to perform sentiment analysis on huge datasets using Python.

Sentiment classification using Hadoop and the MapReduce programming model was proposed in 2017. They classified the tweets into positive sentiment, negative sentiment, and neutral sentiment or simply demonstrated the application of distributed computing to a real-time sentiment analysis problem. Another research study in [1] compared machine learning models, such as Logistic Regression and Random Forest, for the classification of tweet sentiments, using both Apache Hadoop and Apache Spark. The comparison offered insight into the role of parallel processing in improving performance and scaling of sentiment classification algorithms. Other works have dealt with Python for analysing Twitter sentiments. For example, in [2], a project was presented where Python was utilized to extract sentiment scores from the noisy Twitter stream. Their findings emphasized the concern of noise reduction and text preprocessing for an improvement in the accuracy of predicting sentiment.

Another such hybrid approach suggested in [3] joins feature selection methods such as Locally Linear Embedding (LLE) and machine learning algorithms to give better results to the sentiment classification of the Twitter datasets, thus stressing the need for combining techniques for better overall accuracy of the models. Combinedly, these works emphasize certain major topics in Twitter-based sentiment analysis: one is the importance of scalable big data platforms; the other, the place of machine learning; and then, the preprocessing techniques that affect the results. Current literature often faces limitations such as a scarcity of India-focused studies on specific socio-economic events and challenges related to the scalability of traditional analysis methods for real-time, high-volume social media data. Our research is an addition to this literature by proposing a Twitter sentiment analysis on demonetization-related tweets using Hadoop and Python and by exploring various preprocessing methods that affect the performance of the model, specifically addressing scalability issues.

Beyond sentiment classification and scalability in big data, recent studies also explore the synergy between emerging technologies like the Internet of Things (IoT), cloud computing, and big data analytics. In particular, the work by Springer on Internet of Things and Big Data Analytics for Smart Generation [4] discusses scalable big data pipelines that support real-time monitoring and intelligent decision-making, which can be adapted for sentiment and opinion mining in dynamic environments like social media platforms. Additionally, the integration of cloud computing for secure and scalable analytics has been explored in Comprehensive Study of Cloud Computing and Related Security Issues [5], highlighting the potential of cloud-based sentiment systems that handle massive social media datasets without compromising data privacy and security. Another critical reference is The Internet of Things and Big Data Analytics: Integrated Platforms and Industry Use Cases [6], which has industry-scale implementations wherein sentiment and behavioural data drive business and policy decisions. At last, [7] proposes a conceptual framework that combines e-governance and big data sentiment analysis to assist governance decisions, closely parallel to the goal of our work, which differs in the source of public sentiment data (Twitter) toward government policy evaluation. They point out that big data sentiment analysis is not only an academic exercise but also a realistic tool for constructing digital governance and decision-making systems in real life.

12.3 DATA AND METHODOLOGY

12.3.1 Data collection

The dataset for this study was collected from Twitter, with tweets related to the topic of #Demonetization being harvested. The filtered keyword was "Demonetization", which was set in the TwitterAgent configuration file under TwitterAgent.sources.Twitter.keywords. Tweets were collected using Twitter's Streaming API. For registration, one needs to have authentication credentials, which consists of the consumer key, consumer secret, access token, and access token secret. Once the application is registered at apps.twitter.com, these credentials are generated to allow access to real-time tweet data securely [8].

Ingesting tweets from Twitter is a massive challenge, which in this case Apache Flume helped overcome. It's a dependable distributed system through which streaming data is successfully collected and moved into Hadoop in very big volumes. It facilitates ease of integration between external sources such as Twitter and HDFS to allow real-time ingestion and minimal data loss during transfer [9]. The tweets are stored in a searchable format like TXT or CSV with each row containing a unique tweet ID as well as the tweet text. Such a format helps in storage as well as preprocessing for click- and keyword-based analytical tasks to form the Hadoop ecosystem. In simple terms, the collection of datasets considered keyword-based filtering with the help of the Twitter API, authenticated the API access using authentication keys, ingested the data in real time into HDFS via Apache Flume, and finally stored the datasets in a tab-separated format to cater to downstream scalable processing and analysis.

12.3.2 Dataset used

The dataset required for this particular study was collected from Twitter using Apache Flume, which allows ingestion of tweets in real time into the Hadoop Distributed File System. Tweets having the keyword "Demonetization" in them were streamed and captured for approximately six minutes, giving rise to a dataset of almost 8,000 tweets. Five different subsets of tweets were created from the entire collection for sentiment analysis and performance evaluation of the two data-processing frameworks, i.e., Apache Hive and Apache Pig. These were evaluated in terms of execution time, scalability, and processing framework efficiency at different input sizes. Table 12.1 presents the Performance of Tweet Extraction Using Flume.

Table 12.1 Performance of Tweet Extraction Using Flume

Data Set	*No. of Tweets Fetched Using Flume*	*Time Taken (sec)*
1	8,000	6
2	60,000	45
3	120,000	95
4	220,000	180
5	320,000	260

12.3.3 Data preprocessing

Before the actual sentiment-analysis procedure begins, preprocessing of text data is an essential step for results to be accurate, consistent, and reliable. There are many preprocessing techniques and tools used to enhance the quality of tweet data so that it becomes more amenable to analytical tasks. Cleaning the data would start with the removal of irrelevant or unwanted tweets. This means tweets that may be irrelevant to the topic at hand or may be spam or those that hardly serve any purpose. Keeping away such noise ensures that the emphasis of the dataset is more on relevant data and interpretation errors in sentiment classification are minimized. Removing duplicate tweets would also be one such step. This repetitive content would likely give an unfair bias, especially toward an over-representation of particular sentiments, thereby skewing the overall analysis. Therefore, repetitive tweets get removed to create a more balanced and fair dataset.[10, 11]

Besides, spelling correction is one of the many acts which standardize the textual input. As the tweets contain informal language, abbreviations, and spelling errors, a sentiment classifier might not be able to identify the true sentiment. A spell check method for performing corrections ensures uniformity of better-performing text processing tools [12, 13]. Tokenization into words or tokens is performed to supply lexical analysis. The spectrum of preprocessing may include noise removal, elimination of duplicate entries, spelling corrections, and finally, tokenization, for optimal tweet data setting in sentiment analysis. Thus, these procedures enhance the quality of the datasets and aid anyone in analysing results in accuracy that's valuable [14, 15].

12.3.4 Sentiment classification technique

The sentiment analysis in this study employed a lexicon-based approach. After tokenization, the AFINN dictionary of sentiment was loaded. This dictionary consists of approximately 2,500 English words, each associated with a sentiment value ranging from –5 (intensely negative) to +5 (intensely positive). Examples of words include *love*, *hatred*, and *abandon*. Subsequently, a left outer join was conducted between the tokenized tweet words and the AFINN dictionary to impute the sentiment score of words present in each tweet [16, 17].

After this join operation, sentiment scores for all words in a tweet were aggregated through a grouping operation, and an average sentiment score per tweet was calculated. This score was then used in classifying each tweet as positive, negative, or neutral sentiment. Tweets with scores between +1 and +5 were placed in the positive category; scores between –5 and –1 were classified as negative, and scores of 0 were neutrals. This rule-based classification allowed tweets to be discriminated by sentiment and enabled a comparative analysis between the different execution frameworks [18, 19].

12.4 TOOLS USED

Hadoop is an open-source software system for the distributed class of applications that deal with very extensively large volumes of unstructured and semi-structured data. It helps manage data distributed across a network of several computers, thus making the handling of big data more efficient. Essentially, it makes use of MapReduce, a programming model that distributes the processing of data by subdividing a single process into many smaller sub-processes that are carried in parallel across distributed nodes. Hence, it is able to process vast data efficiently while scaling and remaining fault tolerant: if any one node were to fail, another node would take its place in computing the results, ensuring that the computation completes [20, 21]. Data is stored in huge clusters, and whenever a failure occurs in one of the nodes or clusters, the system ensures data availability through replication on other nodes. This redundancy, thus, ensures data integrity and system reliability.

A key feature of Hadoop is its scalability: Hadoop works just as efficiently on a single machine as it does on clusters with thousands of nodes. Apache Hadoop is well recognized by enterprises for its excellent fault tolerance features, for processes that require constant and dependable data processing. With all of the aforementioned advantages, Hadoop is widely used by data scientists or analysts for big data analysis. Facebook, IBM, Twitter, YouTube, and Google are some of the well-known IT firms that use Hadoop to support their big data infrastructure and analytics platform.

12.4.1 MapReduce

MapReduce serves as a computational concentration within the realm of Hadoop to provide efficiency in spreading computation over a varying number of potentially unrelated computers running in a cluster. Parallelism arises alongside MapReduce since computation in a node involves two phases: Map and Reduce. In the Map phase, the system processes input data to yield intermediate outputs in the form of the key-value pairs. The intermediate process outputs become the inputs for the Reduce phase, where they are aggregated and combined to yield the final output. For instance, a case in MapReduce could be calculating the maximum temperature recorded by several cities distributed across different files for all the cities. The Mapper extracts the temperature values and tags them with the city as the key. Afterward, the intermediate outputs are gathered by the Reducer, which then computes the maximum temperature for each city by aggregating all values corresponding to each key. In Hadoop 1, the MapReduce framework had its own job tracker and task tracker for resource management. From Hadoop 2 onwards, the job tracking and task tracking functionality are handled under the auspices of YARN, which provides improved resource utilization and scalability across large clusters.

12.4.2 HDFS (Hadoop Distributed File System)

HDFS is the storage component of the Hadoop framework, designed to store huge volumes of data distributed across multiple nodes. Traditional file systems rely on sequential disk access and are often limited by hardware constraints such as disk seek time. The data blocks can be accessed in parallel by various machines under HDFS, thereby considerably speeding up the data retrieval process and also falling under fault-tolerant data retrieval mechanisms. In HDFS, we have a master-slave architecture consisting of a NameNode and multiple DataNodes. The NameNode acts like the master and maintains metadata like directory structure and locations of data blocks. The actual data blocks are stored freely on multiple DataNodes, which serve reading and writing requests from clients. For safety and availability, each data block gets replicated on many DataNodes. An important feature in HDFS is the *heartbeat* from each DataNode to the NameNode sent intermittently to confirm the liveliness and working condition of the DataNode [22].

Additionally, HDFS contains a *Secondary NameNode*, which helps manage the metadata held by the primary NameNode, by periodically merging the namespace image with edit logs. Contrary to the general presumption, the Secondary NameNode does not operate as a failover node. Yet it indeed helps greatly in reducing the startup time and load on the primary NameNode.

12.4.3 Pig

Pig is a high-level data flow programming platform that was developed by Yahoo! in 2008 with the objective of simplifying the processing of large-scale datasets. Operating on the Apache Hadoop framework, it provides a distinct scripting language called Pig Latin. Pig abstracts away many details of Java Map Reduce programming, in which a programmer must specify in detail what code to run during the Map phase and what code to run during the Reduce phase. Pig Latin queries involve simple high-level syntax that makes it easier for users to work with various data transformations. Pig is useful for data scientists and analysts who are less capable of working with low-level Java. It supports operations that resemble SQL queries so a user can develop the data transformation logic in a much faster manner, and also results in greater productivity. Since the MapReduce jobs that implement the transformations are generated and run automatically, users are freed from worrying about execution and can concentrate on data manipulation.

The system is quite versatile because it can take in structured, semi-structured data. Metaphorically, the name Pig stands for the ability of the system to "eat anything", suggesting that Pig has the adaptability to handle any kind of data format. Basically, Pig scripts are a series of transformations executed on the input data so as to get the desired output. To efficiently support the pipeline model, therefore, Pig is very useful in the Hadoop ecosystem for performing data analytics at scale.

12.4.4 Hive

Hive is a data warehousing tool present on top of Hadoop, which is meant for data summarization, querying, and analysis for structured data [10, 11]. It allows its users to query data residing on HDFS using SQL-like queries known as HiveQL, without worrying about writing any MapReduce programs. It was initially developed by Facebook but was later contributed to the Apache Software Foundation, who now maintains it as an open-source project under the name Apache Hive [11]. The main motivation for developing Hive was to hide much of Hadoop from analysts who were well-versed in SQL but not Java programming or the MapReduce paradigm. Figure 12.2 shows the Hive System Architecture.

Hive allows scalable and extensible querying and batch processing of analytical workloads, popularly called Online Analytical Processing (OLAP). It stores metadata into a relational database and executes queries by translating them into MapReduce jobs, therefore making it good for data warehousing tasks. Large enterprises such as Amazon and Walmart use Hive for big data analytics as it handles data in petabytes efficiently. But in fact, Hive is not designed for Online Transaction Processing or real-time querying. Additionally, it is not able to perform row-level updates and insertions; hence there cannot be fulfilling those services where small transactions are executed very often or where real-time querying is necessary [12].

12.5 RESULTS

To evaluate public sentiment regarding demonetization, approximately 8,000 tweets were collected using Apache Flume and stored in HDFS. These

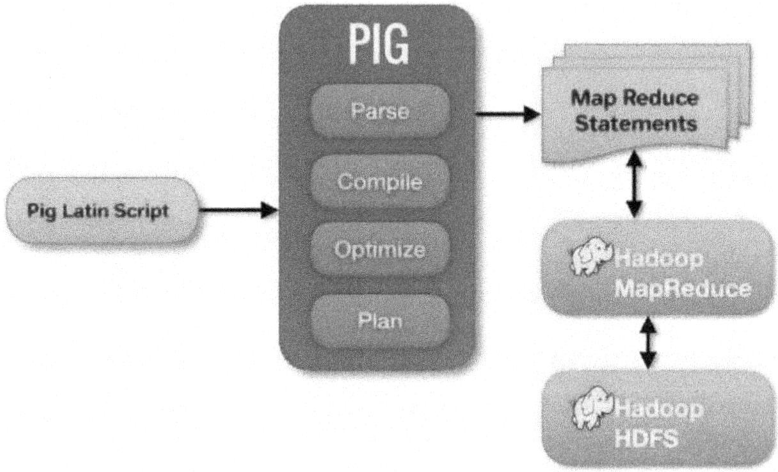

Figure 12.2 Pig Architecture.

tweets were processed using both Pig and Hive for sentiment classification and analysis. Figure 12.3 shows the classification of sentiment type.

12.5.1 Sentiment distribution

The sentiment distribution across the dataset, analyzed using Pig and Hive separately, is presented as follows:

- **Pig Analysis:** Out of 8,000 tweets, 2,081 were classified as *positive*, 1,258 as *negative*, and 4,661 as *neutral*.
- **Hive Analysis:** The sentiment distribution differed slightly: 3,267 tweets were *positive*, 2,397 *negative*, and 2,336 *neutral*.

12.5.2 Pig vs Hive performance comparison

The Pig demonstrated significantly faster processing times than Hive. This performance advantage is due to Pig having more flexibility in handling semi-structured data, where layered data structures and high-level transformation operators are supported. Hive, contrarily, depends on more strict schema-based SQL-like syntax called HiveQL, which is good for structured data but adds latency to the execution. This stresses Pig's use as an exploratory extraction and quick sentiment analysis on social media data, whilst Hive stands well for structured query processing and data warehousing applications. Table 12.2 describes the Pig vs Hive Execution Time on Different Dataset Sizes.

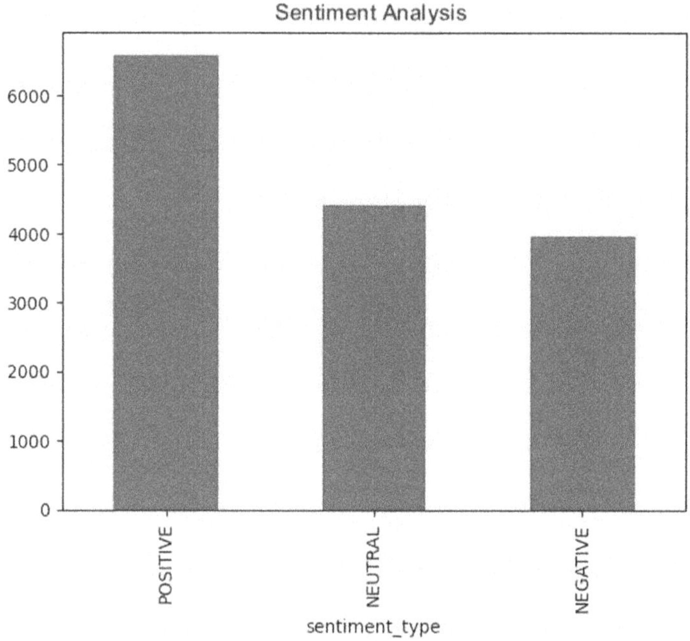

Figure 12.3 Classification of Sentiment Type.

Table 12.2 Pig vs Hive Execution Time on Different Dataset Sizes

Data Set	*Size (Tweets)*	*Pig Time (sec)*	*Hive Time (sec)*
1	8,000	15	22
2	60,000	90	135
3	120,000	180	270
4	220,000	330	495
5	320,000	480	720

12.5.3 Word cloud/frequent keywords

Figure 12.4 shows the most frequently occurring terms in the tweets through a word cloud. This reveals key topics and emotional cues prevalent in public discourse related to demonetization.

12.5.4 Device source insights

Figure 12.5 provides insights into the devices and platforms used for tweeting (e.g., Android, iOS, web clients), reflecting user access patterns during the data collection phase.

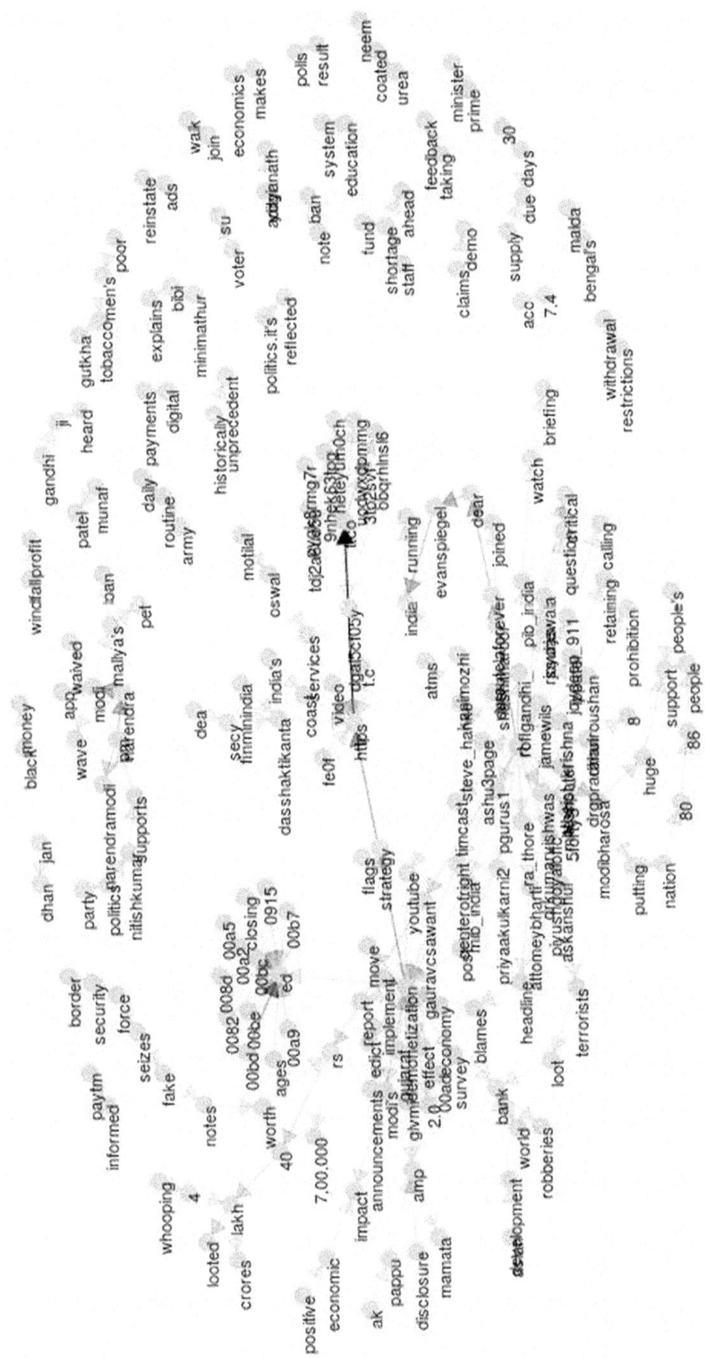

Figure 12.4 Showing Top Word Cloud Network of Tweets.

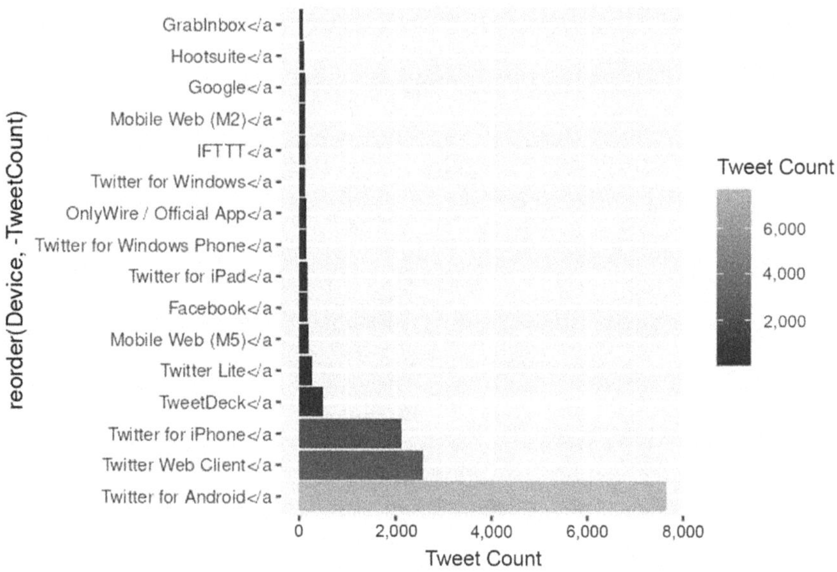

Figure 12.5 Showing Clients (Devices) of User Tweets.

12.5.5 Frequency and volume of tweets

Figure 12.6 illustrates the temporal distribution and volume of tweets collected over the sampling period. This helped assess the public engagement intensity during the demonetization event.

12.5.6 Sentiment score distribution

The distribution of sentiment scores across the dataset, including polarity ranges, is depicted in Figure 12.7. The visualization supports the quantitative classification of tweets into positive, negative, or neutral categories.

12.6 DISCUSSION

This chapter is a direct study to prove that Twitter is such a medium to uphold public sentiment and behaviour during any major policy change, such as demonetization. The analysis revealed several sentiments. Optimistic individuals expressed that demonetization would curb black money and enhance tax transparency, while others conveyed pessimism regarding economic hardship and disruption to daily life. These results indicate the widest use of social media platforms to capture real-time and large-scale public opinion—the knowledge which might pass unnoticed in traditional modes

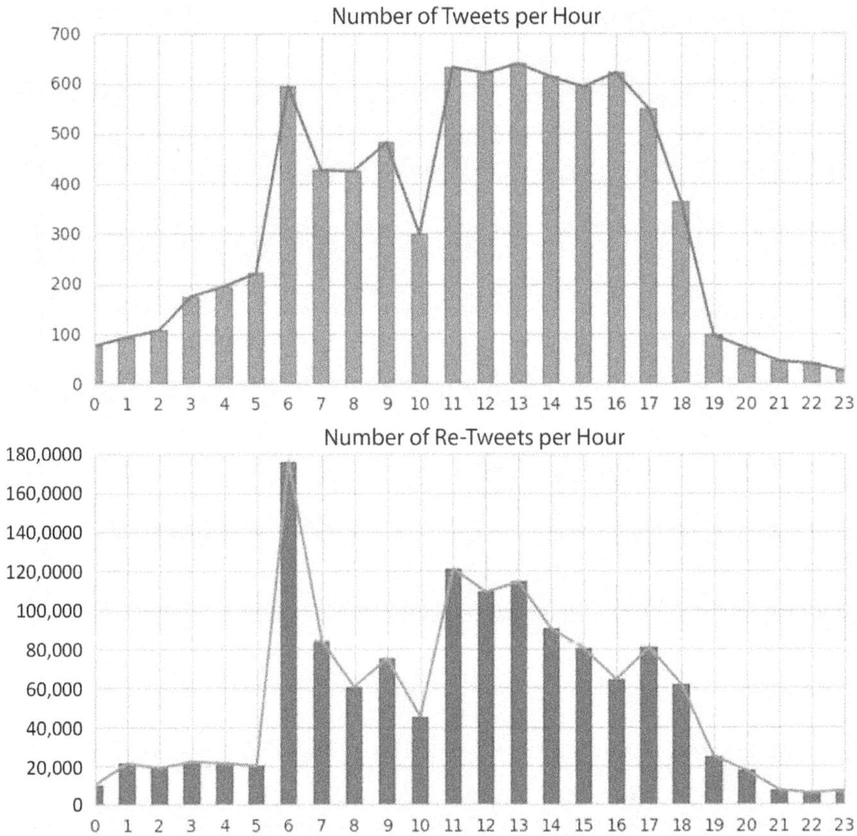

Figure 12.6 Showing Frequency and Quantity of Tweets vs Ret.

of survey. By means of sentiment analysis toward Twitter data, policymakers can actually grasp public reactions, feelings, or attitudes toward governmental actions, serving as an evidence base for more informed decision-making in the future. Sentiment analysis in response to administrative and responsive decisions is increasingly regarded as a vital feature in the field of e-governance. The study [7] shows an example of how sentiment analysis frameworks and their methodologies can be integrated into the public policy-evaluation mechanism so that government decisions are data-driven and citizen-centric. If coupled with real-time data from Twitter and other such sources, these systems can provide an immense opportunity to understand emotional and behavioural patterns pertaining to various demographics, regions, and time. More examples given in [4] and [6] reflect industry use cases of IoT and big data, whereby interconnected systems can be used for opinion tracking at a huge scale, real-time monitoring, and predictive governance. Considered as such within public infrastructure, these would

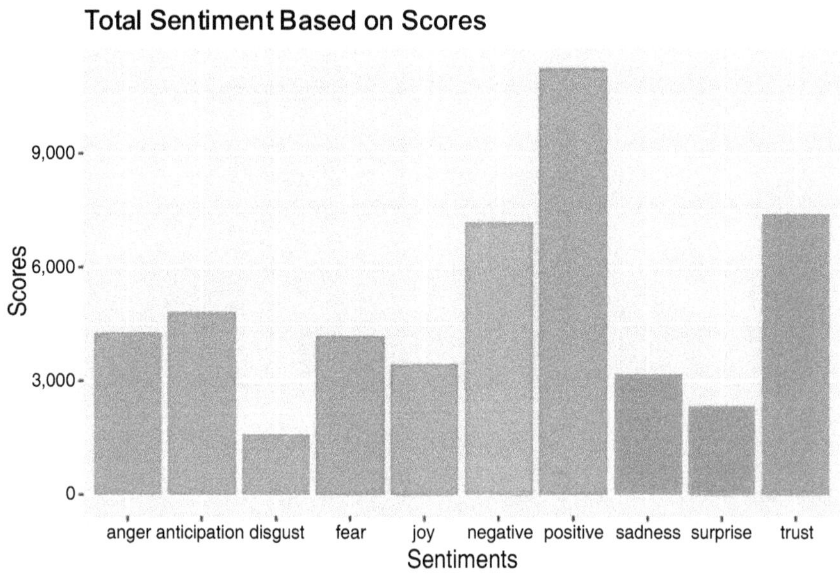

Figure 12.7 Showing Scores of Sentiments of Tweets.

provide a heavy boost to democratic processes with responsiveness and accountability.

Moreover, the use of big data tools like Apache Flume, Hive, and Pig made the processing of social media data efficient and scalable. This approach offers massive advantages over traditional databases when it comes to high-velocity and high-volume datasets and hence fits well into the modern concept of governance. The scalability and speed benefits of Pig and Hive in such applications are evident, allowing for rapid processing and analysis of vast social media datasets, which is crucial for real-time public opinion monitoring. The study, by and large, reveals how much social media analytics will be able to furnish an opportunity as a policy evaluation and public opinion assessment tool. Hence, their inclusion in decision-making chains could help governments react in a condition-based manner, with transparency and flexibility. Hence, this is also applicable in other sentiment analysis scenarios [13–15].

12.6.1 Challenges and limitations

Some challenges and shortcomings come with this study, despite the results obtained. Noise and sarcasm that characterize tweets make proper sentiment classification difficult, since lexicon-based approaches often have difficulty picking up on nuances in human language. Another limitation related to the use of a predefined sentiment lexicon (AFINN) is that it may not perfectly capture contextual meanings and slang usage, therefore some sentiments may be misclassified. Besides, because the collection process is based on keyword filtering, the dataset might not be reflective of the whole array of public opinions, as it surely does not include tweets that do not contain the chosen keyword. Lastly, while both Pig and Hive are powerful tools, Hive was observed to be slower than Pig for processing semi-structured data, primarily due to its more rigid schema requirements and the overhead of translating queries into MapReduce jobs, which can add latency.

12.7 CONCLUSION AND FUTURE WORK

The study basically corroborates the practicality and merits of big data methods for social sentiment analysis of Twitter data concerning the demonetization policy in India. The analysis reveals that there are varying levels within the general sentiment-area in which a majority of tweets are either positive, negative, or neutral. The research posits that the real-time view of the public is very important for policymaking, whereby opinion mining techniques from social media serve as avenues for the dynamic analysis of public acceptance of critical socio-economic events. It considers that Hadoop, Pig, and Hive being successfully applied presage the scalability and efficiency of these tools to process the unlimited number and volume of unstructured data, thus affording a strong platform for policy appraisal and targeted communication.

Further research efforts could extend toward the exploration of advanced NLP models, especially Large Language Models (LLMs) such as GPT-4o, for fine-grained sentiment analysis. These models are able to comprehend context, sarcasm, multilingualism, and emotional tones, the usual features found in social media conversations. Real-time sentiment tracking using streaming data could boost this predictive potential. Incorporating geographic, demographic, and temporal metadata could provide granular insights for designing policies targeted at specific regions or populations. Therefore, a broader analysis considering visual content such as memes, images, and videos on Twitter should be conducted, utilizing multimodal techniques. This will certainly amplify the understanding of public sentiment. With such advanced techniques, precision in polarity analysis will increase, enabling swifter and informed decisions to be made in the digital age.

Other research may involve working with big-data IoT platforms to gather and analyze sentiment data in real time across different channels, beyond simply Twitter, including location-based devices and government dashboards, as suggested by [4] and [6]. Furthermore, cloud-native solutions could be added to sentiment analysis pipelines, as explored in [5], which would lend the analysis greater scalability, real-time querying capabilities, and fault tolerance spread across distributed systems. With such improvements, the system could be transformed from a post-event analysis tool into a proactive governance tool that constantly feeds into strategic planning and citizen engagement efforts.

REFERENCES

[1] S. D. Shetty, "Sentiment analysis, tweet analysis and visualization on big data using Apache Spark and Hadoop," *IOP Conference Series: Materials Science and Engineering*, vol. 1099, no. 1, p. 012002, 2021.

[2] "Python report on Twitter sentiment analysis," *SlideShare*, Dec. 22, 2018. [Online], https://www.slideshare.net/AntaraBhattacharya12/python-report-on-twitter-sentiment-analysis.

[3] T. Zahra, H. Ghous, and I. Hussain, "Sentiment analysis of Twitter dataset using LLE and classification methods," *International Research Journal of Modernization in Engineering Technology and Science*, vol. 3, no. 1, pp. 1151–1164, 2021.

[4] V. E Balas, V. K. Solanki, R. Kumar, & M. Khari, M. (Eds.). (2019). *Internet of things and big data analytics for smart generation* (Vol. 154, p. 309). Heidelberg: Springer.

[5] S. Srinivasa and R. B. K. Tripathy, "Comprehensive study of cloud computing and related security issues," in *Big Data Analytics: Proceedings of CSI 2015*, Singapore: Springer, 2016, pp. 699–707.

[6] P. Raj and A. C. Raman, *The Internet of Things and Big Data Analytics: Integrated Platforms and Industry Use Cases*, CRC Press, 2017.

[7] A. Dwivedi, R. P. Pant, M. Khari, S. Pandey, L. Mohan, & M. Pande, (2019, March). E-governance and big data framework for e-governance and use of sentiment analysis. In *International Conference on Advances in Engineering Science Management & Technology (ICAESMT)-2019, Uttaranchal University, Dehradun, India.*

[8] Twitter Developer Platform, "Using the Twitter API," [Online], https://developer.twitter.com/en/docs [Accessed: July 25, 2025].

[9] Apache Flume, "Flume 1.9.0 user guide," *Apache Software Foundation* [Online], https://flume.apache.org/FlumeUserGuide.html [Accessed: July 25, 2025].

[10] A. Thusoo, J. Sarma, N. Jain, Z. Shao, P. Chakka, N. Zhang, S. Antony, H. Liu, and R. Murthy, "Hive—a warehousing solution over a map-reduce framework," *Proceedings of the VLDB Endowment*, vol. 2, no. 2, pp. 1626–1629, 2009.

[11] Apache Hive, "Apache hive—data warehouse software for reading, writing, and managing large datasets," [Online], https://hive.apache.org/ [Accessed: July 25, 2025].

[12] D. Borthakur, "The Hadoop distributed file system: Architecture and design," *Hadoop Project Website*, 2007. [Online], https://hadoop.apache.org/ [Accessed: July 25, 2025].

[13] V. E. Balas, V. K. Solanki, R. Kumar, and M. Khari (Eds.), *Internet of Things and Big Data Analytics for Smart Generation* (Vol. 154, p. 309), Heidelberg: Springer, 2019.

[14] M. Khari, M. Kumar, and Vaishali, "Comprehensive study of cloud computing and related security issues," in *Big Data Analytics: Proceedings of CSI 2015* (pp. 699–707), Singapore: Springer, 2017.
[15] S. Raj, P. Gupta, and O. Singh, "Working on R programming language for big data: A review," *International Journal of Enhanced Research in Science, Technology & Engineering*, vol. 5, no. 7, 2016. ISSN: 2319-7463.
[16] N. Anusha, G. Divya, and B. Ramya, "Sentiment analysis of Twitter data through big data," *International Journal of Engineering Research & Technology (IJERT)*, vol. 6, no. 6, pp. 307–309, 2017.
[17] R. Dalal, A. Sangwan, and M. Khari, M., "The bibliometrics assessment of opportunistic network protocols & simulation tools," *Telematics and Informatics Reports*, vol. 11, p. 100082, 2023.
[18] R. Dalal and M. Khari, "Empirical analysis of routing protocols in opportunistic network," in *Research in Intelligent and Computing in Engineering: Select Proceedings of RICE 2020*, Singapore: Springer, 2021.
[19] R. Dalal et al., "Facial emotion recognition and detection using convolutional neural networks," in *2023 3rd International Conference on Smart Generation Computing, Communication and Networking (SMART GENCON)*, IEEE, 2023.
[20] R. Dalal et al., "0A comprehensive review on anomaly detection techniques for web data logging," in *International Conference on Information and Communication Technology for Competitive Strategies*, Singapore: Springer Nature, 2023.
[21] A. Dwivedi, R. P. Pant, M. Khari, S. Pandey, L. Mohan, and M. Pande, "E-governance and big data framework for e-governance and use of sentiment analysis," in *International Conference on Advances in Engineering Science Management & Technology (ICAESMT)-2019*, Dehradun, India: Uttaranchal University, March 2019.
[22] P. Raj, T. Poongodi, B. Balusamy, and M. Khari (Eds.), *The Internet of Things and Big Data Analytics: Integrated Platforms and Industry Use Cases*, CRC Press, 2020.

Chapter 13

Deep Transfer Learning in Edge Computing

Enhancing Intelligence at the Network Edge

Asha Kurian, Arunima Banerjee, Swetha C.B., and Daphne Grace Backiam Nathaniel

13.1 INTRODUCTION

Technology has been rapidly evolving in the past few decades, leading to a gradual migration from data processing on small- to mid-sized networks to performing computations of workloads on the cloud. Termed as cloud computing, any demand for resources—storage, processors, databases—are met on the network. With the burgeoning amount of data generated by edge devices like cameras, IoT sensors [1], and other connected devices, there has been an exponential increase in the demand for processing power, bandwidth, swift response times, and advanced computational capabilities. Edge computing is a computational paradigm that strives to move the computation and storage of workloads from centralized servers closer to the data source—to localized or edge servers [2]. For data-intensive applications, the migration translates to faster response times and minimizes the bandwidth needs. Edge computing plays an indispensable role and has seen widespread use in diverse sectors like smart homes, healthcare systems, autonomous vehicles, smart cities and grids, industrial IoT, telecommunications etc. The explosive growth of interconnected devices calls for innovative adaptations to the edge computing architecture to support scalability while maintaining low latency.

Deep learning is a subfield of artificial intelligence that uses machine learning algorithms to learn data patterns to aid in automatic decision-making processes [3]. Generalized architectures called deep neural networks are trained on large datasets, empowering machines to predict outcomes based on previously unseen data. Applications of deep learning—computer vision, natural language processing, and generative AI—have revolutionized various aspects of life. Machines are endowed with intelligence to take informed decisions like humans. The major limitation of these architectures is the high demand for resources and data.

 DOI: 10.1201/9781003635147-13

13.2 MOTIVATION FOR INTEGRATING DEEP TRANSFER LEARNING WITH EDGE COMPUTING

Transfer learning is an approach designed to alleviate the high resource data requirements of the deep models [4]. Transfer learning leverages the power of pre-trained models (models trained to perform a specific task) and fine-tunes them to perform well on downstream tasks. The use of off-the-shelf neural architectures reduces the number of parameters to be learned, reducing the resource requirements.

Embedding the pre-trained model architectures in edge devices offers a feasible solution to move the computational intelligence in proximity to the originating data, at the same time keeping the memory and processor requirements at a minimum.

The primary advantages of integrating deep transfer learning in edge environments can be highlighted as follows:

1. **Resource Downscaling:** Using pre-trained models saves the computational load of training and processing DL models on edge computing devices [5].
2. **Low Latency:** The migration of computational intelligence from the cloud to the edge devices reduces the response rates encountered.
3. **Scalability:** DTL eliminates the need for excessive training, which can be translated into scalable solutions across diverse domains.

13.3 OBJECTIVES OF THE CHAPTER

The chapter highlights the convergence of two interdisciplinary methodologies—edge computing and deep transfer learning [6]. The amalgamation of deep transfer learning within an edge environment has been explored as a potential solution to counter challenges while processing large workloads in a network edge environment—low latency, bandwidth, scalability, and intensive resource demands. The chapter elaborates on the following:

1. Core principles of deep learning
2. Key concepts of transfer learning
3. The fundamental principles of edge computing
4. Applications of deep transfer learning in edge computing

13.4 FUNDAMENTALS OF DEEP LEARNING

The rise of AI, along with innovations in computer vision and natural language processing, led to the unprecedented growth of deep learning architectures capable of taking AI-driven decisions requiring human intelligence. DL has been widely established in a variety of disciplines, including healthcare, medicine, automation, robotics, and finance. The remarkable success of deep neural architectures in decision-making can be attributed to the availability of high-performance computational resources like GPUs and TPUs [7–9] combined with the proliferation of vast amounts of data sources.

The origins of deep architectures can be traced back to neural networks. Neural networks are architectures conceptualized to understand underlying patterns in the data and mimic the way humans assimilate knowledge through interaction with the environment. Neural networks were devised to mimic the learning of the human brain. A deep neural network typically consists of an input layer, one or more hidden layers, and an output layer [10]. The fundamental unit of computation/learning in a deep neural network is a perceptron. The unit is designed to produce an activation for certain combinations of weighted values above a certain threshold. Each layer is built from a single/multiple perceptron. The objective of a neural network is to learn a function, given vector input, to find optimal weight coefficients to model the output. The simple perceptron was designed to find solutions for linear separable problems and performed poorly for more complex problems [11]. This led to the conception of larger architectures (deep neural networks) as most real-life applications are non-linearly separable.

$$Y = \sum w_i x_i + b_i \tag{13.1}$$

Where *xi*'s are the input vectors, *wi*'s represent the corresponding weight parameters [12], *b* is a bias component added to the weighted sum, *Y* is the output vector being modeled. *Y* is a function that causes an activation on weighted inputs. Parameter tuning involves finding the optimal parameters *wi* and *bi*.

The output variable given in Equation (1) is the linear combination of the weighted sum of the input data with a bias factor added to it. The data is propagated forward at each intermediary layer and permeated to the final output layer to produce predictions. The prediction error (measured as a deviation of the model prediction from the expected ground truth) results in the propagation of the error back to the network as modified weights and biases. The process goes on for multiple epochs until the prediction error of the model in comparison to the ground truth is negligible and can be discounted. At each iteration, the weights are modified in a bid to reduce the model error [13].

$$W_{new} = W_{old} - \propto \frac{dJ}{dW} \tag{13.2}$$

Equation (2) represents the weight updation during backpropagation.

Figures 13.1 and 13.2 depict the structure of a simple perceptron and a multi-layer perceptron.

The shortfall of a perception in solving complex problems led to the evolution of "deep learning" architectures constituted from a series of neurons placed in tandem [14]. The weighted vectors of the input layer are passed on to inner layers termed as hidden layers. The weighted inputs from the input layer are passed to hidden layers, where each hidden/computational layer captures finer nuances of the data which a perceptron acting alone is unable

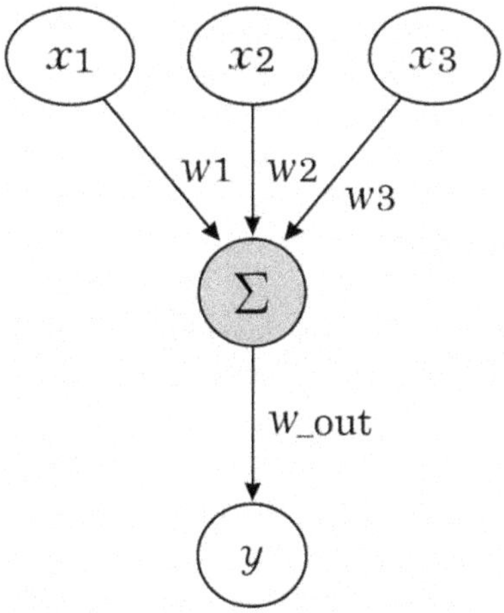

Figure 13.1 Simple Perceptron.

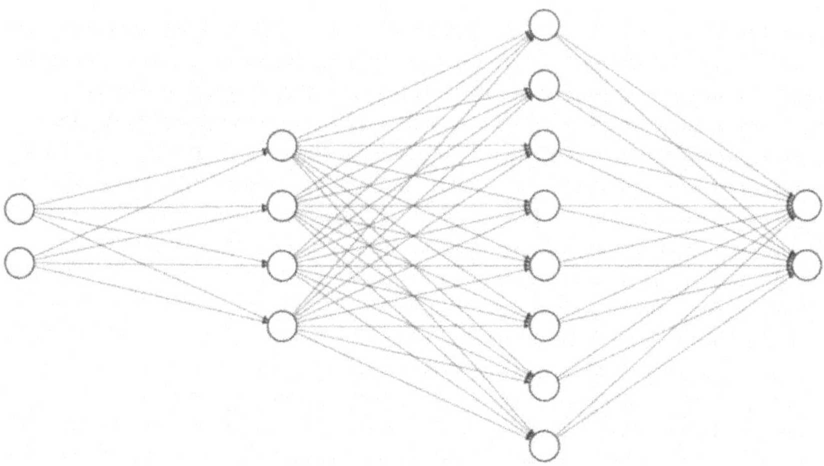

Figure 13.2 Multi-Layer Perceptron.

to capture. Each hidden layer comprises neurons, which are activated for certain combinations of the input. An activation function is responsible for deciding which combination of input results in an activation [15]. As the number of hidden layers increases, the number of tunable parameters also increases significantly.

All these components work together to learn patterns from data and further enable classification, prediction of data—discriminative analysis. After the forward propagation of input and weights, the final layer called the output layer predicts a result. The result could be a classification or prediction task. A procedure called the backpropagation modifies the weights and biases of the inputs to reduce the error of misclassification [16]. The process of forward and back propagation continues until the weights are optimized to perform the classification/recognition problem with minimal/negligible error margin.

13.5 TRANSFER LEARNING

In the traditional setting, machine learning models are adapted to a particular task by training them on large datasets. To generalize, a framework using extensive/available data is time-consuming and resource intensive as it involves the optimization of a substantial number of tunable parameters. Furthermore, machine learning models are based on the assumption that the training and testing samples originate from the same distribution. The parameters of the generalized model are optimized to recognize patterns in the training data. If the data encountered by the model varies from the one that it was trained on, the model is unable to classify/predict optimally.

Transfer learning is a sub-field of deep learning that leverages the power of pre-trained models generalized for a particular task to build models for a related task [17, 18]. The basic idea is to utilize the knowledge gained by a model from large datasets to solve a problem in a domain with limited available data. In a deep architecture, the early layers in the network learn generic features from the data while the higher layers are adapted to learn the task specific features. The transfer learning methodology allows deep network architectures to be reused by freezing the initial layers that can recognize the lower-level features. The higher layers are fine-tuned on the task at hand to learn the task specific features with the limited data available. Using pre-trained models by suspending training their initial layers and retraining the later layers results in huge cost and time savings and improved performance.

As a practical example, consider the integration of transfer learning in medical imaging to detect a disease like pneumonia. Typically, medical image datasets are limited owing to the difficulty in procuring and annotating them [19, 20]. Classification of a disease like pneumonia can be done using any of the available pre-trained models (Inception, ResNet50, MobileNet) by

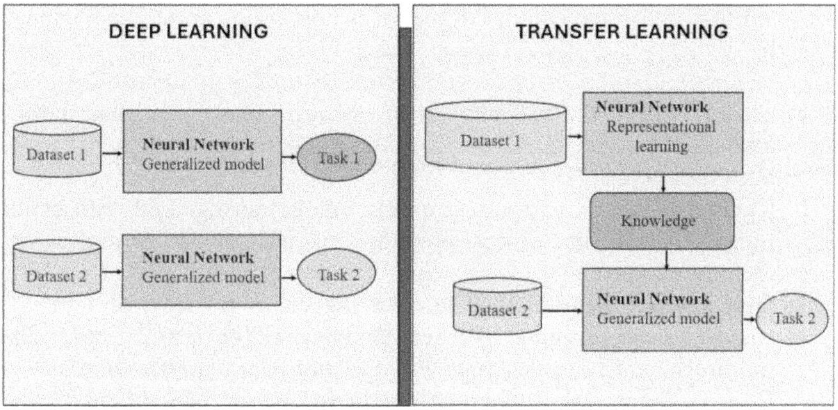

Figure 13.3 Deep Learning vs Transfer Learning.

leveraging the feature extraction blocks of these deep architectures. Retaining the trained feature extraction blocks and retraining the classification head on the small pneumonia dataset, x-ray images can be easily classified as pneumonia vs normal by the fine-tuned model. Figure 13.3 accentuates the differences between deep learning and transfer learning [21, 22].

13.6 EDGE COMPUTING

The advent of edge computing has resulted in a migration of data processing from cloud centric nodes to servers located closer to the source of data origin [23]. This migration has led to faster response times as processing can be handled by computing centers close to the devices that assimilate data. These independent and interconnected devices termed edge devices are hardware systems capable of capturing, processing, and transmitting data to servers/nodes [24]. Edge computing devices like cameras, IoT sensors, surveillance systems, and smart home devices generate an enormous amount of data on a regular basis, with high velocity. Edge computing has found its applications in smart computing, healthcare, and autonomous vehicles. With the proliferation of data, there is growing concern in scaling networks that can handle the processing load, bandwidth, and security.

13.7 EDGE DEVICES

Processing components located at the periphery of the network enabled to collect, process, or relay data are termed as edge devices [25]. These devices can perform computations locally, or they can act as agents to route data for computation on the cloud. Edge devices ubiquitous in our environment include

a. **Sensors within the IoT Ecosystem**—Motion sensors in smart homes and agriculture, temperature, fire, and humidity detectors in industrial automation [26]
b. **Cameras**—Smart cameras capable of capturing and processing images for real-time object detection, image classification, and person identification
c. **Smartphones/Wearables**—The data logged in by smartphones and wearables can be mined by AI enabled devices for personalized health monitoring and running augmented reality-based applications.
d. **Routers/Gateways**—Tasked with preliminary processing of data aggregated from sensors before transmitting them to cloud computing systems

The edge architecture embodies a federated network by relaying cloud services downstream and extending cloud services to the edge layer by integrating edge devices that communicate with the terminal devices capable of performing the processing work/compute of the cloud [27]. The edge computing architecture is depicted in Figure 13.4.

Based on the introduction of processing nodes/data centers between the cloud and the data generating devices, the collaboration between the cloud and edge can be demarcated as three layers—the edge devices layer, an edge layer/fog layer, and a cloud computing layer.

The role of each of the layers in the architecture is described as follows:

a. **Edge Devices Layer**—At the lowest tier of the cloud computing architecture lies the diverse types of interconnected devices capable of

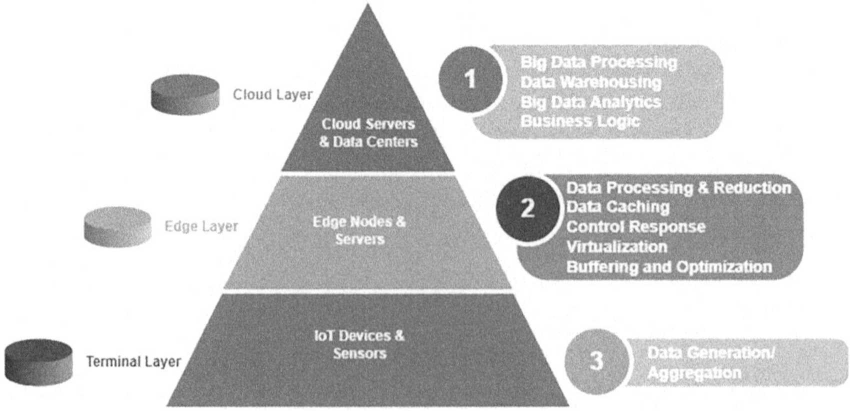

Figure 13.4 Edge Computing Architecture.

producing and consuming data [28]. These devices include mobile devices, IoT devices like sensors, mobile phones, smart phones, cameras, and smart devices. Data in its raw form is generated in this layer and propagated to the next higher level called the fog layer. In the cloud computing architecture, millions of devices in the terminal layer are not invested with any computing power but merely upload the data to the network edge [29].

b. **Fog Computing Layer**—This layer is at the core of the architecture, with processing nodes interspersed between the terminal layer and the cloud clusters. Gateways, routers, base stations, and switches are the primary data processing devices of the edge computing layer. It is an end-to-end layer that permits the access of the terminal devices by cloud computing centers while supporting the real-time computation and analysis of the data uploaded by the terminals. As this layer is closer to the networked devices/point of origin of the data, it results in faster processing than if the data was relayed to the cloud for processing.
c. **Cloud Layer**—The most powerful component of the cloud-edge architecture providing data processing centers is the cloud layer. Advanced storage capabilities along with high-performance cores make the cloud layer essential for tasks requiring high compute capacity and extensive data analysis. The cloud is equipped to store the large amount of data produced at the terminal layer/edge layer and take up the processing tasks which are unable to be handled by the edge layer or tasks that require the integration of global services. The cloud nodes have the authority to dynamically change the deployment of applications, workloads or services on the edge, fog, or cloud to optimize performance.

13.8 INTEGRATING DEEP TRANSFER LEARNING INTO EDGE COMPUTING

Edge devices are severely restricted in their computing capability due to critical factors that affect the processing of the burgeoning data generated by the interconnected devices. These devices need to address the low availability of resources, high computational power, large bandwidth, and the real-time processing needs of the user. Intelligent decision-making can be enabled by integrating deep learning architectures within an edge network [30, 31]. However, the training of deep architectures is resource intensive and time consuming as the models must be fed a lot of data for pattern and representation learning. A significant step in this direction is to embed the edge devices with pre-trained models, leveraging the deep learning capabilities of the model for domain specific tasks. Reusing the weights of models trained on large datasets, followed by fine-tuning on scarce task specific data, permits domain adaptation with minimal resource requirements.

13.9 BRIDGING TRANSFER LEARNING WITH EDGE COMPUTING

Transfer learning helps in faster inference times by re-using models improving response times.

The initial hidden layers of a deep neural network are responsible for feature extraction, and the higher layers are adapted for domain-specific tasks. State-of-the-art models like BERT, ResNet, and Inception have been trained on large datasets for natural language processing and computer vision tasks. The knowledge acquired by the lower layers as a feature learner can be retained by unhinging the pre-penultimate layers from the decision processing layers. In transfer learning, the weights of the pre-penultimate layers are frozen, disabling any incremental training [32, 33]. Re-training the initial layers could lead to overfitting, a condition where the model is overexposed to the training data, and weights are optimized on the training data. The model is unable to perform optimally on generalized or unseen data. Additionally, re-training from scratch is a burden on resources, and time-consuming.

13.10 METHODOLOGY

Integrating transfer learning models into the edge layer for faster processing requires a carefully designed methodology that optimizes both computational efficiency and performance. A detailed methodology and the components required for achieving this integration is given as follows:

13.10.1 Task-specific adaptation of the pre-trained model

Pre-trained model frameworks are chosen specific to the type of application data processed by the edge network. Image-based models like MobileNet, Inception, and ResNet are critically acclaimed vision models [34, 35]. Common text-based models like BERT [36], RoBERTa, and ULMFit are state-of-the-art language processing models capable of handling complex textual representational tasks. The base model is adapted to the edge application by fine-tuning for the edge specific task. Fine-tuning the model can be performed on the cloud to utilize the high-performance data processing centers [37]. Alternatively, this can be achieved on the terminal layer as well. Data adaptation can also be achieved remotely, and the trained weights sent back to the edge node.

The myriad types of data generated from the various data sources (IoT devices, sensors, smartphones) need to undergo on-device pre-processing before being redirected as model input. In addition, data augmentation techniques are implemented to increase the quantity and variety of data perceived by the model.

13.10.2 Model optimization

Abstracting lightweight models is achieved using various strategies. The primary technique to achieve a lightweight model is parameter pruning. On pruning, the neural paths result in a smaller model without compromising the inferential speed or accuracy. Knowledge distillation is a process where a larger parameter intensive model acting as a "teacher" is used to train a smaller "student" that has imbibed the knowledge of the teacher [38]. The student is trained to retain analytical power with reduced computational cost.

13.10.3 Model update mechanisms

Edge-Based Learning: Robust integration of the edge network demands an incremental learning mechanism adept at adapting the model with local data, reducing a need for constant retraining in the cloud.

Federated Learning: In federated learning networks, model updates are shared to the cloud server. This obviates the need to transfer local data to an edge network. Along with increased security and privacy of indigenous data, federated learning reduces the communication overhead required for model feedback and updations.

13.10.4 Security and privacy protocols

To keep the data safe in transit, security protocols are an essential component of an edge network. Prevalent protocols for data encryption include End-to-End Encryption, and Homomorphic encryption. Secure Multi-Party Computation provides protection for multiple parties in communication, and Trusted Execution Environments ensure hardware-based security features. Additional techniques like federated learning ensure that confidential data does not travel beyond the edge of the device.

13.10.5 Model deployment

The lightweight models are deployed on the edge using frameworks designed to optimize model inference at the edge to handle real-time demands—ONNX or OpenVINO are popular frameworks. These frameworks are equipped with specialized hardware (TPUs, hardware accelerators) to accelerate inference times [39]. The model is integrated into an inference engine that can counter the edge device's hardware limitations. NVIDIA Jetson and Coral from Google use specialized hardware accelerators designed to optimize inference times.

The architecture of the proposed model is given in Figure 13.5

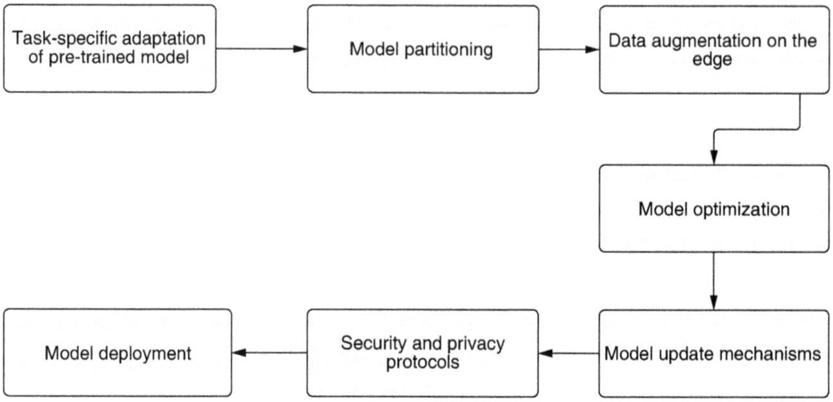

Figure 13.5 Architecture of the Proposed Model.

13.11 EXAMPLE SCENARIO: SMART TRAFFIC MONITORING SYSTEM

Imagine a smart city that is struggling with traffic congestion and frequent accidents. To address these challenges, the city officials decided to implement an AI-powered smart traffic monitoring system. This system integrates deep transfer learning within an edge computing framework to ensure real-time traffic analysis without overwhelming cloud resources. The components can be elaborated as:

a. **Cloud Server**: The process begins in the cloud, where a high-performance computer server trains an advanced object detection model such as YOLO (You Only Look Once). This model learns to identify vehicles, pedestrians, and road conditions by analyzing a vast dataset of urban traffic patterns. Once trained, the optimized model is prepared for deployment across the city's network of edge devices.
b. **Edge Gateway**: Downloads the model and the edge gateway acts as a bridge, sends the trained model to traffic cameras, and manages periodic updates. This ensures the camera stays up to date with the latest improvements.
c. **Edge Devices (Traffic Cameras)**: Each traffic camera runs an optimized YOLO model, detecting congestion, accidents, and rule violations locally. Instead of sending raw footage, it transmits only critical alterations and summarizes data to the cloud, reducing bandwidth usage.

d. **Cloud Feedback:** The cloud receives alerts, processes data trends, and helps authorities respond in real time.
   It also fine-tunes the model with new traffic patterns and pushes the updates back to edge devices. And lastly, deploy updated models to improve performance.

13.12 FUTURE DIRECTIONS AND EMERGING TRENDS

The integration of deep transfer learning with edge computing represents a significant shift in how AI-driven systems process and analyze data. By leveraging deep transfer architectures, which have already demonstrated remarkable success in fields like healthcare, automation, and finance and combining them with the efficiency of edge computing, we are moving towards a future where intelligent decision-making happens closer to the source of data generation.

Traditionally, deep learning models thrive on vast computational resources, typically relying on cloud architecture. However, as real-time applications such as smart surveillance, autonomous vehicles, and IoT-based healthcare monitoring continue to grow, the need for low latency and high-speed processing becomes critical. This is where edge computing bridges the gap, enabling deep learning to function efficiently without excessive reliance on centralized cloud systems.

The fundamental principles of deep learning, particularly the role of neural networks, perceptron, and multi-layer architectures, lay the groundwork of power AI models. However, training models from scratch for every task is neither practical nor scalable. Transfer learning addresses this challenge by repurposing pre-trained models, significantly reducing the computational burden while maintaining high accuracy. When implemented within an edge environment, transfer learning enhances performance, making intelligent systems more adaptive, responsive, and resource efficient.

The architecture of edge computing plays a crucial role in this transformation. By distributing workloads across three layers of edge devices, fog computing, and cloud infrastructure, the system can balance efficiency with processing power. Edge devices like IoT sensors, smart cameras, and smartphones act as the first line of data collection. Fog computing ensures real-time processing closer to the sources, reducing delays. Meanwhile, the cloud layer, with its advanced capabilities, handles large scale processing and manages applications deployment.

This combination of deep transfer learning and edge computing is not just a technological upgrade but a necessity. With the explosion of data, relying solely on centralized cloud systems is impractical. Instead, edge driven deep learning enables faster, smarter, and more scalable solutions across industries by reducing latency, optimizing resource utilization, and ensuring real-time data processing closer to the source.

Despite these advancements, challenges such as resource constraints and security concerns still need to be addressed. However, as deep learning continues to evolve within the edge environment, its role in enabling efficient, scalable, and intelligent computing solutions will only grow stronger.

REFERENCES

[1] Lan, L., R. Shi, B. Wang, and L. Zhang. "An IoT unified access platform for heterogeneity sensing devices based on edge computing." *IEEE Access* 7 (2019): 44199–44211. https://doi.org/10.1109/ACCESS.2019.2908684.

[2] Shi, W., J. Cao, Q. Zhang, Y. Li, and L. Xu. "Edge computing: Vision and challenges." *IEEE Internet of Things Journal* 3.5 (October 2016): 637–646. https://doi.org/10.1109/JIOT.2016.2579198.

[3] Rao, T. Venkat Narayana, et al. "Reliance on artificial intelligence, machine learning and deep learning in the era of industry 4.0." *Smart Healthcare System Design: Security and Privacy Aspects* (2022): 281–299.

[4] Alzubaidi, Laith, et al. "Novel transfer learning approach for medical imaging with limited labeled data." *Cancers* 13.7 (2021): 1590.

[5] Alajlan, Norah N., and Dina M. Ibrahim. "TinyML: Enabling of inference deep learning models on ultra-low-power IoT edge devices for AI applications." *Micromachines* 13.6 (2022): 851.

[6] Dec, Grzegorz, et al. "Role of academics in transferring knowledge and skills on artificial intelligence, internet of things and edge computing." *Sensors* 22.7 (2022): 2496.

[7] Armoni, Marco. (2023). Tensor processing units (TPU): A technical analysis and their impact on artificial intelligence. *Tech4Future Information Technology Report*.

[8] Chopra, Bhuvi. "Enhancing machine learning performance: The role of GPU-based AI compute architectures." *Journal of Knowledge Learning and Science Technology* (2024): 29–42. ISSN: 2959-6386.

[9] Raju, M., et al. "Architectures of high-performance computing systems for machine learning workloads." In *Integrating Machine Learning Into HPC-Based Simulations and Analytics*. IGI Global Scientific Publishing, 2025, 435–460.

[10] Sivaram, Abhishek, Laya Das, and Venkat Venkatasubramanian. "Hidden representations in deep neural networks: Part 1. Classification problems." *Computers & Chemical Engineering* 134 (2020): 106669.

[11] Elizondo, David. "The linear separability problem: Some testing methods." IEEE Transactions on Neural Networks 17.2 (2006): 330–344.

[12] Shieh, Ching-Sung, and Chin-Teng Lin. "A vector neural network for emitter identification." *IEEE Transactions on Antennas and Propagation* 50.8 (2002): 1120–1127.

[13] Bonaccorso, Giuseppe. *Machine Learning Algorithms: Popular Algorithms for Data Science and Machine Learning*. Packt Publishing Ltd, 2018.

[14] Riedmiller, Martin, and A. Lernen. "Multi layer perceptron." *Machine Learning Lab Special Lecture (University of Freiburg)* 24 (2014).

[15] Singh, Jaswinder, and Rajdeep Banerjee. "A study on single and multi-layer perceptron neural network." In *2019 3rd International Conference on Computing Methodologies and Communication (ICCMC)*. IEEE, 2019.
[16] Kramer, Mark A., and J. A. Leonard. "Diagnosis using backpropagation neural networks—analysis and criticism." *Computers & Chemical Engineering* 14.12 (1990): 1323–1338.
[17] Pan, Sinno Jialin. "Transfer learning." *Learning* 21 (2020): 1–2.
[18] Torrey, Lisa, and Jude Shavlik. "Transfer learning." In *Handbook of Research on Machine Learning Applications and Trends: Algorithms, Methods, and Techniques*. IGI Global, 2010, 242–264.
[19] Jahanian, Mojtaba, et al. "MedTransCluster: Transfer learning for deep medical image clustering." *Intelligence-Based Medicine* 9 (2024): 100139.
[20] Mwaniki, Paul M. *Using Transfer Learning to Leverage Large Un-labelled Datasets to Improve Classification Models in Cases with Small-Labelled Datasets: Application to Paediatric Diagnostic and Prognostic Models*. Diss. University of Nairobi, 2023.
[21] Dutta, K., P. Gupta, and D. Bajaj. "Robo-net: A novel reinforced walking biped design using an augmented random search approach." In *2024 International Conference on Augmented Reality, Intelligent Systems, and Industrial Automation (ARIIA)*. IEEE, December 2024, 1–9.
[22] Dutta, K., and P. Gupta. "Reso-net: Generic image resolution enhancement using convolutional autoencoders." In *International Conference on Intelligent Systems and Machine Learning*. Cham, Switzerland: Springer Nature, December 2022, 298–308.
[23] Varghese, Blesson, et al. "Challenges and opportunities in edge computing." In *2016 IEEE International Conference on Smart Cloud (SmartCloud)*. IEEE, 2016.
[24] Satyanarayanan, Mahadev. "The emergence of edge computing." Computer 50.1 (2017): 30–39.
[25] Young, Roger, Sheila Fallon, and Paul Jacob. "An architecture for intelligent data processing on iot edge devices." In *2017 UKSim-AMSS 19th International Conference on Computer Modelling & Simulation (UKSim)*. IEEE, 2017.
[26] McGibney, Alan, Alejandro Esquiva Rodríguez, and Susan Rea. "Managing wireless sensor networks within IoT ecosystems." In *2015 IEEE 2nd World Forum on Internet of Things (WF-IoT)*. IEEE, 2015.
[27] Fernández, Cristian Martín, Manuel Díaz Rodríguez, and Bartolomé Rubio Muñoz. "An edge computing architecture in the internet of things." In *2018 IEEE 21st International Symposium on Real-Time Distributed Computing (ISORC)*. IEEE, 2018.
[28] Talebkhah, Marieh, et al. "Edge computing: Architecture, applications and future perspectives." in *2020 IEEE 2nd International Conference on Artificial Intelligence in Engineering and Technology (IICAIET)*. IEEE, 2020.
[29] Tsai, Wei-Tek, Xin Sun, and Janaka Balasooriya. "Service-oriented cloud computing architecture." In *2010 Seventh International Conference on Information Technology: New Generations*. IEEE, 2010.
[30] Jang, Ingook, et al. "Knowledge transfer for on-device deep reinforcement learning in resource constrained edge computing systems." *IEEE Access* 8 (2020): 146588–146597.
[31] Sufian, Abu, Changsheng You, and Mianxiong Dong. "A deep transfer learning-based edge computing method for home health monitoring." In

2021 55th Annual Conference on Information Sciences and Systems (CISS). IEEE, 2021.

[32] Xiang, Evan Wei, et al. "Bridging domains using world wide knowledge for transfer learning." *IEEE Transactions on Knowledge and Data Engineering* 22.6 (2010): 770–783.

[33] Liu, Zhengchun, et al. "Bridging data center AI systems with edge computing for actionable information retrieval." In *2021 3rd Annual Workshop on Extreme-Scale Experiment-in-the-Loop Computing (XLOOP)*. IEEE, 2021.

[34] Cao, C., Zhang, Y., Yu, Y., Lv, Q., Min, L., & Zhang, Y. (2024, October). Task-adapter: Task-specific adaptation of image models for few-shot action recognition. In *Proceedings of the 32nd ACM International Conference on Multimedia* (pp. 9038–9047).

[35] Li, Junlong, et al. "Task-specific objectives of pre-trained language models for dialogue adaptation." *arXiv Preprint* (2020). arXiv:2009.04984.

[36] Shu, Xin. "BERT and RoBERTa for sarcasm detection: Optimizing performance through advanced fine-tuning." *Applied and Computational Engineering* 97 (2024): 1–11.

[37] Wang, Yu-Xiong, Deva Ramanan, and Martial Hebert. (2017). Growing a brain: Fine-tuning by increasing model capacity. In *Proceedings of the IEEE conference on computer vision and pattern recognition* (pp. 2471–2480).

[38] Srinivas, Suraj, and R. Venkatesh Babu. "Data-free parameter pruning for deep neural networks." *arXiv Preprint* (2015). arXiv:1507.06149.

[39] Kljucaric, Luke, and Alan D. George. "Deep learning inferencing with high-performance hardware accelerators." *ACM Transactions on Intelligent Systems and Technology* 14.4 (2023): 1–25.

Chapter 14

A Unified Framework for Trustworthy and Explainable Edge AI (TE2AI)

Enhancing Transparency, Security, and Reliability in Decentralized Intelligence

Neha Sharma, Urvashi Sugandh, Arvind Panwar, Nitin Jain, Shakir Khan, and Sunita Yadav

14.1 INTRODUCTION

The AI paradigm is shifting from the classical centralized cloud computing processing to decentralized edge computing. Edge AI is widely deployed near the edge, where the data is generated, as chip-based implementation on devices such as sensors, smartphones, autonomous vehicles, and industrial robots [1]. The paradigm shift brings abundant benefits such as lower latency, less networking bandwidth, better privacy for local traffic, improved operational independence, disconnected, and soon [2]. Applications range from critical areas such as real-time patient monitoring in the medical domain, predictive maintenance within smart factories, intelligent traffic management in smart cities, and responsive control in autonomous robotics [3, 4].

While these benefits are promising, the broad acceptance and social acceptance of edge AI depend largely on trust in such systems. AI trustworthiness is a multi-faceted concept that includes credibility, robustness, security, privacy, fairness, and explainability [5]. These challenges are exacerbated in edge AI because of several intrinsic characteristics of edge AI. Edge end devices are usually resource-limited; they do not possess much computational power, memory, and energy, making deploying complex AI models and heavy security and explainability algorithms particularly challenging [6]. In addition, the edge data is often heterogeneous and distributed over many devices which complicates model generalization, robust performance, and explanation coherence [7]. The situation will be worsened by the physical exposure and control of the edge computing devices in untrusted environments, which in turn renders new attack surfaces that can be exploited for several purposes including data exfiltration, model poisoning, and denial of service [8]. Although edge processing achieves data privacy by performing

DOI: 10.1201/9781003635147-14

edge computing, and reducing the cloud data transmission, local processing of private data still requires strong privacy-preserving abilities against the natural privacy risk [9]. The lack of interpretability among AI models is perhaps one of the most substantial obstacles; most powerful AI models function like 'black boxes' and their decision process is hard to interpret. Failure to achieve this transparency is a formidable barrier to trust, particularly on mission-critical applications, which require accountability [10].

Explainable AI has gained a lot of importance as a way to open the black box and provide reasons why AI algorithms make specific decisions [11]. However, classical XAI solutions, which are frequently targeted at centralized systems, may not be directly capable or effective for resource-limited and distributed edge AI ecosystems. Moreover, explainability is not enough by itself; this must be combined with other aspects of trustworthiness, security, and privacy for a truly reliable edge AI system.

This chapter is focused on the imperative of a comprehensive approach to creating Trustworthy and Explainable edge AI. However, the current state of the art typically only tackles these properties in separate silos, leaving an open space for the holistic conceptualization of frameworks that consistently incorporate trustworthiness and explainability across the scope of edge AI systems' lifecycle. To address this void, in this chapter, we present the Trustworthy and Explainable edge AI (TE2AI) framework. This framework establishes a scaffold for creating edge AI systems with intelligence and efficiency, as well as visibility, security, contactability, and reliability.

The premises of the chapter are the principles that have been developed in various fields. It utilizes techniques for model explanation and justification from Explainable AI (XAI) [11, 12]. It is based on decentralized computation and collaborative learning principles and maintains data privacy, which forms the basics of the DS and Federated Learning (FL) work [13]. The framework has been secured by the application of state-of-the-art security mechanisms as evidenced in cyber security, for resource-constrained and distributed environments. Moreover, it also combines Privacy-Enhancing Technologies (PETs) like differential privacy and homomorphic encryption to ensure the privacy of sensitive data and adopts models of computational trust and reputation systems from trust management literature [14]. We believe that by consolidating these foundations, the TE2AI conceptual framework can serve as an overall roadmap that guides the development of edge AI for the betterment of society. The following sections will cover a critical review of related work, the proposed TE2AI framework, and implications along with directions for future work.

14.2 LITERATURE REVIEW

Ensuring that such edge AI systems can be trusted and are capable of explaining their reasoning requires joint efforts from a variety of related but distinct research domains. This section provides a critical review of previous conceptual and theoretical contributions in edge AI, XAI, AI trustworthiness, and security and privacy in distributed systems, in terms of important progress and ongoing limitations that justify the TE2AI framework proposed herein.

14.2.1 Edge AI: evolution and challenges

Edge computing has transformed from early content delivery networks into a sophisticated paradigm capable of increasingly sophisticated AI computations at the network edge [15]. The literature discusses different architectures for edge AI, such as device-centric, edge-server-centric, and hybrid cloud-edge models [16]. Important research in edge AI includes model compression and optimization, such as pruning and quantization, which is critical for resource-constrained devices [17]. There are also active areas of research around efficient data-driven management and processing at the edge and frameworks for deploying and managing edge AI applications, including, but not limited to Kubernetes for edge [18].

Yet as edge AI systems become more autonomous and embedded in mission-critical applications the issues of their trustworthiness and comprehensibility are reaching new dimensions. Though a lot of research is done on cost and performance of latency and energy, less integrated attention is paid to the aspects of trust, security, and explainability [19]. The intrinsically heterogeneous nature of edge devices makes edge intelligence in the presence of adversarial attacks and dynamic networks noncompliant with traditional AI governance frameworks [20].

14.2.2 Explainable AI (XAI) in centralized and edge contexts

Explainable AI is a set of methods and approaches aimed at making AI decisions interpretable by humans. These methods fall under general categories. Some are conditioned to certain models, tailored to specific model classes like interpreting tree-based models or visualizing Convolutional Neural Network (CNN) filters. Others are model-agnostic and may be used for any black-box model. Well-known model-agnostic methods are the Local Interpretable Model-Agnostic Explanations (LIME) [21], which locally approximates the black-box model by an interpretable one, and the Shapley Additive explanations (SHAP) [22] that leverage game theory to define a feature importance metric. Moreover, the XAI approaches can also be characterized according to whether the post-hoc explanations are given to the models after the model was trained that based on inherently transparent models, for example, linear regression or decision trees.

These techniques have greatly contributed to make AI transparent in centralized contexts, but they are not directly applicable to edge AI. The high computational complexity of many XAI methods, particularly model-agnostic ones, has become a barrier for resource-constrained edge devices [23]. To derive explanations, training data or model internals are typically needed, however, it may be impossible and undesirable in the distributed edge due to privacy or proprietary reasons [24]. Edge AI often assumes that there are several distributed models, and generating coherent global explanations from local device-specific explanations is challenging, especially without compromising privacy [25]. Additionally, edge applications typically require real-time decisions and therefore real-time explanations, however many XAI approaches cannot offer this functionality [26].

Recently, work has been done on the 'edgeX AI' or lighter version of XAI for edge devices in particular [27]. This ranges from adopting techniques of existing XAI to new low-overhead techniques and FL-based cooperation for explanation (generation) [28]. But a holistic approach of XAI with the more general trustworthiness at the edge is still immature.

14.2.3 Trustworthy AI: dimensions and frameworks

Trustworthy AI is a broad concept that encompasses multiple critical dimensions. These include the **reliability and robustness** of an AI system, referring to its ability to perform consistently and accurately, even under noisy conditions or adversarial attacks [29]. **Security** is another vital dimension, focusing on protecting AI systems from threats that could compromise their integrity, availability, or confidentiality. **Privacy** ensures that sensitive data used or generated by AI systems is protected according to ethical and legal standards. **Fairness and bias mitigation** aim to prevent AI systems from making discriminatory decisions or perpetuating societal biases [30]. **Accountability and governance** establish clear lines of responsibility for AI system behavior and ensure compliance with relevant regulations [31]. Finally, **transparency and explainability**, as previously discussed, are crucial for making AI decisions understandable.

Several frameworks and guidelines for trustworthy AI have been proposed by academic institutions, industry consortia, and governmental bodies, such as the EU AI Act and the NIST AI Risk Management Framework [32, 33]. These frameworks typically outline principles and best practices for developing and deploying AI responsibly. However, most are high-level and do not offer specific architectural guidance for implementing these principles in complex, distributed environments like edge AI. Furthermore, the interplay and potential trade-offs between these dimensions, for instance, between privacy and explainability, are critical considerations that need to be addressed systematically in system design [34].

14.2.4 Security and privacy in edge AI

Security and privacy are also crucial in edge AI as the data is usually sensitive, and the edge environments are typically vulnerable [35]. Security challenges such as secure device authentication, secure communication protocols, immunity to personal devices, as well as protection of AI models running at the edge from tampering and malware [36] are critical. Assisted by various advanced algorithmic technologies, privacy directly stems from the ethical aggregation, secure processing, and responsible distribution of personal and private data from edge devices.

Work has also been performed in the context of edge AI security into lightweight cryptographic protocols, intrusion detection systems custom-designed for edge networks, and secure enclave technology for the construction of trusted execution environments [37, 38]. For privacy-preserving, one popular solution is FL, which enables models to be trained jointly on distributed data without sharing the raw data, and can further improve the data protection [39]. Other PETs, including differential privacy, homomorphic encryption, and secure multi-party computations are also explored for use in edge AI, the issue being the computational overhead that they require can be too heavy in resource-constrained environments [40, 41].

Although a lot has been achieved on individual security and privacy approaches, the development of an overall trustworthy AI platform comprising these along with a robust explainability component still constitutes an open problem. One open question is, e.g., how to integrate privacy-preserving FL with techniques for generating meaningful explanations without the risk of unintentionally leaking sensitive information from local data sources [42, 43].

14.2.5 Gaps in existing literature

This literature review also highlights a number of important voids that have been identified as objectives to be filled by the proposed TE2AI framework. First, there is no clear integrated solution; most of the work that is available focuses only on one part, e.g., XAI for edge or security protocols for edge devices, and no comprehensive XBIA is given for edge AI. Second, although there exist high-level principles for trustworthy AI, it is an urgent necessity to operationalize trustworthiness in concrete architectural models that order these principles, within the specificities of the edge AI (i.e., constraints of edge hardware and edge input data, etc.). Third, the conflicts and synergies between various aspects of trustworthiness that may exist between different types of trustworthiness, such as trade-offs between privacy and explainability or security and performance, have not yet been fully understood or systematically managed in the design of edge AI, calling for balance of such trade-offs. Fourth, dependability and explainability have been implemented as additional functions instead of taking them into account throughout the system lifecycle; the lifecycle perspective from design and development to deployment and operation should be acquired. Lastly, XAI is the pursuit of human understanding but the wide-ranging human-centric dimensions from how users interface and trust edge AI systems to ways in which they have control over these require more explicit focus in framework design.

Our TE2AI framework will address these gaps conceptually by providing a modular, unified, and adaptive architecture for creating edge AI systems that are provably trustworthy and inherently explainable.

14.3 PROPOSED TE2AI FRAMEWORK: TRUSTWORTHY AND EXPLAINABLE EDGE AI

The TE2AI conceptual framework is a conceptual architecture specifically defined to systematically and structurally integrate trustworthiness and explainability into edge AI systems. It offers a layered, modular map model that meets specific requirements of edge computing, such as the resource constraints, data distribution, and security holes, as well as real-time, human-readable intelligence.

14.3.1 Conceptual underpinnings and design principles

The TE2AI framework has been designed following a set of core design principles to enable the development of robust and understandable edge AI. One key principle is the intrinsic part of the trustworthiness and explainability in the system design, which means the trustworthiness and model explanation are not second class citizens and they do not appear as afterthoughts but as integrated parts of the system design, which includes security, privacy, reliability, fairness, and transparency, as a holistic integration. The modular design of the framework facilitates the flexible implementation of the system according to specific application needs and device capabilities, since the modules can be mixed and matched in different ways according to the specific requirements of the application. Functions are layered from the device itself down to the edge network, and even potentially the cloud, resulting in a clean separation of concerns and processing hierarchy. Given the dynamic nature of edge environments, it is important to be adaptive and context-aware, by supporting mechanisms for monitoring the context (enabling continuous adaptation to the changing environment, device capabilities, as well as the threat landscape). Privacy-by-design is another building block, with built-in PETs to safeguard the privacy of sensitive data during training and inference as well as when generating explanations. Due to edge devices' limited resources, we make the components on-device light and efficient for low computations. Ultimately, the human factor is put in front, aiming at producing explanations that are human-understandable, interpretable for a wide range of stakeholders, and also enabling feedback and control.

14.3.2 Core components of the TE2AI framework

This TE2AI architecture is proposed as a multi-tier architecture with a set of artifacts running across different layers, including the edge devices, the edge servers (if they exist), and possibly involved with the central cloud for orchestration or global model aggregation.

Multiple key components function on the edge Device Layer. Secure AI Execution Environment (SAIEE) provides a secure foundation for executing AI models by isolating sensitive components through ARM TrustZone or lightweight containers, which helps protect data, preserve model integrity, and safeguard critical modules from tampering. This is to make sure that the AI model and its trust and explainability mechanisms function correctly. To support this, the Lightweight explainable AI Module (L-XAIM) provides local, real-time explanations to the predictions of the AI model directly on the edge device. It operates computationally cheaper XAI methods, like simplified rule-based explanations or feature importance scores from lightweight surrogate models, to offer immediate transparency at the decision point. The Local Trust Assessor (LTA) is continuously observing the behavior of the local AI model and operational context of the device together performance of the model, status of the device, and characteristics of the input data to anticipate any reliability or security issues. In order to protect privacy-sensitive information, the Privacy Preserving Data Handler (PPDH) controls local data in a privacy-preserving manner and may apply operations to local data to make it minimally sensitive or for anonymization before data processing and/or publishing data in an aggregated form.

In some exemplary implementations, a second optional edge Network/ Aggregation Layer, which may be distributed or located on edge servers, enables more widespread coordination. If FL or explanation is used, the Federated Explanation Aggregator (FEA) can aggregate local explanation results among several edge devices to obtain relatively global or archetype explanations without direct access to raw data on the devices, also using secure aggregation protocols. Additionally, the Federated Trust Coordinator (FTC) consolidates local trust assessments from LTAs as well as network data to construct an account of system trustworthiness, utilizing, e.g., reputation systems or federated anomaly detection to detect systemic threats and coordinate responses. The security posture is tracked by and, if necessary, instigated by the Adaptive Security Orchestrator (ASO) which orchestrates and adapts security policies and countermeasures across the system based on FTCs, threats, and trust assessments, pushing out updates and coordinating incident response.

A few Cross-Cutting Components apply to these layers. Trustworthy Model Repository & Updater (TMRU) securely provides the repository, versioning, and distribution of AI model updates on the edge device and TE2AI component updates, using cryptographic signing and secure channels, to avoid the installation of malicious or broken components. The Explanation Interface & Delivery Module (EIDM) packages and delivers explanations, both local and aggregated, to a wide range of stakeholders in a human-readable and context-appropriate fashion for various modalities and levels of depth. Lastly, the Ethical & Fairness Monitor (EFM), with possible cloud support for more computationally expensive analyses, monitors AI model behavior for potential biases and fairness concerns, particularly from aggregated data or models, using fairness metrics and bias detection algorithms to foster responsible AI.

14.3.3 Operational dynamics of TE2AI

The TE2AI framework is a persistent and evolving process. At first, data are ingested in the edge devices and, subsequently, passed to the PPDH to perform the first privacy-preserving actions. The AI model does inference in a safe environment of the SAIEE to make predictions or decisions. At the same time, the L-XAIM produces a local explanation for this output and the LTA evaluates the behavior and device state of the model to calculate a local trust level. The EIDM subsequently serves an appropriate local explanation to the user or consuming application.

Model updates (or summaries) and trust evaluations at different edge networks or the aggregation layer federated operations will be securely exchanged in the form of local explainability and trust evaluations (e.g., encrypted model updates or anonymous extracts) with the FEA and FTC. They do then feed into pooled insights and coordinated systemwide responses. For example, the FEA may use a generic explanation model to explain patterns of distrust or anomalies. ASO constantly discovers threats and sees attacks based on FTC and other threat intelligence and rolls out the suitable, pre-emptive security countermeasures through the distributed system. The newly signed or validated model and component updates from TMRU are then securely delivered to edge devices to preserve high-level performance, to counter new threats, or to enable evolving features. The entire system is under continuous monitoring as feedback enables AI models, explanation strategies, trust thresholds, and security countermeasures to be adjusted, while the EFM supervise the checked for fairness issues, to ensure the system evolves responsibly.

14.3.4 Architecture diagram

14.3.4.1 The TE2AI framework: layers and components

The TE2AI framework is realized as a multi-layered model that aims to unify the concepts of trustworthiness and explainability in the context of edge AI applications. It consists of three main working tiers: the Cloud/Orchestration Layer, the edge Network/Aggregation Layer, and the edge Device Layer, along with a transversal module for explanation delivery. Figure 14.1 shows the **TE2AI Framework.**

14.3.4.2 Cloud/Orchestration layer

This top layer is the 'central brain' or the control and management center of the entire TE2AI framework. It oversees globally the distributed edge

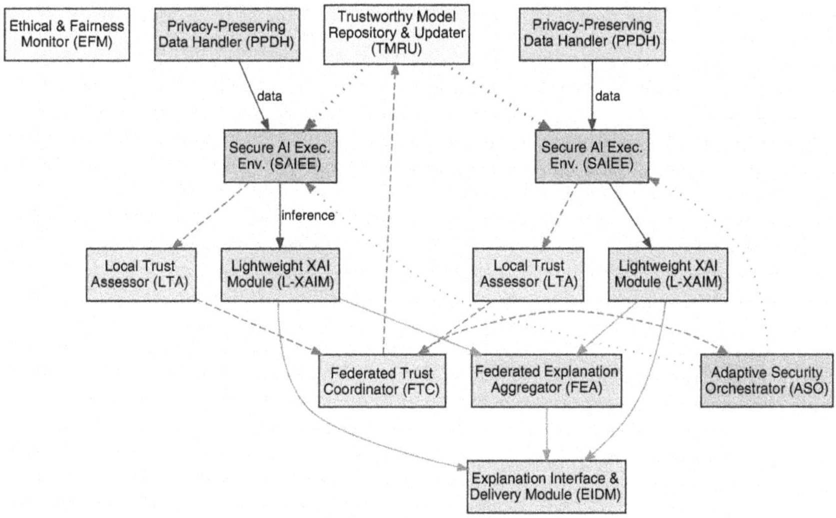

Figure 14.1 TE2AI Framework.

AI system, managing models lifecycles and enforcing high-level ethical and security policies.

- **Ethical & Fairness Monitor (EFM):** This is a key block responsible for safeguarding the AI models launched in the framework with respect to pre-set ethical rules and fairness demands. It dynamically processes FEA pooled explanation data, FTC trust reports, and TMRU model behavior to identify and flag potential bias, disparate adverse impact, or ethical non-compliance. The EFM may raise red flags and alert alarms and propose retraining models and policy changes for the ethical robustness of the edge AIs.
- **Trustworthy Model Repository & Updater (TMRU):** TMRU acts as a safe model repository for AIs and the related trustworthiness describing metadata. It is the controller of the AI model lifecycle, from the initial validation of the model, its versioning, and secure distribution to edge devices. The TMRU accepts trust feedback from the FTC and fairness feedback from the EFM which are used to measure model performance and trustworthiness. It manages device-side updates (or rollbacks) of the model so that only validated and trusted models are up and running.

14.3.4.3 Edge network/aggregation layer

Sitting between the cloud and the individual edge devices, this middle layer manages coordination, aggregation of the insights from multiple edge devices, and distributed intelligence services. It is instrumental in supporting FL and security as well as trust management at a local or network level.

- **Federated Explanation Aggregator (FEA):** The FEA is in charge of collecting local explanations produced by the L-XAIMs of individual edge devices. It uses methods to aggregate these local interpretations into higher-level or summary insights about the aggregate behaviour of the AI models over a set of devices, altogether without raw data ever leaving the edge. These collected explanations are transmitted to EIDM for user visualization and to EFM for ethical recording.
- **Federated Trust Coordinator (FTC):** The FTC gathers the local trust evaluations that the LTAs may share from multiple edge devices. It creates a more holistic perspective on the reliability of both devices and models that fall under its aegis. Such a component can look around in search for declining trust patterns, facilitate a joint effort among trust violations, provide the TMRU with aggregated trust information for model updates and the ASO with trust information for adaptive security mechanisms. It also provides trust scores to the EFM for a holistic ethical inspection.
- **Adaptive Security Orchestrator (ASO):** The ASO centrally orchestrates and dynamically enforces security polices in edge network and devices. It gets trust information of the FTC and the security posture update from the SAIEE. With this information and global policies from the cloud, the ASO can modulate security settings, deploy countermeasures in response to identified threats, quarantine infected devices, and fortify the entire system. It also collects security instance to the EFM to be audited.

14.3.4.4 Edge device layer

This layer is the distributed edge fleet, where data is created and processed locally and AI allocations are made. Components at this level are lightweight and efficient; behave in a resource-constrained style; are privacy-preserving, secure, and locally explainable; and can assess trust.

- **Privacy-Preserving Data Handler (PPDH):** PPDH is the initial access interface for raw data in the edge device. Its goal is to enforce PETs such as anonymization and differential privacy or homomorphic encryption (when applicable) before the data is used by the AI model. This guarantees that sensitive data is safeguarded across the data life cycle within the device as well as during any required (privacy-preserving) communication.
- **Secure environment to execute AI (SAIEE):** This SAIEE offers a secure sandboxed space to execute AI models on the edge device. It leverages

technologies such as Trusted Execution Environments (TEEs), secure boot, and runtime integrity monitoring to protect the AI model against tampering, unauthorized access, and reverse engineering. It keeps watching for the operational performance of the AI model for security and trust evaluation and then reports to the LTA and ASO.

- **Lightweight XAI Module:** The Memo-Lite is in charge of producing local explanations about the predictions or decisions taken by the AI model being executed at the edge device. Since edge devices have resource constraints, in this chapter we make use of lightweight explainability techniques (such as simplified LIME, SHAP, and/or rule-based explanations) that are suitable for edge settings. The aforementioned local explanations are given to the user through the EIDM and then submitted to the FEA for aggregation.
- **Local Trust Assessor (LTA):** The LTA constantly checks the AI model behaviour, and the operational status of the edge device itself, and computes a score of local trust. It includes aspects such as model prediction certainty, data drift, security integrity as reported by SAIEE, and compliance with anticipated operational parameters. Local trust assessment is used for local decision-making and submitted to the FTC for federated trust coordination.

14.3.4.5 Cross-cutting component

Though not a traditional layer, it sits outside layers and serves an essential purpose bridging layers.

- **Explanation Interface & Delivery Module (EIDM):** The EIDM is the central place for collecting, formatting, and delivering explanations to the end-users or interested parties. It accepts local explanations from L-XAIMs (for immediate, device-specific views) and also holistic explanations from the FEXA component (for a general appreciation). The EIDM provides the aforementioned explanations in a human understandable format, based upon the desires or requirements of the particular user/application, increasing the transparency and user trust in the edge AI system.
  This multi-layered architecture of specialized building blocks will form the basis to develop a resilient and flexible TE2AI framework to handle the inherent multidisciplinary problems in developing the trustworthy explainable AI systems at the edge.

14.4 DISCUSSION

The TE2AI model presents a holistic structure that embeds trustworthiness and explainability across the edge AI system. The theoretical contributions, anticipated effects, how it fills out gaps in the literature existent and its links with existing theories of this section are emphasized.

14.4.1 Theoretical contributions and implications

The TE2AI framework contributes to several theoretical aspects. It takes a step forward in research on integrated trust/explanation models for distributed systems. In contrast to existing theories, which usually keep trust and explainability disjoint targeted to centralized architectures, TE2AI defines a new synthesis. It formalizes how explanations at various levels and trust dependencies from different regions can be federated and aggregated to construct global understandings of the behavior of a distributed AI system and its reliability, thus extending distributed consensus and collective intelligence theorizations to AI governance. Indeed, TE2AI treats trustworthiness as a dynamic attribute rather than an immutable characteristic where trustworthiness has to be managed over its life-cycle, in AI. This suggests a move towards a theoretical framework of continuous assurance and adaptation, where trust metrics and explainability mechanisms co-evolve with the AI model and its operational context.

From the edge's perspective, the framework also paves the way towards context-aware explainability. EdgeXAI further asserts that edge explanations must be designed for device constraints, user requirements, and operation demand in real time, and expects new XAI theories to explicitly account for these context factors, instead of being purely model centric. Importantly, TE2AI draws attention to the privacy-explainability intersection, and through that to the emergent theory of Privacy-Preserving Explainable AI (PPXAI). By proposing architectural structures such as the PPDH and the FEA, it stimulates research on approaches like secure multi-party computation for XAI and differentially private explanation generation to provide useful explanations while avoiding the leakage of sensitive information.

14.4.2 Addressing gaps and expected benefits

The TE2AI model is designed to fill these gaps in the literature review. It adopts an integrated model and holistic architecture, instead of isolating model and architecture artifacts, to obtain a more reliable and robust edge AI systems. It facilitates the operationalization of trustworthiness by abstract reasoning into a concrete architecture with identifiable components and their interactions, which will support a practical implementation in edge AI. Although the framework does not directly address all trade-offs among dimensions such as privacy and explainability, it offers an organizing principle with which such trade-offs can be navigated and parameterized, such as between explanation fidelity and computational cost in the L-XAIM.

The framework follows a lifecycle approach and includes TMRU, LTA, FTC, and ASO, which collectively provide a trust and security management attitude set that is applicable throughout the deployment, operations, and upgrade. Furthermore, its human-centered design, especially through the EIDM module, prompts the provision of valuable explanations and strengthened transparency, which help fulfill user goals and public trust.

It is expected that the introduction or modification of the TE2AI concept will have a number of advantages. These benefits come from greater user confidence and the use of more transparent and credible systems, along with the integration of security into trust assessment mechanisms for developing more robust and secure systems. Additionally, debugging and uplift augmentations are supported by explainability, allowing developers to understand why models fail and making debugging and maintenance easier. Also, organizations can use demonstrable trustworthiness and explainability to help ensure that their AI systems comply with regulations. In the end, granting more complex but reliable local intelligence, TE2AI may offer power for edge devices and open new possibilities for building up autonomous edge systems.

14.4.3 Synergy with established theories

The TE2AI framework is rooted in, and extends, a number of established theories, making its design based on rich scientific knowledge. It is consistent with socio-technical systems theory in that the social factors, such as human trust and comprehension, are seen to interact with the technical elements of edge AI systems. In addition, the focus of the framework on continuous monitoring (LTA: Long-term adaptation, FTC: Feedback control) and adaptation (ASO: Adaptation scenario, TMRU: Trust management) should be justified with Control Theory, since they will guide how to preserve stability and performance, such as trustworthiness, for edge AI systems. The production and transmission of explanations can be analyzed from the perspective of Information Theories: how efficiently one can encode and transmit information about the AI decision reaching, minimizing noise and maximizing clarity. In addition, insights from Game Theory can be relevant in establishing relationships among possibly adversarial parties in the edge environment and the provision of secure approaches in the ASO: noteworthy XAI techniques such as SHAP, indeed, have their roots in game-theoretic principles [22]. The goal of TE2AI is to be a principled statement for design of future generation of edge AI, as opposed to an ad hoc library of components.

14.5 IMPLICATIONS AND APPLICATIONS

The TE2AI framework has significant implications for both academic research and practical applications across various domains where edge AI is being deployed or considered.

14.5.1 Academic implications

The framing of TE2AI suggests several stimulating streams of research for future academics. A critical gap in the field is the creation of new low-power XAI methods optimized for resource-constrained edge devices, and that must find the right trade-off between interpretability and computational efficiency [44]. This could include much more work on highly compressed surrogate models, incremental explanation, or even hardware-assisted XAI. Equally important is the research on federated and privacy-preserving XAI, aiming at the development of secure and efficient protocols to generate global or archetype explanations based on distributed local explanations without leaking data privacy, and it would be interesting to explore the possible synergy between FL, secure multi-party computation, differential privacy, and XAI [45, 46].

Standardized metrics and benchmarks for trustworthy edge AI are also needed by which systems can be objectively assessed; the TE2AI framework can be used as a model for evaluating systems in terms of explainability, security, robustness, and fairness. The LTA and FTC components, for example, also require adaptive trust management systems for edge AI where trust scores may evolve according to observed behavior, environmental context, and user feedback [47]. In addition, investigating formal verification techniques for verifying security properties of the components, such as the SAIEE or the correctness of the explanation generation by the L-XAIM under some particular assumptions, would greatly increase the trust in the framework. Human-Computer Interaction (HCI) for edge XAI is another critical domain, which calls for studying optimal explanation interfaces (EIDM) for different users and diverse edge application contexts, not limited to how users view and respond to explanations emanating from the edge [48]. Lastly, the EFM lens signals the importance of further developing ethical framings for governing edge AI that can tackle more nuanced matters of distributed accountability and bias in federated settings in cities in the 21st century.

14.5.2 Practical applications and case scenarios

TE2AI can be fit into numerous practical edge AI applications to improve the trustworthiness and explainability of these systems, in turn producing a more trustworthy and user-acceptable system.

In an autonomous car, for example, if the vehicle made a life-or-death decision (e.g., emergency brake activation), the L-XAIM system could provide real-time justifications to the driver, or record them for later accident analysis. The LTA would keep track of the integrity of sensor data and model confidence and the ASO would prevent cyber-attacks. This application would result in improved safety, driver confidence, and traceable decision-making.

For healthcare IOT (HIOT), think about wearable sensors that are monitoring a patient's vitals and reach out to an edge device that can predict a potential health crisis. The L-XAIM might elucidate the predictive factors to the patient or caregivers, the PPDH secure patient data confidentiality, and the LTA monitor sensor trustworthiness. It would lead to better patient outcomes thanks to clear, timely alerts and effective remote monitoring.

In Industrial IoT and smart manufacturing, the edge AI systems for predictive maintenance or production optimization, e.g., L-XAIM, could explain why a machine is scheduled for maintenance [49]. The SAIEE would safeguard proprietary AI models, and the FTC could consolidate trust information across several machines to discover systemic problems, leading to fewer operating hours and more transparency in operations.

At the level of smart cities, edge AI could control traffic or help reduce energy use. Through the EIDM, explanations about traffic guidance or energy saving behaviors could be given to human citizens or operators. It would safeguard critical infrastructure for the ASO, and for the EFM it could be used to analyze the fairness of allocating resources, thereby allowing for more efficient city operations and increased trust from the city's population [50].

Even in brokering personalized edge devices like smartphones, where an assistant is generating recommendations, the L-XAIM could do the explanation, the PPDH preserve the privacy of the user, and the user would be better able to understand and control the AI's behavior, leading to improved personalization and trust [51–55].

By offering a structured way to embed trust and transparency, the TE2AI framework can contribute to unlocking the full potential of such and other edge AI systems via their responsible development and deployment.

14.6 CONCLUSION AND FUTURE DIRECTIONS

The fast evolution of edge AI brings huge opportunities and significant challenges of trust, security, privacy, and transparency. In this chapter, we have presented the TE2AI framework, a newly proposed conceptual architecture that comprehensively tackles these challenges. TE2AI proposes a modular, layered strategy for integrating trust-building mechanisms and explanations at every stage of the life cycle of edge AI systems, spanning from standalone devices to networked setups. The main value of the framework is the focus on the close integration of trustworthiness dimensions as opposed to trustworthiness in general and explainability, its consideration of resource-constrained constraints at the edge environment, and support for federated approaches. By describing elements and their interactions, TE2AI provides a blueprint for edge AI systems that are not only intelligent, but also verifiably secure, capable of explanation, and reliable. Theoretical implications include developing more sophisticated integrated trust-explanation models and context-aware XAI, while the practical implications range from increased user adoption, improved system robustness, and eased regulatory compliance in a variety of applications such as autonomous vehicles, healthcare, and smart cities.

Despite its comprehensive conceptual basis, significant research and development work is needed to catch up with the potentially provided by TE2AI. Next steps will be to further explore and empirically validate and prototype the framework in real or simulated edge AI settings to observe performance, overhead, and user experience. TE2AI interfaces and protocols will need to be standardized to facilitate interoperability and wide acceptance. Investigating co-design of hardware and TE2AI software (e.g., special-purpose XAI accelerators, improved TEEs) can significantly enhance efficiency and security. Finally, it would also be interesting to study how the advanced AI systems we are using in TE2AI can also be applied to the components of TE2AI, e.g., reinforcement learning for the ASO. The EFM should be extended to more completely support the diverse types of fairness in decentralized scenarios. There is a need for longitudinal studies regarding trust and explainability to better understand how user trust develops in systems developed around such frameworks. Lastly, building a full-fledged TE2AI toolkit as an open source package would help expedite the R&D end and make it more comfortable to utilize these important features as they develop as part of edge AI.

In summary, the path to fully reliable and explainable edge AI is still ongoing. The potential for this journey is the framework provided by TE2AI, which fosters a paradigm shift from closed edge AI to openness with trust, based on transparency, trustworthiness, security, and reliability in the design of edge AI systems. By encouraging further study and collaboration in this area, we can enable the transformational benefits of edge AI in an accountable, ethical manner, creating a future in which distributed intelligence benefits both individuals and society while protected by the core values of trust and accountability.

REFERENCES

[1] W. Shi, J. Cao, Q. Zhang, Y. Li, and L. Xu, "Edge computing: Vision and challenges," *IEEE Internet of Things Journal*, vol. 3, no. 5, pp. 637–646, Oct. 2016.
[2] M. Satyanarayanan, "The emergence of edge computing," *Computer*, vol. 50, no. 1, pp. 30–39, Jan. 2017.
[3] S. M. R. Islam, D. Kwak, M. H. Kabir, M. Hossain, and K. S. Kwak, "The internet of things for health care: A comprehensive survey," *IEEE Access*, vol. 3, pp. 678–708, 2015.
[4] Y. Lu and L. Xu, "Internet of Things (IoT) cybersecurity research: A review of current research topics," *IEEE Internet of Things Journal*, vol. 6, no. 2, pp. 2103–2115, Apr. 2019.

[5] J. Chen and X. Ran, "Deep learning with edge computing: A review," *Proceedings of the IEEE*, vol. 107, no. 8, pp. 1655–1674, Aug. 2019.

[6] S. H. Reddy, H. Bathini, V. N. Ajmeera, R. S. Marella, T. V. V. Kumar, and M. Khari, "Startup unicorn success prediction using ensemble machine learning algorithm," in *Lecture Notes in Computer Science (Including Subseries Lecture Notes in Artificial Intelligence and Lecture Notes in Bioinformatics)*, vol. 14532. LNCS, 2024, pp. 330–338. https://doi.org/10.1007/978-3-031-53830-8_34.

[7] A. Agarwal, M. Khari, and R. Singh, "Detection of DDOS attack using deep learning model in cloud storage application," *Wireless Personal Communications*, vol. 127, no. 1, pp. 419–439, 2022. https://doi.org/10.1007/s11277-021-08271-z.

[8] R. G. Crespo, *Concepts and Real-Time Applications of Deep Learning*. Springer, 2021. [Online]. https://link.springer.com/10.1007/978-3-030-76167-7.

[9] M. Dayal, A. Chawla, M. Khari, and A. N. Mahajan, "Artificial intelligence-based smart packet filter," *Lecture Notes in Networks and Systems*, vol. 421, pp. 791–801, 2023. https://doi.org/10.1007/978-981-19-1142-2_62.

[10] S. Srivastava, G. Chaudhary, and C. Shukla, "Text-independent speaker recognition using deep learning," *EAI/Springer Innovations in Communication and Computing*, pp. 41–51, 2021. https://doi.org/10.1007/978-3-030-76167-7_2.

[11] A. Adadi and M. Berrada, "Peeking inside the black-box: A survey on explainable artificial intelligence (XAI)," *IEEE Access*, vol. 6, pp. 52138–52160, 2018.

[12] D. Gunning, M. Stefik, J. Choi, T. Miller, S. Stumpf, and G. Z. Yang, "XAI—explainable artificial intelligence," *Science Robotics*, vol. 4, no. 37, p. eaaw9368, 2019.

[13] R. Guidotti, A. Monreale, S. Ruggieri, F. Turini, F. Giannotti, and D. Pedreschi, "A survey of methods for explaining black box models," *ACM Computing Surveys*, vol. 51, no. 5, art. 93, pp. 1–42, Aug. 2018.

[14] H. B. McMahan, E. Moore, D. Ramage, S. Hampson, and B. Agüera y Arcas, "Communication-efficient learning of deep networks from decentralized data," in Proceedings of the 20th International Conference on Artificial Intelligence and Statistics (AISTATS), *Proceedings of Machine Learning Research (PMLR)*, vol. 54, 2017, pp. 1273–1282.

[15] A. Josang, R. Ismail, and C. Boyd, "A survey of trust and reputation systems for online service provision," *Decision Support Systems*, vol. 43, no. 2, pp. 618–644, Mar. 2007.

[16] Y. C. Hu, M. Patel, D. Sabella, N. Sprecher, and V. Young, "Mobile edge computing—A key technology towards 5G," ETSI White Paper, No. 11, Sep. 2015.

[17] S. Khan and L. AlSuwaidan, "Agricultural monitoring system in video surveillance object detection using feature extraction and classification by deep learning techniques," *Computers & Electrical Engineering*, vol. 102, p. 108201, 2022. https://doi.org/10.1016/j.compeleceng.2022.108201.

[18] S. Han, H. Mao, and W. J. Dally, "Deep compression: Compressing deep neural networks with pruning, trained quantization and Huffman coding," in *Proceedings of the ICLR*, 2016.

[19] R. Li Wang, Y. Zhang, and S. Ci, "KubeEdge: Kubernetes native edge computing framework," in *Proceedings of the 2019 IEEE/ACM Symposium on Edge Computing (SEC)*, 2019, pp. 276–276.

[20] A. V. Dastjerdi and R. Buyya, "Fog computing: Helping the Internet of Things realize its potential," *Computer*, vol. 49, no. 8, pp. 112–116, Aug. 2016.

[21] F. Bonomi, R. A. Milito, J. Zhu, and S. Addepalli, "Fog computing and its role in the internet of things," in *Proceedings of the First Edition of the ACM MCC Workshop on Mobile Cloud Computing (MCC '12)*, Helsinki, Finland, Aug. 2012, pp. 13–16, doi: 10.1145/2342509.2342513.

[22] M. T. Ribeiro, S. Singh, and C. Guestrin, "Why should I trust you? Explaining the predictions of any classifier," in *Proceedings of the 22nd ACM SIGKDD International Conference on Knowledge Discovery and Data Mining (KDD '16)*, San Francisco, CA, USA, Aug. 2016, pp. 1135–1144, doi: 10.1145/2939672.2939778.

[23] S. M. Lundberg and S.-I. Lee, "A unified approach to interpreting model predictions," in *Advances in Neural Information Processing Systems 30 (NeurIPS 2017)*, I. Guyon, U. V. Luxburg, S. Bengio, H. Wallach, R. Fergus, S. Vishwanathan, and R. Garnett, Eds., Curran Associates, Inc., 2017, pp. 4765–4774.

[24] A. U. Haq et al., "DEBCM: Deep learning-based enhanced breast invasive Ductal Carcinoma classification model in IoMT healthcare systems," *IEEE Journal of Biomedical and Health Informatics*, vol. 28, no. 3, pp. 1207–1217, 2024. https://doi.org/10.1109/JBHI.2022.3228577.

[25] S. Ahmad et al., "Deep learning enabled disease diagnosis for secure internet of medical things," *Computers, Materials & Continua*, vol. 73, no. 1, pp. 965–979, 2022. https://doi.org/10.32604/cmc.2022.025760.

[26] P. Kairouz et al., "Advances and open problems in federated learning," *Foundations and Trends® in Machine Learning*, vol. 14, no. 1–2, pp. 1–210, 2021.

[27] M. Vilone and L. Longo, "Explainable artificial intelligence: A systematic review," *ACM Computing Surveys*, vol. 53, no. 5, art. 97, pp. 1–42, Sep. 2020.

[28] S. Khan et al., "Manufacturing industry based on dynamic soft sensors in integrated with feature representation and classification using fuzzy logic and deep learning architecture," *International Journal of Advanced Manufacturing Technology*, vol. 128, no. 7–8, pp. 2885–2897, 2023. https://doi.org/10.1007/s00170-023-11602-y.

[29] Q. Li, Z. Wen, Z. Wu, S. Hu, N. Wang, Y. Li, X. Liu, and B. He, "A Survey on Federated Learning Systems: Vision, Hype and Reality for Data Privacy and Protection," *IEEE Transactions on Knowledge and Data Engineering*, vol. 35, no. 4, pp. 3347–3366, Apr. 2023, doi: 10.1109/TKDE.2021.3124599.

[30] P. Madry, A. Makelov, L. Schmidt, D. Tsipras, and A. Vladu, "Towards deep learning models resistant to adversarial attacks," in *Proceedings of the 6th International Conference on Learning Representations (ICLR 2018)*, Vancouver, BC, Canada, 2018.

[31] N. Mehrabi, F. Morstatter, N. Saxena, K. Lerman, and A. Galstyan, "A survey on bias and fairness in machine learning," *ACM Computing Surveys*, vol. 54, no. 6, art. 115, pp. 1–35, Jul. 2021.

[32] B. C. M. Mittelstadt, "Principles alone cannot guarantee ethical AI," *Nature Machine Intelligence*, vol. 1, no. 11, pp. 501–507, 2019.

[33] R. Yousef, S. Khan, G. Gupta, B. M. Albahlal, S. A. Alajlan, and A. Ali, "Bridged-U-Net-ASPP-EVO and deep learning optimization for brain tumor segmentation," *Diagnostics*, vol. 13, no. 16, p. 2633, 2023. https://doi.org/10.3390/diagnostics13162633.

[34] National Institute of Standards and Technology (NIST), "AI risk management framework," Jan. 2023. [Online]. Available: https://www.nist.gov/itl/ai-risk-management-framework.

[35] S. Wachter, B. Mittelstadt, and L. Floridi, "Why a right to explanation of automated decision-making does not exist in the general data protection regulation," *International Data Privacy Law*, vol. 7, no. 2, pp. 76–99, May 2017.
[36] S. Khan, C. Viswanathasarma, K. Sekaran, K. Joshi, C. K. Roy, and M. Tiwari, "Incorporating deep learning methodologies into the creation of healthcare systems," in *2023 International Conference on Artificial Intelligence and Smart Communication, AISC 2023*, 2023, pp. 994–998. https://doi.org/10.1109/AISC56616.2023.10085651.
[37] A. Saboor et al., "DDFC: Deep learning approach for deep feature extraction and classification of brain tumors using magnetic resonance imaging in e-healthcare system," *Scientific Reports*, vol. 14, no. 1, p. 6425, 2024. https://doi.org/10.1038/s41598-024-56983-6.
[38] A. U. Haq, J. P. Li, S. Khan, M. A. Alshara, R. M. Alotaibi, and C. B. Mawuli, "DACBT: Deep learning approach for classification of brain tumors using MRI data in IoT healthcare environment," *Scientific Reports*, vol. 12, no. 1, p. 15331, 2022. https://doi.org/10.1038/s41598-022-19465-1.
[39] S. Vimal, M. Khari, N. Dey, R. G. Crespo, and Y. Harold Robinson, "Enhanced resource allocation in mobile edge computing using reinforcement learning based MOACO algorithm for IIOT," *Computer Communications*, vol. 151, pp. 355–364, 2020. https://doi.org/10.1016/j.comcom.2020.01.018.
[40] K. Kumar, M. Khari, and A. Kumar, "Comprehensive method of distributed denial of service detection by artificial neural network," in *Wireless Communication with Artificial Intelligence*. CRC Press, 2022, pp. 27–43. https://doi.org/10.1201/9781003230526-2.
[41] M. S. Pillai, G. Chaudhary, M. Khari, and R. G. Crespo, "Real-time image enhancement for an automatic automobile accident detection through CCTV using deep learning," *Soft Computing*, vol. 25, no. 18, pp. 11929–11940, 2021. https://doi.org/10.1007/s00500-021-05576-w.
[42] R. Yousef, G. Gupta, N. Yousef, and M. Khari, "A holistic overview of deep learning approach in medical imaging," *Multimedia Systems*, vol. 28, no. 3, pp. 881–914, 2022. https://doi.org/10.1007/s00530-021-00884-5.
[43] A. Verma and M. Khari, "Vision-based accident anticipation and detection using deep learning," *IEEE Instrumentation & Measurement Magazine*, vol. 27, no. 3, pp. 22–29, 2024. https://doi.org/10.1109/MIM.2024.10505198.
[44] M. Mishra, A. K. Yadav, B. D. Mazumdar, P. K. Gupta, A. Panwar, and S. Bharadwaj, "Recognition and evaluation of cutaneous condition through assorted artificial intelligence reliant algorithms," *International Journal of Information Technology*, pp. 1–13, 2025. https://doi.org/10.1007/s41870-025-02483-3.
[45] T. Singh, A. Panwar, N. Sharma, U. Sugandh, and M. Singh, "Leveraging blockchain technology for secure and efficient smart city applications," in *Lecture Notes in Networks and Systems*, 2025, pp. 619–633. https://doi.org/10.1007/978-981-97-8329-8_45.
[46] P. S. Lamba et al., "Detection of driver drowsiness using adaptive eye characteristic ratio for enhanced road safety," *IEEE Access*, 2025. https://doi.org/10.1109/ACCESS.2025.3566981.
[47] N. Sharma, U. Sugandh, J. Agarwal, A. Panwar, and P. Gaba, "Federated learning on wheels: A decentralized approach to privacy-enhanced data collection in internet of vehicles," in *Federated Learning Based Intelligent Systems to*

Handle Issues and Challenges in IoVs (Part 2). Bentham Science Publishers, 2025, pp. 1–33.

[48] P. Gaba, U. Sugandh, A. Panwar, and M. Kumar, "Beyond cash and cards: Exploring the potential of blockchain for electronic payments in the internet of vehicles," in *Federated Learning Based Intelligent Systems to Handle Issues and Challenges in IoVs (Part 2)*. Bentham Science Publishers, 2025, pp. 34–70.

[49] A. Panwar, P. Gaba, U. Sugandh, and N. Bohra, "Breaking the centralization barrier: Exploring decentralized federated learning for vehicle number plate recognition in IoV," in *Federated Learning Based Intelligent Systems to Handle Issues and Challenges in IoVs (Part 2)*. Bentham Science Publishers, 2025, pp. 102–139.

[50] N. Sharma, A. Panwar, U. Sugandh, V. Jain, and K. Singh Kaswan, "Digital twins in healthcare: Applications, challenges, and future directions," in *2024 4th International Conference on Advancement in Electronics and Communication Engineering, AECE 2024*, 2024, pp. 1401–1406. https://doi.org/10.1109/AECE62803.2024.10911658.

[51] T. Singh, M. Singh, U. Sugandh, and A. Panwar, "Feature extraction in image processing using medical scans: Detailed survey," in *2nd IEEE International Conference on IoT, Communication and Automation Technology, ICICAT 2024*, 2024, pp. 1587–1592. https://doi.org/10.1109/ICICAT62666.2024.10923105.

[52] P. S. Lamba, A. Jain, H. Taneja, P. Priyadarshi, A. Panwar, and A. K. Dubey, "Ensemble sentiment analysis using Bi-LSTM and CNN," in *2024 4th International Conference on Advancement in Electronics and Communication Engineering, AECE 2024*, 2024, pp. 575–579. https://doi.org/10.1109/AECE62803.2024.10911211.

[53] P. Gupta, A. N. Mathur, K. Kathuria, R. Chandak, and S. Sangal, "Recommender system based on fuzzy C-means," in *International Conference on Next Generation Computing Technologies*. Singapore: Springer, Oct. 2017, pp. 3–18.

[54] S. N. Mohanty, J. M. Chatterjee, S. Jain, A. A. Elngar, and P. Gupta (Eds.), *Recommender System with Machine Learning and Artificial Intelligence: Practical Tools and Applications in Medical, Agricultural and Other Industries*. John Wiley & Sons, 2020.

[55] K. Dutta, R. Lenka, P. Gupta, A. Goel, and J. V. N. Ramesh, "Swift diagnose: A high-performance shallow convolutional neural network for rapid and reliable SARS-COV-2 induced pneumonia detection," *EAI Endorsed Transactions on Pervasive Health & Technology*, vol. 10, no. 1, 2024.

Chapter 15

Security and Privacy Considerations in Edge AI

Trilok Bhardwaj, Sandhya Malikar Patel, Poonam Arora, and Manju Khari

15.1 INTRODUCTION

The rise of edge AI, which integrated the power of AI with edge computing, has transformed intelligent computation to occur directly on edge devices [1], along with the processing of data locally at the device itself. With this integration significant benefits can be achieved, like reduced latency, improved bandwidth efficiency, and a better user experience. However, it also presents new challenges related to security and privacy due to localised processing of sensitive user data in a decentralised environment. Since edge AI involves sensitive data processing locally, the chances of physical tampering, adversarial attacks, malware exploitation, and eavesdropping are naturally evident. Due to the nascent distributed environment and varied functional characteristics of edge devices, there are significant challenges in forming a cohesive security framework. This complexity enforces the necessity for implementing robust strategies on fundamental security measures. In addition to the aforementioned, secure communication protocols need to be deployed to ensure that data transmitted between devices remains confidential and tamper-proof. This will not only fortify the overall security posture in edge computing environments but also will introduce trust among its users.

Privacy concerns become more prominent as personal and sensitive information, like health records, biometric information, and location data, etc., is being generated at edge devices and is frequently collected and processed locally at the device itself for further communication in decentralised environments. This raises the need for compliance with regulations such as General Data Protection Regulation (GDPR), which is a European Union law governing the processing and sharing of personal data; Health Insurance Portability and Accountability Act (HIPAA), a set of rules designed to protect the privacy and security of patients' health information; and DPDP Act 2023, which is an Indian law applicable to the processing of digital personal data within India. This is applicable to and encompasses data collected online or offline which is further digitised for use in various smart applications. These regulations need to be enforced on healthcare providers, smart

DOI: 10.1201/9781003644323-15

city developers, smart transportation entities, etc. To mitigate risks arising from local data collection and processing, techniques like data anonymisation, access control, and privacy-preserving computation have become essential. Furthermore, due to the unique challenge posed by machine learning attacks, capabilities checks and balances need to be incorporated by continuous monitoring and regular enhancements in local model capabilities.

There are many emerging technologies available which can be deployed at local edge devices and in the architecture which promise solutions to the mentioned challenges. Hardware-based approaches, federated learning, and trusted execution environments (TEEs) can enhance privacy and reduce exposure to breaches by doing decentralised model training without sharing raw data. In order to safeguard sensitive personal information, differential privacy mechanisms can be deployed which ensure that individual data remain indistinguishable within larger datasets. In this chapter we put efforts into exploring some of the fundamental security and privacy considerations to be deployed in edge AI. We mainly emphasise in the coming sections the importance of a multi-layered approach to protect against diverse threats. By deploying the discussed security measures in this chapter, edge AI can achieve its full potential and will definitely offer transformative benefits across all verticals deploying AI architecture, along with safeguarding user data and fostering trust in smart decentralised systems.

In this chapter, an in-depth exploration of the critical security and privacy issues associated with edge AI is explained in detail. The chapter begins by defining essential terms related to artificial intelligence, which allow us to clearly understand the unique characteristics that set "edge AI" apart from traditional AI.

Our main concern and focus in this chapter is for the reader to get answers for some important aspects of edge AI with respect to security concerns and why privacy is the most crucial consideration in designing applications for edge AI.

Also, authors address various parameters which influence both privacy and security concerns in edge AI along with their appropriate mitigation strategies while implementing them in real-world applications.

15.2 INTRODUCTION TO AI

AI, by core definition, is the development of computer systems that can perform tasks typically requiring human intelligence [2]. These tasks include learning, reasoning, problem-solving, understanding natural language, recognising patterns, and making decisions. The idea of artificial intelligence is far older than many realise. Its origin can be traced back to 1950 [3], when Alan Turing introduced the groundbreaking Turing test, a pivotal moment in the quest to understand machine intelligence. Fast forward to the 1960s, and we see the emergence of ELIZA [4], the first chatbot program, which paved the way for future developments in AI. By 1977, IBM Deep Blue [5] made history as a revolutionary chess computer that not only challenged but also defeated a world chess champion in two out of six games, with one victory for the champion and three games ending in draws. In the year 2000, Professor Cynthia Breazeal developed the first robot that could simulate human emotions with its face, which included eyes, eyebrows, ears, and a mouth. It was called Kismet [6]. Eventually In 2011, Siri was released as the first popular digital assistant by Apple [7]. Elon Musk and some others founded OpenAI in 2015 [8] with a goal to benefit humanity. In the year 2016, Hanson Robotics created a humanoid robot named Sophia, who became known as the first "robot citizen" and was the first robot created with a realistic human appearance and the ability to see and replicate emotions, as well as to communicate [9]. Subsequently, in the year 2021, OpenAI developed DALL-E, which can process and understand images enough to produce accurate captions, moving AI one step closer to understanding the visual world [10]. This remarkable journey highlights the long-standing ambition to create machines capable of human-like thinking and decision-making.

At its core, AI is about creating machines that simulate behavioural aspects of human nature. AI has dramatically transformed countless industries, including healthcare and finance, by significantly enhancing efficiency and accuracy in a wide range of tasks. However, it is essential to note that not all AI systems are created equal [11]; they can be categorised into two primary types: Narrow AI and General AI. Narrow AI (also known as weak AI) is designed to perform specific tasks—such as voice recognition, image classification, or playing a board game. General AI, also known as Strong AI, is the next step in the future direction of AI technology. It possesses the ability to understand, learn, and apply knowledge across a wide range of tasks at a level equal to or beyond human intelligence. While general AI remains a theoretical goal, narrow AI is already widely used in technologies such as Siri, Google Search, autonomous vehicles, and recommendation systems on platforms like Netflix or Amazon.

The main critical actors controlling AI are machine learning (ML), deep learning, and natural language processing (NLP) [11]. In ML, systems improve their performance over a period of time on a particular domain by constant learning from data, and no separate explicit programming is required for the system. On the other hand, in deep learning, a neural network of many deep layers is used to learn complex patterns which are particularly useful in image processing and speech recognition applications. NLP trains machines so that they can interact using human language, understand it, and reply accordingly, e.g., like chatbots and virtual assistants.

AI's applications span across the health sector in diagnosing diseases from medical images and predicting patient health in real time. In the financial domain, fraudulent transaction detection systems and algorithmic trading can be easily carried out using AI tools. Apart from these in the manufacturing industry and smart transport applications, AI can improve quality control and predictive maintenance by interpreting sensory data and making split-second driving decisions.

So, from this angle it seems clear that induction of AI has enormous benefits, yet AI has many significant concerns that raise eyebrows, like ethical, social, and economic concerns. Among all the most important is bias, i.e., how much truthful data is being used to train AI models. This means if an AI system is trained on biased data with malign intention, it can produce biased outcomes which can reinforce inequalities, especially in areas like smart homes, quality control, and law enforcement. The second most vital concern is privacy. Since large datasets are being used to train any AI systems, which, of course, sometimes involve sensitive personal information to function effectively, this can raise crucial trustworthiness and social security aspects of the application. As a result of all the points already discussed, not only government organisations and researchers but also the private sector are working tirelessly to develop fully trustworthy and secured AI frameworks and regulations to guide the responsible use of AI for the benefit of human beings and society.

In conclusion, the future of AI relies not only on technological advancements but also on our ethical intentions to use artificial intelligence in developing human-centric designs that reshape the modern world.

15.3 INTRODUCTION TO EDGE AI

Edge AI represents a vertical of AI where processing of data takes place close to its source of generation, responsible for significant advancement of intelligent systems. Since AI models operate locally in edge AI, it enables faster, more private, and context-aware decision-making [12] capability in the AI application. With the constant evolution in technology, edge AI will play a central role in deciding the future of smart devices, smart industries, and AI-based services. Now in edge AI algorithms are directly deployed on smart devices like IoT sensors, drones, smartphones, and embedded systems, thus removing dependability on centralised cloud servers for processing. This localised processing of data close to its source provides power to normal AI with the advantages of edge computing. As a result, edge AI will offer faster decision-making capability in the deployed system, along with reduced latency, the highest bandwidth efficiency, and enhanced privacy in comparison to traditional AI. As in traditional AI applications, there is a huge dependency on cloud computing, where data is collected from devices and, after preprocessing, sent to cloud data centres for final processing, and then results are sent back to the devices. Though a normal AI model has vast computational resources, its drawbacks, like latency, bandwidth consumption, and privacy concerns, make it inferior when compared with the power of edge AI.

15.3.1 Main benefits of edge AI

The concept of edge Intelligence enhances decentralised computing by integrating artificial intelligence, data analytics, and localised processing. This integration contributes to the development of adaptive and responsive networks [13, 14]. Edge Intelligence is crucial for minimising reliance on centralised data processing, leading to the creation of more efficient and resilient systems. Edge AI revolutionises data processing on devices, offering numerous advantages as follows:

- **Low Latency:** The most crucial aspect for which edge AI is gaining a boom is the capability to process data locally so that decisions can be made in a few seconds. This enables real-time-dependent applications like financial transactions, autonomous vehicles, industrial automation, and medical monitoring devices to function in real time with faster speed.
- **Bandwidth Efficiency:** Due to the highest availability of bandwidth, data which is to be analysed on the device is stored locally on the device, and the information essential for the application is transmitted to the cloud, thereby reducing network congestion. Implementation of this also results in lowering operational costs.
- **Privacy and Security:** Since data resides locally on the device and is processed at the device only, it minimises data exposure, thus reducing the risk of interception during transmission. This feature becomes crucial, especially for applications where sensitive personal information, such as facial recognition or health monitoring parameters, is processed.
- **Offline Functionality:** As most of the edge AI devices do not essentially require constant internet connectivity, it enables edge AI to be deployed in remote areas or in applications where connectivity is intermittent, such as in drones or mobile devices.

15.3.2 Applications of edge AI

Edge computing is an innovative paradigm that transfers traditional cloud-based network computing capabilities from centralised data centres to end-user devices and local area networks [15]. This development fundamentally extends the cloud computing architecture to the periphery of the network, facilitating a diverse range of new services and applications for end users. The proliferation of Internet of Things (IoT) applications across various sectors has resulted in a remarkable increase in the number of Internet-enabled devices, rising from millions to billions [16]. This shift presents significant opportunities for enhancing operational efficiency and fostering technological advancement in an increasingly interconnected environment.

Edge AI is widely deployed across various sectors:

- **Smart Homes:** Devices like smart speakers, thermal sensors, motion sensors, and security cameras use AI at the edge to recognise voices, detect motion, or optimise energy use.
- **Healthcare:** Wearable devices analyse biometric data in real time to detect abnormalities and provide instant alerts.
- **Manufacturing:** Edge AI helps in predictive maintenance, quality control, and anomaly detection by analysing machine data on-site.
- **Retail:** Smart shelves, customer behaviour analysis, and automated checkouts leverage edge AI to enhance efficiency and customer experience.
- **Transportation:** Autonomous vehicles depend in real time on edge AI to process traffic light and environment sensor data and make split-second driving decisions.

15.3.3 Challenges and considerations

Many challenges are also associated with edge AI despite lots of advantages:

- **Hardware Limitations:** One of the prime factors which limits the full functionality of edge devices is processing power, memory, and energy compared to cloud servers. AI models must be optimised to run efficiently on edge devices with limited hardware technologies and their associated constrained environments.
- **Model Updates and Scalability:** To keep all the edge devices updated with the latest patches and software updates, it requires updating AI models across a decentralised network of edge devices. This can be complex and resource-intensive.
- **Security Risks:** Edge devices are often physically accessible and more vulnerable to tampering, making robust security measures essential.
- **Data Management:** Ensuring consistent data quality and handling data retention responsibly are crucial for privacy and compliance.

15.4 COMPARISON BETWEEN EDGE AI AND AI

AI is a new fieldthat targets building smart machines and autonomous systems which can independently perform tasks that typically require human intelligence. The tasks of nature can be of any kind, like self-learning and reasoning over a problem domain, problem-solving, making perceptions about a real-time application, and machine language understanding. Traditional AI, especially in recent years, has relied heavily on cloud computing, where large datasets are processed in centralised data centres with powerful servers. Edge AI, by contrast, represents a newer approach [1, 17] where AI models are deployed directly on edge devices—such as smartphones, sensors, drones, or embedded systems—so that data can be processed locally, at or near the source of generation.

This difference in architecture and deployment results in several key distinctions between traditional AI and edge AI.

15.4.1 Location and data processing

In traditional AI-based systems, raw data from devices is sent to remote servers (the cloud) for processing. These servers analyse the data and send results back to the device. In contrast, edge AI performs computation on the edge device itself, often without needing to communicate with the cloud. This reduces the dependency on internet connectivity and allows for faster response times.

15.4.2 Latency and real-time processing

One of the most critical differences is latency, because data must travel to the cloud and back in traditional AI systems. However, that methodology introduces a delay, which can be problematic for real-time-based applications that require real-time processing for making decisions in applications. Edge AI, by processing data locally, offers ultra-low latency, enabling immediate decision-making without delays caused by network communication.

15.4.3 Bandwidth and connectivity

Traditional AI models consume considerable **bandwidth** due to the continuous transmission of large datasets to and from the cloud. Such an approach can be inefficient, especially in bandwidth-constrained environments. Edge AI minimises this need by analysing data on-site and only sending essential information or summaries to the cloud. This makes edge AI more suitable for remote or rural areas with limited internet access.

15.4.4 Privacy and security

In edge AI, due to localised processing of data closer to its source of generation, the user gets enhanced privacy, thus reducing the chance of exposure of sensitive user information during transmission. However, traditional AI systems often transmit the user's personal or sensitive data directly to the cloud, making privacy concerns more significant. Due to the traditional features of normal AI, data breaches or unauthorised access to centralised cloud servers can always compromise user privacy. In edge AI, since the data always remain on the device, the chance of breaches is less prominent. This localised data handling is especially important for applications in healthcare, surveillance, and personal devices.

15.4.5 Resource requirements

In traditional AI systems, massive computational power from cloud servers is required to run large and complex models. However, in edge AI, within the **limited processing power, storage capacity, and energy constraints, processing can be carried out locally over**edge devices. That enforces the use of techniques like model pruning or quantisation to generate lightweight or compressed AI models.

15.4.6 Use cases

On the basis of use cases, traditional AI will be more advantageous for applications and systems that require deep insights from large-scale data. A few examples of such applications can be natural language translation, search engines, or large-scale recommendation systems. However, in edge AI, the use cases will focus more on localised, real-time tasks such as facial recognition on smartphones, predictive maintenance in factories, and object detection in autonomous drones.

15.4.7 Conclusion

Both traditional AI and edge AI aim toward making machines intelligent, though their architectures, applications, and constraints differ significantly. It can be concluded from the preceding facts that edge AI complements traditional AI by doing more private computing at the edge node of the network. However, these two approaches will continue to coexist, often working together in hybrid mode.

15.5 MAJOR SECURITY CONCERNS IN EDGE AI

15.5.1 Physical vulnerability of devices

Edge AI starts by deploying AI models directly on edge devices such as smartphones, drones, and IoT devices, with real-time data processing at the edge node of the network, thereby removing dependency on centralised cloud infrastructure. This decentralisation approach enhances speed, privacy, and bandwidth efficiency, but it also exposes edge devices to physical vulnerabilities that will raise serious security and reliability concerns in the near future. One of the most critical physical vulnerabilities among them is physical tampering. Since in decentralised hostile environments edge devices are often deployed, that will make them vulnerable to direct physical access by malicious actors. Attackers can have unauthorised access and may open the device to extract memory chips and do reverse engineering with the components or even manipulate circuitry for their malign intentions. This type of intrusion can result in the extraction of sensitive data or proprietary AI models, ultimately compromising intellectual property and end-user privacy. Threats like side-channel attacks also augment the physical tampering of edge devices due to decentralised hostile environments. As in this kind of attack, generally indirect physical emissions from edge devices, such as electromagnetic radiation and power consumption, can be exploited to infer sensitive information about the AI models or the data being processed at that particular edge device deployed in the system. These attacks are more dangerous because they can be carried out very easily without needing to physically tamper with the device.

Another major drawback of edge AI devices is hardware degradation due to environmental parameters like extreme weather, fluctuating temperatures, humidity, or exposure to dust and moisture. This can significantly degrade performance over time, due to which components like sensors, storage devices, and microcontrollers may fail and can produce inaccurate results, which may in turn lead to system breakdowns. In addition to such issues, the supply chain also possesses physical vulnerability, as many edge devices are built using components sourced globally, and this can introduce malicious modifications during manufacturing at the hardware level itself. For example, any attackers can embed persistent malware or spy modules that are difficult to detect at the hardware manufacturing itself, thus posing long-term security risks. On a similar line, device theft from a manufacturing plant allows the attacker to gain access to edge AI devices and to bypass encryption, clone AI models, or spoof device behaviour for malicious purposes.

15.5.2 Insecure data storage

As we know, edge AI processes and analyses data directly on local devices. Devices can be integrated or smart, like IoT sensors, mobile phones, drones, and robotic machines. It provides advantages of reduced latency, high bandwidth, and high real-time decision-making capability.

As data is processed locally, there are significant security concerns with respect to insecure data storage, weak encryption, improper key management, secured key storage hardware, data lifecycle management, and firmware vulnerabilities. As in edge devices, data is often stored locally, and there are no extensive protections along with weak encryption, so these devices become attractive targets for malicious actors. Weak encryption of sensitive data, such as user identity, location, health information, or operational logs, in plain text stored in edge devices may result in breaches of personal privacy or industrial espionage or manipulation of critical systems if an attacker gains physical or remote access to the device.

Improper key management and poor handling of cryptographic keys increase the vulnerability of data at rest. Attackers can extract these keys and decrypt stored data with minimal effort. Furthermore, due to a lack of secure key storage hardware, such as Trusted Platform Modules (TPMs), edge devices become more vulnerable to tampering or reverse engineering. Along with these weak authentication mechanisms, using default company-set passwords can provide attackers access to sensitive data, thereby allowing them to modify configurations or inject malicious code into the device firmware. This will make the whole smart system vulnerable, and this is particularly more concerning in the case of shared or publicly accessible devices, such as smart kiosks or connected medical devices. Another major issue with data storage is in respect of the absence of data lifecycle management and outdated or unpatched software and firmware vulnerabilities. These weaknesses can provide backdoors for attackers to extract or manipulate historically stored information on the devices which are deployed in environments with limited IT oversight.

15.5.3 Insecure communication channels

Edge AI brings AI capabilities closer to the source of data generation with reduced latency and less dependency on centralised cloud systems. It also introduces new security concerns in respect of insecure communication channels. By insecure communication channels we mean that communication between edge devices happens over public or unsecured networks, making them highly vulnerable to attacks, which compromises data confidentiality, integrity, and availability and becomes suspicious.

Due to a lack of strong encryption (like TLS 1.3 or AES-256) or the use of outdated and weak encryption protocols in communication, which is a major critical vulnerability in edge device communication, transmitting sensitive data like personal information or model outputs over weakly encrypted Wi-Fi networks, Bluetooth, or cellular networks gave attackers easy and free hands to launch man-in-the-middle (MITM) attacks, intercepting and thereby altering data in transit, leading to data leaks, privacy breaches, or manipulation of AI decisions [18]. Even when encryption is used, poor implementation can render it ineffective, as some devices may still support insecure protocols like HTTP, FTP, or outdated SSL versions, which pave the way for skilled attackers to impersonate trusted entities or decrypt traffic using known keys. Another problem in edge AI communication is poor mutual authentication, like when devices authenticate to centralised servers but do not verify the server's authenticity and legitimacy in return. This type of poorly implemented single-side verification model gives attackers an advantage in spoofing endpoints and gaining unauthorised access to the communication channel. Apart from the aforementioned vulnerability, replay attacks, packet injection, and device-to-device (D2D) communication are also common in insecure communication setups. Attackers can capture legitimate data packets and resend them later to manipulate system behaviour or gain unauthorised access by injecting malicious commands or falsified data into the network and thus corrupting AI model behaviour, particularly in edge AI systems that rely on continuous sensor data or command feedback. Poor D2D communication where multiple edge devices communicate locally allows one compromised device to infect or eavesdrop on others in the network.

15.5.4 Model theft and reverse engineering

The decentralised approach, edge AI, brings ML models to function locally on the edge device of the network, thereby allowing data to be processed locally on devices. It introduces critical security risks like model theft and reverse engineering. These attacks target the intellectual property embedded in the AI models.

Model theft, sometimes also called model extraction, involves replicating a ML model by either directly accessing it or indirectly querying it to infer its behaviour [19] if physical security or storage protections are weak on edge devices. It becomes easy for attackers to gain access to model files. Once access is obtained, the model can be copied, redistributed, or reconfigured without authorisation, thus compromising the owner's intellectual property and violating data protection regulations. Especially if the models are stored unencrypted or if cryptographic protections are poorly implemented, attackers can disassemble devices and retrieve model files from local storage.

On the same line, reverse engineering takes this threat further by analysing the model's architecture, weights, and behaviour to understand how it works [20], thereby enabling attackers to identify vulnerabilities within the model. In critical applications like facial recognition, industrial automation, or predictive maintenance, reverse engineering can allow attackers to develop countermeasures or spoof system behaviour.

API-based model extraction attacks can also be used where attackers can send a large number of queries to the device for remote access or inference and use the outputs to train a surrogate model that closely mimics the original model. If proper access controls are not in place, then an API-based model extraction attack does not require direct access to the device's internal components and even can be performed remotely.

15.5.5 Adversarial attacks on AI models

In adversarial attacks, attackers inject inputs that are typically small and imperceptible to the human eye to deceive AI models. This procedure is done to manipulate input data in such a way that the AI model makes incorrect predictions. As more and more AI models are being integrated into the real-world systems, mitigating adversarial attacks has become a vital area of research and development.

The most prevalent types of adversarial attacks are White-box attacks (attacker has full knowledge of the model's architecture, weights, and parameters) and Black-box attacks (attacker has no direct access to the model but can observe its inputs and outputs) [21].

The second category of adversarial attacks is the transferability attack, in which examples are generated for one model in such a way that they appear to be effective against different models trained on similar tasks. This tactic allows attackers the capability of executing the scope and feasibility of attacks in real-world scenarios.

Adversarial attacks are not limited to image data but also target NLP models, speech recognition systems, and time-series data used in financial or health monitoring systems. The implications of adversarial attacks are serious [21,22]. They can lead to safety hazards, financial loss, privacy breaches, and loss of trust in AI systems. In critical infrastructure, adversarial inputs can be used to bypass security systems or cause automation failures.

15.5.6 Malicious or unauthorised software updates

Edge AI systems operate at the edge node of decentralised networks with real-time data processing on local devices. These devices rely on frequent software and firmware updates to patch vulnerabilities, improve AI model performance, and add new functionalities. However, this update process introduces a significant security risk of malicious or unauthorised software updates. If attackers can manipulate the update mechanism, they can compromise the entire device and potentially the larger network it connects to.

Malicious or unauthorised updates refer to the injection or installation of unverified, tampered, or intentionally harmful software onto edge devices. Any attackers, by exploiting insecure update channels and compromising update servers, can force users to download rogue updates, making edge devices vulnerable to such attacks, which often operate autonomously. Though it is ensured in the secure systems that software updates are signed using cryptographic keys, and before any installation of updates, these signatures are verified. If signature verification is not carried out before installation, then attackers can replace legitimate updates with malicious versions. Once these updates are installed, they can disable security features, exfiltrate sensitive data, install backdoors, and even take control of the device.

Another kind of update attack is MITM attacks if the update process does not use secure communication protocols [23, 24]. For example, if updates are transmitted over unencrypted channels (like HTTP instead of HTTPS), attackers can intercept and modify update files in transit.

15.5.7 Lack of standardised security protocols

As edge AI technology is growing rapidly across sectors such as healthcare, transportation, manufacturing, and smart cities with the deployment of AI models directly on local edge devices, one major challenge remains unresolved: the lack of standardised security protocols. While in centralised cloud systems, security measures are relatively mature and standardised, the edge computing landscape is still fragmented and needs new formulation in terms of standards for different sectors. We all know that edge AI systems are using heterogeneous hardware platforms and a wide variety of operating systems to design application-specific AI frameworks. This heterogeneity enforces problems in applying consistent security policies across different devices, leaving them vulnerable to cyber and physical attacks.

Due to a lack of standardisation protocol, some edge devices may use strong encryption like TLS 1.3 [18] and AES-256 [25], while others still rely on outdated or proprietary methods [26] that are easier to break. Similarly, authentication mechanisms can vary widely from secure multi-factor systems to simple password-based schemes that are easily exploitable.

Due to a lack of standardised security protocols, when each device follows a different update process, it becomes challenging for administrators to maintain a secure, consistent state across the entire edge infrastructure. This inconsistency leads to unpatched vulnerabilities, misconfigurations, and an increased attack surface. In addition to the that, there is the absence of agreed-upon standards for secure data transmission and device identification between different vendors, thus leading to implementations of custom or ad hoc solutions which are often poorly tested and may contain security flaws. This fragmentation also hinders large-scale adoption of security best practices.

The lack of standardisation also complicates compliance and regulatory oversight. In sectors such as healthcare and finance, data protection is governed by strict laws like HIPAA, GDPR, and the DPDP Act [27–29]. Ensuring compliance becomes much harder when devices do not adhere to consistent security protocols.

15.5.8 Resource constraints

Resource constraints are the most significant challenges in edge AI. As in edge AI, models and algorithms are deployed on devices at the edge of the network. This paradigm enables data to be processed locally, with reduced latency, conserving bandwidth, and enhancing privacy. In edge AI devices have limited computational power, memory, storage capacity, and energy availability, unlike powerful cloud servers or data centres in cloud computing or traditional AI environments.

These features of edge AI restrict the complexity and scale of AI models which can be run on such devices, as extensive processing capabilities and memory resources are required in most common traditional deep learning models, like large convolutional neural networks (CNNs), which most edge devices simply do not possess. Since the edge devices are not designed for intensive computation, as they are equipped with low-power CPUs or microcontrollers or AI accelerators, running complex AI inference tasks locally can lead to performance bottlenecks, slow responses, or system overheating. Edge device local memory capacity and energy efficiency also make it difficult to load and execute large AI models, thereby restricting the ability to handle large volumes of input data, and due to sustained inference workloads, the batteries of local edge devices such as drones, mobile robots, and wearable sensors can be drained rapidly. These limitations require that lightweight AI models and energy-efficient hardware need to be developed for effective implementations of AI systems.

15.5.9 Insider threats and supply chain attacks

Edge networks are more susceptible to cyber threats because most of the focus in edge AI security is carried out on external attacks; however, insider threats and supply chain attacks represent some of the most dangerous and difficult-to-detect vulnerabilities. These threats raise concerns about the data integrity, system availability, and trustworthiness of AI systems operating at the edge.

Insider Threats: These are the threat vectors which are associated with individuals working inside an organisation. These actors also have authorised access to systems and data, either maliciously or unintentionally. They can exploit their access to tamper with AI models, leak sensitive data, or sabotage devices, altering configurations, installing backdoors, or introducing vulnerabilities without immediate detection.

Supply Chain Attacks: When deploying edge AI, applications rely on various third-party libraries, firmware, hardware modules, and cloud services. In these scenarios, supply chain attacks target the development, manufacturing, or distribution of hardware and software components. A compromise in any part of this chain can infect the final product before it reaches the end user. These attacks are especially dangerous because they exploit trusted relationships.

15.5.10 Data integrity attacks

The most crucial thing in edge AI is keeping data integrity intact. By data integrity we mean that information is accurate, consistent, and trustworthy. As in edge AI, data is collected and processed locally at the edge of a network; thus, maintaining data integrity is both critically important and uniquely challenging.

Attackers always try to attack the accuracy and trustworthiness of data and AI models. If they are successful, then it can lead to incorrect decisions, security breaches, or even physical harm in safety-critical applications. Since in edge AI data is going through a life cycle from collection, local processing via AI models, and sometimes synchronisation with cloud systems, a data integrity attack can occur at any of these stages. These attacks are more concerning, as the nature of the devices in edge environments is decentralised with limited resources and a lack of centralised monitoring. Data poisoning, data replay, and injection attacks are common forms of data integrity attacks. An attacker deliberately feeds false or misleading data into an AI model during its training through compromised sensors or network channels, leading to a corrupted AI model that makes systematically flawed predictions. If these attacks are doing real-time data manipulation by tampering with live data streams before they reach the AI model, they become a major threat to the whole system, as such manipulations can disrupt normal operations and lead to system malfunctions with eroded user trust and can cause physical damage or loss of life in critical systems.

15.5.11 Lack of centralised oversight

Edge AI enables real-time decision-making and reduced dependence on centralised cloud systems. However, one of the major challenges with this decentralised approach is the lack of centralised oversight, which can lead to various security, operational, and management issues. Edge devices may be deployed in various environments, which creates a complex, fragmented system that is difficult to monitor and manage from a central point. This lack of centralised monitoring arising from the lack of centralised oversight creates issues in ensuring consistent security across all devices in an edge AI network. Without centralised monitoring, vulnerabilities on individual devices may go unnoticed for long periods, leaving them open to cyberattacks. Further, edge AI systems often require frequent updates to address emerging security threats, so without centralised oversight, coordinating these updates across a fleet of devices can become a cumbersome process, leaving outdated or insecure software and exacerbating the risk of exploitation. Also, it becomes difficult to ensure that all devices are operating the same version of an AI model if there is a lack of central monitoring or it is missing completely.

15.5.12 Physical access risks

Since edge AI devices are deployed in a variety of environments, security of edge devices is often compromised, as edge AI devices are deployed autonomously and sometimes without constant supervision. When attackers gain physical access to them, they try to bypass traditional cybersecurity defences and cause extensive damage to both the devices and the networks. Physical access to edge AI devices provides attackers the opportunity to directly tamper with the hardware and software, which can have serious consequences like hardware modifications and installing malicious components and implants that could enable attackers to intercept or alter data, disable security mechanisms, or even cause physical damage to the device. Many edge devices strongly rely on firmware updates, and by tampering with the firmware, an attacker after getting physical access could replace the firmware with malicious code, leading to backdoors, data leaks, or system failures. This could result in the disruption of vital services, such as surveillance, traffic management, or even healthcare monitoring in critical infrastructure applications.

15.5.13 Access and privilege escalation

Vulnerabilities like unauthorised access and privilege escalation become prominent when an attacker gains access to edge AI devices or systems without the proper permissions. They are inherently present in edge AI systems due to the decentralised nature of the deployment of devices. These devices can be physically exposed in public spaces, factories, or remote locations, making it easier for attackers to interact with them directly or exploit network vulnerabilities to gain unauthorised access. The major effect of unauthorised access in edge AI can be severe, resulting in loss of privacy to the disruption of critical infrastructure or services. Similarly, once an attacker gains unauthorised access to an edge AI device or system, they may attempt privilege escalation, which is the process of gaining higher levels of access or control within the system. In edge AI systems, privilege escalation often takes the form of exploiting flaws in software, misconfigurations, or weak access controls to escalate from a low-level user account to an administrator-level account with more control over the device or system. Privilege escalation can occur by exploiting software vulnerabilities, misconfigured access controls, and weak separation of duties.

Privilege escalation in edge AI systems can have serious consequences. An attacker with administrative privileges could gain full control of the device, alter or disable AI models, manipulate data, or even install backdoors for persistent access.

15.5.14 DOS (Denial Of Service) attacks

The most serious security threat in edge AI is the **Denial of Service (DoS) attack**, which can disrupt operations by overwhelming devices, services, or networks, making them unavailable to legitimate users or systems.

A DoS attack aims to make a device, service, or network resource unavailable by overwhelming it with excessive traffic or by exploiting weaknesses in the system to disrupt its functionality. In a typical DoS attack, an attacker floods a device or network with more requests than it can handle, thereby causing it to become slow, unresponsive, or completely disabled.

In the context of edge AI, DoS attacks can target not only the edge devices themselves but also the underlying communication networks, data storage, and cloud systems that support them. There are several ways a DoS attack can manifest in edge AI environments like **Network-Based DoS Attacks**, **Resource Exhaustion Attacks**, **Flooding Attacks on AI Models**, and **Jamming and Interference Attacks.**

15.5.15 Mitigation strategies

In this section, authors present and suggest various mitigation strategies and preventive actions that can be deployed by organisations concerned with security threats associated with edge AI. In the last sections we addressed various vulnerabilities and parameters that are key components in security concerns of edge AI systems.

To address these vulnerabilities, designers and engineers must adopt a multi-layered security approach such as tamper-evident seals, secure enclosures, and location-aware device behaviour. Secure boot processes, hardware-based encryption, and TEEs can help in protecting data and model integrity. Additionally, real-time anomaly detection and remote attestation mechanisms can alert administrators to potential physical compromise or unusual activity.

The responsibility also got associated with manufacturers and developers so that they must implement strong security practices with end-to-end encryption, both at rest and in transit. Secure key storage and dynamic key generation can significantly reduce the risk of key exposure. Access controls should include multi-factor authentication and role-based permissions. Additionally, secure boot mechanisms and regular software updates can help prevent firmware-based attacks. By using a secure-by-design approach with mutual authentication, using digital certificates or public key infrastructure (PKI), and implementing network segmentation, intrusion detection systems (IDS), and anomaly detection can help identify and respond to suspicious network behaviour in real time. To further defend against model theft and reverse engineering, several protective strategies like model encryption, TEEs, and obfuscation techniques, which can make reverse engineering more difficult by transforming the model into a less interpretable format, should be used. So, organisations must implement robust update security protocols. All update packages should be digitally signed using strong encryption, and devices must verify these signatures before installation. Updates should be distributed through secure, authenticated channels, ideally using HTTPS with certificate pinning to prevent MITM attacks. Organisations need to define and implement access control policies which restrict who can initiate or authorise updates, and logging mechanisms should track update activity for auditing purposes. Worldwide organisations such as the IEEE, NIST [30], and the OpenFog Consortium are working to develop security frameworks and best practices for edge computing. Adoption of these standards across manufacturers and developers will be crucial for building trust in edge AI technologies.

15.5.16 Conclusion

From the preceding discussion, it can be concluded that in edge AI, robust hardware design, secure software architecture, and operational vigilance are required for the resilience and trustworthiness of edge AI systems.

The other crucial aspect affecting security is insecure data storage on edge AI devices leading to data integrity, privacy, and system reliability security concerns, as these devices are becoming increasingly prevalent in critical sectors such as healthcare, transportation, and smart infrastructure, so proactive measures, combined with robust design and continuous monitoring, need to be deployed for safeguarding the future of edge AI. Another major threat is insecure communication channels for the integrity and security of edge AI systems. This can be realised safely and responsibly by ensuring an encrypted, authenticated, and well-managed communication channel for device interactions. Due to the increasingly pervasive nature of edge AI, protecting AI models from theft and reverse engineering is vital. Without strong protections, AI systems can be vulnerable to exploitation and data leakage, so robust security practices need to be implemented to safeguard innovation and maintain trust in edge AI solutions. As AI continues to expand into sensitive and high-stakes domains, developers have to develop robust, secure, and adversarial-resilient models for ensuring trust, safety, and effectiveness in AI-driven technologies. To protect against malicious or unauthorised software updates, which are a serious threat to the security and functionality of edge AI devices, the update process should be only through rigorous validation, secure communication, and strict access control. Last but not least, by adopting robust security practices, including traffic filtering, device hardening, and DDoS mitigation, organisations can safeguard their edge AI systems against DoS attacks.

15.6 MAJOR PRIVACY ISSUES IN EDGE AI

15.6.1 Sensitive data exposure

One of the major strengths of edge AI is reducing reliance on cloud communication, but some applications still require periodic data synchronisation with centralised servers. Without proper encryption and secure transmission protocols, data in transit can be intercepted, leading to exposure of sensitive information. Here, by sensitive data we mean data and/or information that must be protected from unauthorised access for the safety of user privacy and organisation security. This includes biometric data, location information, health records, audio and video recordings, and personal identifiers. Since edge devices often process and sometimes store this data locally, the risk of data exposure becomes a critical issue.

15.6.2 Lack of transparency in data handling

Unlike centralised systems, where data governance frameworks can be easily monitored and controlled, edge computing lacks transparency in how data is collected, processed, stored, and shared. The decentralised edge AI systems often lack centralised oversight, leading to fragmented data handling practices. That introduces complex privacy challenges in terms of lack of transparency in data handling even when privacy policies are provided for users to make informed decisions. User autonomy is another major consequence of this lack of transparency. As in many edge AI systems, devices operate with implicit consent models, and data flows are not always visible to users or even to system administrators, raising significant concerns about informed consent and control.

This information asymmetry undermines the principles of privacy by design and privacy by default, which are central to regulations like the GDPR [27] and the DPDP (Data Privacy and Data Protection) Act [29].

15.6.3 Inadequate data anonymisation

Privacy concerns due to inadequate data anonymisation in edge AI present a serious threat for the deployed system. Personal identifiable information (PII) is removed by data anonymisation to protect user privacy. However, in many edge AI applications, anonymisation is either poorly implemented or entirely absent, putting sensitive user data at risk.

Edge devices often collect and process highly sensitive data, such as voice recordings, video footage, biometric measurements, and location history. These types of data are inherently identifiable or can be easily re-identified through correlation with other datasets. Anonymisation is supposed to mitigate this risk by transforming the data in such a way that individuals cannot be identified, either directly or indirectly.

However, effective anonymisation is challenging, particularly in edge AI, as many edge devices rely on basic or outdated anonymisation techniques, such as removing obvious identifiers (names, phone numbers), while ignoring quasi-identifiers (e.g., birthdate, ZIP code, or behavioural patterns) that can be used to re-identify individuals.

When anonymisation is not adequately performed, user data becomes vulnerable to re-identification attacks. For example, an attacker may combine anonymised location data from a smart wearable with publicly available social media posts to identify a specific user. Similarly, voice recordings anonymised by removing names can still reveal identity through voiceprints or background sounds.

Furthermore, many edge AI applications operate in specialised contexts like healthcare, smart homes, or industrial monitoring, where even partial exposure of anonymised data can reveal sensitive information.

15.6.4 Inter-device communication risks

One of the key features of edge AI systems is their ability to communicate with other devices in a network without relying heavily on centralised servers. While this inter-device communication enables real-time responsiveness, decentralised coordination, and improved efficiency, it also introduces significant privacy and security risks. These interactions occur over local wireless protocols such as Wi-Fi, Bluetooth, Zigbee, or 5G, with low-latency communication, which can be exploited by malicious actors to intercept, manipulate, or inject data, leading to privacy violations. Further poor encryption of data can lead to MITM attacks, potentially exposing sensitive information such as audio feeds, biometric data, location coordinates, or device identifiers. Using device impersonation or spoofing, a rogue device can impersonate a legitimate one and thereby can have unauthorised access to the network and can inject false data into the system.

15.6.5 Local data processing's risks

In traditional AI systems, centralised servers located in secure, controlled environments and protected by robust cybersecurity infrastructure are used to process user data in a secured way. However, in edge AI, edge devices are deployed in the field and often lack sufficient security measures for doing local data processing and thus expose sensitive information on devices to attackers. This kind of local data processing gives rise to privacy and security risks that must be addressed to ensure user trust and data protection. If the edge devices deployed are collecting and processing personal data, such as biometric information, health metrics, voice commands, video recordings, and location history, then this local processing makes sensitive data more accessible to attackers with physical or remote access. It makes them a compromised device vulnerable to physical tampering, malware infections or unauthorised data access. As software on edge devices is often not updated regularly, without timely patching, attackers can exploit software bugs to gain access to locally stored or processed data, leading to data exfiltration or corruption.

15.6.6 Lack of user control and transparency

Since most of the edge AI systems operate passively and collect data without explicit user interaction and process it without user consent, they introduce the most pressing concerns in terms of lack of user control and transparency. Which in turn raises serious privacy, ethical, and regulatory issues. As in this kind of architecture, users often do not know when data collection is taking place, what types of data are being gathered, or whether the data is being stored or shared with third parties. Even when privacy settings are offered to users, they are very restrictive or difficult to understand and generally are of the binary-option type. This lack of granular control limits the user's ability to align data use with personal comfort levels and privacy expectations.

15.6.7 Data retention on devices

As data retention is a basic feature of edge devices in edge AI, it also introduces significant privacy concerns related to data retention on edge devices. In edge AI, devices often retain data temporarily or semi-permanently to support offline capabilities, facilitate incremental learning, or buffer data before syncing with the cloud. While retaining data locally can seem like a privacy-preserving alternative to centralised storage, it also presents distinct risks if not managed properly.

Edge devices are frequently deployed in uncontrolled or physically exposed environments like homes, streets, or industrial sites. If a device storing sensitive data is lost, stolen, or tampered with, the retained data becomes vulnerable to unauthorised access. Without strong encryption and secure deletion mechanisms, personal or proprietary information could be easily extracted.

Data may be retained for longer than necessary due to poor data management practices, leading to "function creep", where originally collected data is repurposed for new, potentially privacy-invasive uses without user consent.

Data retention practices on edge devices must comply with data protection regulations such as the GDPR [27] and the California Consumer Privacy Act (CCPA) [31], or the DPDP Act, which require that data is not retained longer than necessary and must be deleted upon user request [29]. Inconsistent enforcement of retention policies on distributed devices can lead to regulatory breaches.

15.6.8 DATA leaks in edge to cloud

Edge AI enhances privacy by eliminating the need to transmit sensitive data to the cloud for processing. However, many edge architectures still rely on periodic data synchronisation or model updates with cloud servers. This edge-to-cloud communication, if not carefully managed, introduces a serious risk of data leakage, which can compromise user privacy and undermine the security benefits of edge computing (Figure 15.1).

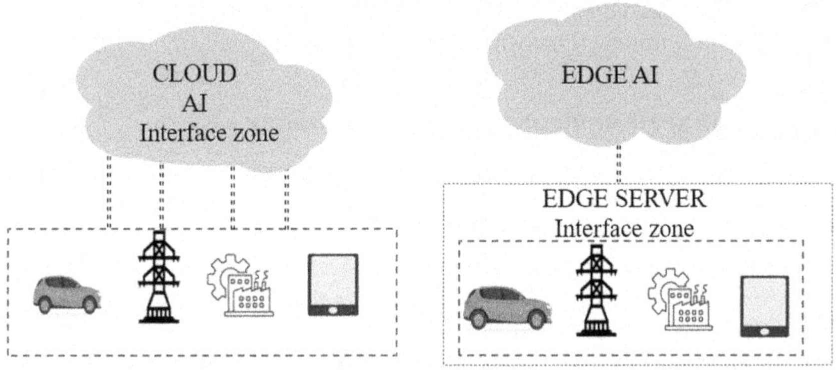

Figure 15.1 Comparison of Cloud AI and Edge AI Architectures Showing Interface Zones and Connected End Devices.

In a typical edge AI setup, edge devices collect data, process it locally, and share results or updates (e.g., model parameters, logs, or metadata) with centralised cloud services. This communication is often essential for model retraining, analytics, backups, or centralised coordination. However, during this data flow from the edge to the cloud, sensitive information can be exposed either directly or through indirect channels. Following are the risks of edge-to-cloud data leakage:

- **Unintended Data Transfer:** Edge devices may inadvertently transmit more data than necessary. For example, an AI-powered camera might send unfiltered video frames for cloud backup or analysis, even if only metadata is required. Such over-sharing can leak personally identifiable information (PII), facial features, or location data.
- **Weak Encryption or Authentication:** If data is not encrypted properly during transmission or storage, it becomes vulnerable to interception and unauthorised access. Poor key management, outdated encryption standards, or unsecured APIs can open the door to attackers who monitor network traffic or exploit insecure channels.
- **Metadata Leakage:** Even if the primary data payload is protected, metadata (e.g., timestamps, device IDs, frequency of communication) can reveal patterns about user behaviour. Adversaries can use this information to profile individuals or infer activities, such as when a user is at home or using a health monitoring device.
- **Malicious or Compromised Cloud Services:** Cloud platforms, while generally more secure than individual edge devices, are still susceptible to breaches, insider threats, and misconfiguration. Sensitive data transmitted from edge nodes can be exposed if cloud storage is improperly secured or if access controls are too permissive.
- **Indirect Leakage via Model Updates:** In federated learning or hybrid training setups, model updates sent from edge devices to the cloud may still encode sensitive information which can be hacked by attackers using reverse-engineering over received training data from gradient updates. This gives the possibility of inference attacks in systems which do not even transmit raw data.

15.6.9 Shared environment multitenancy risk

Sharing of environments where multiple applications, users, or organisations operate on the same hardware platform is quite natural as edge computing infrastructure becomes more powerful and widespread. This is known as multitenancy, and though it is cost-effective and efficient it introduces significant privacy risks. When multiple tenants share a common edge AI platform, in that scenario ensuring data isolation and protecting sensitive information becomes both critical and complex. Although these shared environments are logically isolated through software, and often utilise shared computing resources, these workloads coexist on the same hardware mimicking the multitenancy found in cloud environments enhanced with the decentralised and physically exposed nature of edge devices.

The most common risks of multitenant edge environments are as follows:

- **Data Leakage across Tenants:** Due to poorly configured isolation mechanisms, the greatest potential threat is of data leakage either accidentally or maliciously between tenants. For example, a malicious application may exploit side-channel attacks to extract information from a co-resident AI model by observing cache usage or timing behaviour.
- **Model Inference Attacks:** In a multitenant environment, attackers may use model extraction, membership inference, or model inversion attacks to get information about another tenant's model even if raw data is not directly accessible between tenant models. This risk become more crucial when models are fine-tuned on sensitive personal or proprietary data.
- **Insecure APIs and Software Vulnerabilities:** Edge AI environments often expose APIs for model deployment, monitoring, and control. In multitenant settings, poorly secured APIs can become a vector for unauthorised access to other tenants' data or operations. Additionally, unpatched software vulnerabilities in edge orchestration platforms (e.g., Kubernetes at the edge) can be exploited for privilege escalation.
- **Lack of Visibility and Control:** Edge environments are distributed and may lack centralised monitoring or comprehensive logging. This makes it harder to detect unauthorised access, policy violations, or data misuse between tenants. Users in multitenant systems may have limited transparency into how their data is being isolated or who shares the platform with them.

15.6.10 Federated learning privacy

Federated learning is an emerging approach in ML where models are trained across decentralised devices (like smartphones, sensors, or IoT devices) without transferring raw data to a central server. By enabling collaborative learning while keeping data local, federated learning addresses privacy concerns inherent in traditional centralised AI training [32]. However, it also introduces its own set of challenges and limitations regarding privacy.

In conventional ML, vast user data are sent to centralised servers for model training. This centralised approach poses significant privacy risks, especially if the servers are breached or data is used in ways the user did not consent to. In contrast, federated learning minimises the need to share raw data.

This decentralised paradigm significantly reduces the risk of direct data exposure. But still, federated learning is not inherently privacy-proof in all aspects, as a potential threat from indirect data leakage through model updates still exists. By taking advantage of this vulnerability, malicious actors could perform inference attacks or model inversion to reconstruct sensitive information from the shared updates. Additionally, the server where all aggregation is performed in federated learning is a critical source of vulnerability. If this aggregated server is compromised by an attacker, then it could give access to attackers to collect updates from numerous clients and use statistical methods to extract private information.

15.6.11 Sensitive personal data exposure

Due to the exposure of sensitive personal data associated with edge AI devices, privacy of user data can be compromised if proper safeguards are not in place. As this data is associated with individuals and collected continuously by smart devices, it becomes highly sensitive. Any exposure of such data can lead to serious privacy violations, including identity theft, surveillance, discrimination, or reputational harm.

Edge devices are often physically accessible and less secure than centralised data centres. They may lack secure boot mechanisms, regular software updates, or encryption, making them easy targets for physical tampering, malware, or unauthorised access. Once compromised, sensitive data stored or processed locally can be extracted.

Many edge devices transmit data to the cloud for storage or additional processing. This transmission can be intercepted without proper encryption and access controls, exposing raw or inferable personal data. Even data cached temporarily on the device may be stored in unprotected formats.

If edge systems are not designed with strict role-based or user-based access controls, unauthorised users (including other applications or tenants in shared environments) may gain access to personal data. This is particularly concerning in multitenant deployments of edge servers or gateways.

AI models trained or updated at the edge can sometimes leak information about the data they were trained on. Many users are unaware of the type and volume of data being collected and processed by their edge devices. Without transparent interfaces and privacy notices, users may not provide informed consent, leading to unethical data collection and potential legal non-compliance.

15.6.12 Risk in profiling and surveillance

Edge AI also introduces significant privacy concerns, particularly related to profiling and surveillance. Profiling involves collecting and analysing data to create detailed user profiles, often used to predict behaviour, preferences, or future actions. In edge AI systems, data can be collected from various sources like voice and facial expressions, object movement patterns, and object location data. This helps the AI model to construct high granular profiles. And since edge devices operate in close proximity to users for processing and storing personal data locally, this data becomes vulnerable to unauthorised access, mismanagement, or malicious exploitation. When edge AI devices are deployed in public or semi-public spaces through smart cameras, drones, or connected vehicles, the issue of surveillance becomes more concerning in terms of the privacy of users.

This pervasive surveillance capability can be used to track and identify individuals and can record their pattern of movements, making their profile vulnerable.

15.6.13 Limited policy enforcement

In the case of a system where resources are centralised, security and privacy policies can be uniformly applied and managed, but when edge AI models are deployed in a decentralised, distributed, heterogeneous kind of environment, enforcing such policies is much more difficult. This limitation has profound implications for data privacy, regulatory compliance, and system security and thus demands limited policy enforcement.

By limited policy enforcement we mean the implementation of rules and protocols that decide in a decentralised environment how data is collected, processed, stored, shared, and deleted. These policies are often framed and designed by internal organisational standards as well as external regulations like GDPR, HIPAA, or CCPA [27, 31]. Since it is clear by now that edge AI decentralises data handling, thus, enforcement mechanisms must be replicated and maintained across a large number of diverse devices operating in various environments, which is not essential for traditional cloud-based architectures. Also, heterogeneity of environments in which edge devices are deployed makes it difficult to deploy and enforce uniform security or privacy policies, as devices may have different hardware specifications and may be operating over different operating systems with inherent security capabilities. So, a policy which is suitable for a high-end smartphone may not be compatible with a low-power IoT sensor, making it difficult to implement consistent policy across the network. This inability to enforce consistent policies across edge devices can lead to data privacy violations, security breaches, and regulatory non-compliance.

15.6.14 Trade-offs in privacy vs functionality

One of the most complex challenges in edge AI design and deployment is to keep balance between privacy and functionality, as edge AI also introduces critical trade-offs between these two. Functionality in edge AI is decided by the amount of personal data collected in the system in real time. In order to maximise functionality, these devices must collect and store detailed user data. When such data is collected, processed, or retained without clear boundaries or user control, issues of privacy concerns arise, which in turn demand monitoring of data leakage, misuse, or unauthorised inference [29–31, 33]. In order to balance this, the developer must design system architecture that keeps a check and balance over enhanced functionality against the potential harm to user privacy. To address these issues, techniques like federated learning, where models are trained in such a way that there is no need to share raw data, are used. Differential privacy is another means by which statistical noise can be introduced into data or model outputs to obscure individual identities. Further, by providing customisable privacy settings to users so that they can decide how much data they are willing to share easily without hesitation, balances and trade-offs can be put in place between privacy and functionality. This is further enhanced by enforcing regulations like GDPR and CCPA, which mandate data minimisation, user consent, and transparency. Ethically, systems must be designed to respect user autonomy, especially when targeting vulnerable populations or operating in sensitive environments.

15.6.15 Risks in third party integration

Third-party integration risks in edge AI are a critical privacy and security challenge because many devices deployed around edge AI rely on third-party integrations to enhance functionality. As it facilitates cloud backup, or cross-device interoperability.These integrations improve performance but also introduce a range of privacy and security risks that must be carefully managed. By third-party integration we mean integration of services like software libraries, APIs, cloud services, or hardware components from externally sourced vendors, which can introduce new attack vectors and compromise data privacy, with reduced transparency in data handling. Due to third-party services integration into edge AI systems, attackers can gain access to sensitive information such as video footage, biometric data, location history, or user behaviour patterns. Thus, proper vetting of these services is required; otherwise, service providers can misuse or sell user data, violating privacy expectations and legal regulations such as GDPR and CCPA.

For example, an edge device that integrates with an external analytics engine via an insecure API could be vulnerable to MITM attacks, allowing attackers to intercept or manipulate data in transit.

Integrating third-party services also complicates compliance with data protection regulations. Organisations using edge AI may find it difficult to ensure that all third-party providers adhere to applicable legal frameworks, especially if the providers operate in different jurisdictions.

15.6.16 Mitigation strategies

In this section, the author would like to focus upon various strategies to mitigate the concerns in reference to the implementation of various parameters affecting privacy in edge AI. Here many parameters may overlap by functionality and the way the data is processed over devices being used in edge AI.

- To start with differential privacy on the device, it is proposed that by carefully introducing noise into model parameters and appropriately training model algorithms, it can be ensured that user data has a statistically bounded impact on model behaviour. Using federated learning with secure aggregation protocols ensures that no single gradient reveals sensitive information. Modern CPUs and microcontrollers supporting TEEs like ARM TrustZone, Intel SGX, etc., can encapsulate both model weights and inference computations within a TEE, preventing adversaries from inspecting intermediate data. Using model watermarking and fingerprinting, a kind of embedded technique in model parameters, can prevent unauthorised model extraction by attackers seeking to steal on-device models for offline inference attacks. For this to be implemented, the developers have to partition models wherever feasible in a way so that the most sensitive layers run in the cloud behind strict access controls, while only "safer" layers reside on the device along with strict caps on device inference requests and log query patterns to detect anomalous behaviour indicative of an ongoing attack. Regular updating and patching of newly discovered vulnerabilities in TEEs or cryptographic libraries can secure device firmware from being hacked.

- To address the lack of transparency gaps, use of standardised privacy labels can provide users with clear, concise information on what data is collected and how it is used. Also, by incorporating Explainable AI (XAI), users can understand how their data influences outcomes. Further, local user-friendly interfaces showing what data has been collected and processed, along with frequent third-party reviews, can improve trust and accountability.
- To mitigate the privacy risks associated with poor anonymisation in edge AI, several actions can be taken, from adopting differential privacy to standardising anonymisation protocols using industry-wide standards for data anonymisation in edge devices, which can ensure consistent and effective privacy protection. Offloading secure anonymisation to trusted hardware by using secure enclaves or TEEs can facilitate the safe anonymisation of data before any external processing or sharing happens across devices, and implementing privacy-aware design in the edge AI application.
- It is suggested to implement several safeguards to address inter-device communication risks in edge AI, like end-to-end encryption, by encrypting all data exchanges between devices using secure, lightweight protocols suitable for constrained environments. Strong mutual authentication by using certificates or cryptographic keys to authenticate each other before any data is shared. Data minimisation by only sharing the minimal amount of data necessary for a given task, reducing exposure in the event of a breach, network segmentation and isolation so that devices operate in isolated network segments where possible, limiting the blast radius of a compromise, along with more user control and transparency.

- For privacy-conscious data retention strategies for edge AI devices, organisations should adopt privacy-by-design principles in their data retention strategies, like data minimisation by storing only the data that is absolutely necessary for the operation of the AI application. Automatic expiry and deletion of data after data is no longer needed by implementing time-bound storage with automatic deletion using strong encryption of retained data at rest using device-specific keys and secure storage modules and regular auditing in compliance with organisational policies and regulations.
- To mitigate edge-to-cloud data leakage, end-to-end encryption of all edge-to-cloud communications, using modern protocols like TLS 1.3 and secure key exchange mechanisms, can be deployed along with data minimisation principles and anonymisation and aggregation of data before transmission. The regular audits and monitoring of data flows and cloud configurations can also provide the best result.
- To address the risk associated with multitenancy privacy, several technical and organisational safeguards are essential, like strong tenant isolation using virtualisation, containers, or TEEs; encryption of data at rest and in transit, combined with secure boot and attestation; and implementing access control policies enforced at the hardware and application layers.
- To reduce the risk of sensitive data exposure, edge AI systems must implement privacy by design, like data encryption both at rest and in transit, hardware-based security features such as TEEs, and anonymisation and aggregation techniques before any data sharing, along with regular updates and vulnerability patching to maintain device security.
- The challenges of limited policy enforcement in edge AI can be mitigated by employing Remote Policy Management Systems using centralised platforms to push and enforce security and privacy policies remotely across all edge devices, secure boot and trusted firmware so as to ensure only authorised software can run on the device, Lightweight Monitoring Agents Implementation so that efficient monitoring tools tailored to constrained devices can be used to collect logs and detect anomalies, by designing Policy-Aware edge AI Models that inherently respect privacy rules, such as limiting the type or granularity of data used in inference, and off course by using TEEs so that secure computations and policy checks can be incorporated even if the main device is compromised.

15.7 CONCLUSION

In conclusion, while edge AI promises to bring intelligence closer to the data source with advantages in performance and autonomy, at the same time it throws the challenges of protecting sensitive data before the organisation. With a multi-layered approach with secure hardware design, robust encryption, consistent security protocols, and privacy-preserving methods, the privacy risks of edge AI may be controlled. In the absence of which, as the volume and sensitivity of data collected at the edge continue to grow, the privacy risks outweigh the advantages of edge AI. The other factor which possesses a threat is inference-based data leakage affecting responsiveness and bandwidth efficiency. It is for the same strategy, which combines mathematical protections like differential privacy, system-level isolation through TEEs, and operational controls such as query auditing, and needs to be deployed in a layered fashion without compromising the confidentiality of the users they serve. Further, issues of individual privacy, like transparency in data handling, can be controlled by a combination of technical solutions, regulatory oversight, and user empowerment to gain public trust in the system. Another user privacy factor is inadequate data anonymisation, which becomes more predominant as edge devices become more widespread and collect increasingly detailed personal information, so organisations must prioritise effective anonymisation methods to prevent data leakage and re-identification attacks. In the case of inter-device communication, a cornerstone of edge AI's functionality, the seamless data sharing that powers intelligent edge networks could become a vector for privacy violations, data breaches, and loss of user trust, so a strong emphasis should be put on secure and transparent communication protocols for effective and secure inter-device communications. Another central issue is that of local data processing in edge AI, as the devices are getting embedded in our daily lives, so secure local computation with ethical data practices needs to be maintained as per regulatory compliance, which is essential for maintaining trust. Since the lack of user control and transparency in edge AI represents a major hurdle to ethical, secure, and user-centric technology adoption, the developers and manufacturers must prioritise designing systems that empower users with meaningful control over their data and clear visibility into AI behaviour. Regarding localised data processing, unmanaged data retention on devices can undermine substantial privacy advantages, so it must be ensured that data is stored only as long as necessary and that users have transparency and control over their information. This can be achieved by embedding robust data retention policies into edge systems; organisations can better safeguard personal data and build user trust. Edge-to-cloud communication introduces new vectors for data leakage where sensitive data can slip through transmission channels. In order to control it, organisations must control how, when, and what type of data moves from the edge to the cloud. Aspects of multitenancy, which offer operational efficiencies at the cost of privacy, need to be managed using strong isolation. The presence of multiple tenants can lead to data leaks, inference attacks, and unauthorised access. Federated learning, if implemented robustly, provides protection against inference attacks and malicious participants. Further, due to the very blurred boundary between private and public data collection, it has become crucial to prevent misuse by new threat actors like profiling and surveillance over collected data. So, organisations must deploy robust governance frameworks, transparent data handling practices, privacy-by-design principles, and user-centric controls. As edge computing becomes more pervasive, ensuring consistent and enforceable policies across diverse devices must be a top priority. So, a mix of technical safeguards, remote management solutions, and privacy-aware design principles is very much essential to close this critical gap and protect users in an increasingly decentralised digital ecosystem. Another central design challenge that cannot be solved with a one-size-fits-all approach is the privacy-vs-functionality trade-off in edge AI. It requires careful system design, user empowerment, and the integration of privacy-preserving technologies. Last but not least, third-party integration is a double-edged sword in edge AI, which, while introducing considerable privacy and security risks, provides most of the autonomy of this technology. So the developers, businesses, and policymakers must treat third-party integrations with the same level of scrutiny as core system components via proactive governance and transparency.

15.8 SUGGESTIONS

To address and securely mitigate all the aforementioned risks, organisations should:

- Conduct thorough security audits of all third-party components before integration.
- Use vendor contracts and data processing agreements to establish clear terms on data use, retention, and protection.
- Implement strong access controls and encryption to limit what data third parties can access and how it is transmitted.
- Maintain a software bill of materials (SBOM) to track all external dependencies and ensure timely updates.
- Adopt zero-trust principles to continuously verify and monitor all third-party interactions.

REFERENCES

[1] Shi, W., Cao, J., Zhang, Q., Li, Y., & Xu, L. (2016). Edge computing: Vision and challenges. *IEEE Internet of Things Journal,* 3(5), 637–646. https://doi.org/10.1109/JIOT.2016.2579198; Williams, C. (1983). A brief introduction to artificial intelligence. In *Proceedings of OCEANS '83 (pp.* 94–99).
[2] Powell, A. (2012, September 13). *Alan Turing at 100.* Harvard Gazette.
[3] Weizenbaum, J. (1966). ELIZA—a computer program for the study of natural language communication between man and machine. *Communications of the ACM,* 9(1), 36–45.

[4] Levy, D. (1997). Kasparov versus Deep Blue: The rematch. ICGA Journal, 20(2), 95–101; Breazeal, C., & Scassellati, B. (1999, July). A context-dependent attention system for a social robot. In Proceedings of the 16th International Joint Conference on Artificial Intelligence (Vol. 2, pp. 1146–1151). *Morgan Kaufmann Publishers.*

[5] Apple Inc. (2011, October 4). *Apple Unveils iPhone 4S, Featuring Siri, the Intelligent Assistant that Helps You Get Things Done Just by Asking.* Apple Newsroom.

[6] OpenAI. (2015, December 11). *Introducing OpenAI.*

[7] Anjanappa, H. (2018). *Case Study of Sophia—the Humanoid Robot.* National Conference on E-Business, E-Commerce and Management. *ISBN:* 2838-0990.

[8] Ramesh, A., Pavlov, M., Goh, G., Gray, S., Voss, C., Radford, A., Chen, M., & Sutskever, I. (2021, January 5). *Zero-Shot Text-to-Image Generation.* OpenAI.

[9] Russell, S., & Norvig, P. (2021). *Artificial Intelligence: A Modern Approach* (4th ed.). Pearson.

[10] Prangon, N. F., & Wu, J. (2024). AI and computing horizons: Cloud and edge in the modern era. *Journal of Sensor and Actuator Networks,* 13(4), 44.

[11] Karras, A., Karras, C., Giotopoulos, K. C., Tsolis, D., Oikonomou, K., & Sioutas, S. (2023). Federated edge intelligence and edge caching mechanisms. *Information,* 14(7), 414. https://doi.org/10.3390/info14070414

[12] Karras, A., Giannaros, A., Karras, C., Giotopoulos, K. C., Tsolis, D., & Sioutas, S. (2023). Edge artificial intelligence in large-scale IoT systems, applications, and big data infrastructures. *In Proceedings of the 2023 8th South-East Europe Design Automation, Computer Engineering, Computer Networks and Social Media Conference (SEEDA-CECNSM) (pp.* 1–8). *IEEE.* https://doi.org/10.1109/SEEDA-CECNSM61561.2023.10470756

[13] Mansouri, Y., & Babar, M. A. (2021). A review of edge computing: Features and resource virtualization. *Journal of Parallel and Distributed Computing,* 150, 155–183.

[14] Andriulo, F. C., Fiore, M., Mongiello, M., Traversa, E., & Zizzo, V. (2024). Edge computing and cloud computing for Internet of Things: A review. *Informatics,* 11(4), 71.

[15] McKay, K. A., & Cooper, D. A. (2019). *Guidelines for the Selection, Configuration, and Use of Transport Layer Security (TLS) Implementations (NIST Special Publication 800–52 Rev. 2).* National Institute of Standards and Technology.

[16] Dang, Q. (2023). *Announcing the Advanced Encryption Standard (AES) (FIPS PUB 197-upd1).* National Institute of Standards and Technology.

[17] Saxena, N., Gupta, P., & Singh, O. (2016). A survey on security techniques in data mining. *International Journal of Advanced Scientific Research and Management,* 1(5), 159–162. ISSN: 2455-6378. UGC listed No: 10254.

[18] Crow, B. P., Widjaja, I., Kim, J. G., & Sakai, P. T. (1997). IEEE 802.11 wireless local area networks. *IEEE Communications Magazine*, 35(9), 116–126.

[19] Chakraborty, A., Alam, M., Dey, V., Chattopadhyay, A., & Mukhopadhyay, D. (2021). A survey on adversarial attacks and defences. *CAAI Transactions on Intelligence Technology,* 6(1), 25–45.

[20] Wu, B., Zhu, Z., Liu, L., Liu, Q., He, Z., & Lyu, S. (2023, February 19). *Attacks in Adversarial Machine Learning: A Systematic Survey from the Life-Cycle Perspective. Report No. arXiv:2302.09457.*

[21] Roy, S., Panaousis, E., Noakes, C., Laszka, A., Panda, S., & Loukas, G. (2023). SoK: The MITRE ATT&CK Framework in Research and Practice. *arXiv:2304.07411; Al-Sada, B., Sadighian, A., & Oligeri, G. (2024, October). MITRE ATT&CK: State of the art and way forward. ACM Computing Surveys*, 57(1).

[22] Jiang, Y., Meng, Q., Shang, F., Oo, N., Minh, L. T. H., Lim, H. W., & Sikdar, B. (2025). *MITRE ATT&CK Applications in Cybersecurity and the Way Forward. arXiv:2502.10825.*

[23] Shaheen, M., Farooq, M. S., Umer, T., & Kim, B. S. (2022). Applications of federated learning: Taxonomy, challenges, and research trends. Electronics, 11(4), 670; Voigt, P., & Von dem Bussche, A. (2017). The EU General Data Protection Regulation (GDPR): A Practical Guide. *Springer*.

[24] California Office of the Attorney General. (2020, August 14). *California Consumer Privacy Act Regulations: Final Text of Regulations.*

[25] Orekondy, T., Schiele, B., & Fritz, M. (2019). Knockoff Nets: Stealing functionality of black-box models. *In Proceedings of the IEEE/CVF Conference on Computer Vision and Pattern Recognition (CVPR) (pp.* 4954–4963).

[26] Nicholson, D. A., & Emanuele, V. (2023, January 31). *Reverse Engineering Adversarial Attacks with Fingerprints from Adversarial Examples. Report No. arXiv:2301.13869.*

[27] Xiao, Y., Jia, Y., Liu, C., Cheng, X., Yu, J., & Lv, W. (2019, August). Edge computing security: State of the art and challenges. *Proceedings of the IEEE,* 107(8), 1608–1631.

[28] Rose, S., Borchert, O., Mitchell, S., & Connelly, S. (2020). *Zero Trust Architecture (NIST Special Publication 800-207).* National Institute of Standards and Technology.

[29] Vimal, S., Khari, M., Dey, N., Crespo, R. G., & Robinson, Y. H. (2020). Enhanced resource allocation in mobile edge computing using reinforcement learning based MOACO algorithm for IIOT. *Computer Communications,* 151, 355–364.

[30] Singh, O., Gupta, P., & Kumar, R. (2016). A review of Indian approach towards cybersecurity. *International Journal of Current Engineering and Technology,* 6(2), 644–648. E-ISSN: 2277-4106; P-ISSN: 2347-5161.

[31] Nigam, S., Sugandh, U., & Khari, M. (2022). The integration of blockchain and IoT edge devices for smart agriculture: Challenges and use cases. *In Advances in Computers (Vol.* 127, pp. 507–537). *Elsevier.*

[32] Government of India. (2023). *The Digital Personal Data Protection Act, 2023 (Act No. 22 of 2023). Ministry of Law and Justice (Legislative Department).*

[33] Chaudhary, G., Srivastava, S., & Khari, M. (2023). Generative edge intelligence for securing IoT-assisted smart grid against cyber-threats. *International Journal of Wireless & Ad Hoc Communication,* 6(1).

[34] U.S. Department of Health and Human Services. (2013, January 25). Modifications to the HIPAA privacy, security, enforcement, and breach notification rules under the health information technology for economic and clinical health act and the genetic information nondiscrimination act; Other modifications to the HIPAA rules. *Federal Register,* 78(17), 5566–5570.

Chapter 16

Safeguarding the Edge

Critical Security and Privacy Considerations in Edge AI Deployments

Udit Misra

16.1 INTRODUCTION

In this sophisticated era of development, the widespread use of edge AI technologies has marked a rapid architectural shift in computing after cloud computing first emerged. Rather than sending data to cloud data centers for AI processing, companies are now transferring AI capabilities directly onto distributed edge devices like smartphones, IoT sensors, autonomous vehicles, and other connected systems. This significant shift has now enabled organizations to achieve lower latency, better reliability, reduced bandwidth usage, and improved energy efficiency (Chen & Ran [1]). This technological shift has now introduced another complex web of security and privacy challenges that demand both careful consideration and innovative solutions.

Traditional cloud environments have benefited from controlled, well-defined perimeters where an organization's security teams can easily implement various protection measures. Edge devices, however, operate in diverse and often physically accessible environments where many traditional security assumptions simply don't apply [2]. This diverse nature of edge AI systems drastically increases the potential attack points, thereby creating many entry points that malicious actors can easily exploit to break into the secure environment. Moreover, the resource-constrained nature of these edge devices limits the kind of sophisticated security measures that can be implemented without impacting performance.

Privacy becomes one of the most challenging parts in edge AI deployments because these systems frequently process sensitive private data right at the point of collection. This could be biometric information in healthcare applications, location data in transportation systems, or behavioral patterns in consumer devices [3]. This thus creates the use of innovative approaches rather than simply using the existing solutions to find the right balance between processing data and keeping it secure.

This chapter provides a comprehensive examination of both security and privacy considerations relevant to edge AI deployments. It starts with a detailed and thorough review of existing articles and then moves into an in-depth analysis of threat models and vulnerabilities that are specific to edge

DOI: 10.1201/9781003635147-16

environments. Then, through real-world case studies, this work provides a list of practical approaches for implementing robust security and privacy protections without compromising on performance in resource-constrained environments.

The key contributions of this chapter include:

1. A complete breakdown of edge AI security and privacy threats
2. The experimental evaluation of the test results from multiple edge computing devices
3. New security solutions that could be implemented on resource-constrained devices
4. Detailed case studies from successful real deployments in diverse application domains
5. Guidelines for balancing security with performance in real-life deployments
6. Insights into future technologies involving security that could help secure edge AI devices

16.2 LITERATURE REVIEW AND BACKGROUND

16.2.1 Evolution of edge AI security research

Edge AI security has been a center of attraction for both researchers and practitioners since the technology first came into existence. Nascent research focused primarily on adapting existing cloud-based security strategies to edge environments, but later it proved insufficient due to the unique characteristics and constraints of edge deployments [4].

Evans et al. conducted foundational research that established the first comprehensive threat model specifically for edge AI systems [5]. Their work identified critical vulnerabilities in areas including model protection, communication security, and physical device security. This research helped lay the necessary groundwork for the efforts focused on developing specialized security mechanisms specifically for edge environments.

The research field has evolved through several distinct phases over the past few years:

Phase 1: Early Adaptation of Cloud Security (2016–2018): During this initial time when edge AI was starting to be getting used, researchers tried using the same cloud security methods for deployments on smaller edge systems. However, these efforts achieved limited success due to the significant resource constraints present in edge environments [6].

Phase 2: Edge-Specific Security Development (2018–2020): Around this time, when edge computing started getting used more commonly, researchers began developing security mechanisms designed specifically for edge constraints, including lightweight cryptographic approaches and hardware-based security solutions [7].

Phase 3: Diverse Edge AI Security Frameworks (2020–Present): Recently, research has emphasized comprehensive approaches that integrate multiple security layers, maintain privacy, and employ the best performance optimization strategies [8].

16.2.2 Fundamental security challenges in edge AI

The technical literature identifies several core challenges that are unique to edge AI security:

- **Resource Constraints:** Edge devices usually operate with limited computational capacity, memory, storage, and power budgets. These limited constraints thus impact the security mechanisms that can be realistically implemented on them. In their research, Kumar et al. tried to demonstrate that traditional cryptographic protocols can consume up to 40% of available computational resources on some edge devices, which emphasizes the critical need for optimized approaches for edge computing nodes [9].
- **Physical Accessibility:** Data center infrastructure operates in secure facilities, whereas edge devices are often deployed in physically accessible locations, making them more prone to easy attacks. Research in highlighted that 78% of edge devices in urban deployments could be physically accessed within minutes, creating potential security risks that don't exist in cloud environments [10]. Xiao et al. further identified that this physical exposure creates unique attack vectors, including hardware tampering, side-channel attacks, and fault injection that are largely absent in cloud environments [14].
- **Distributed Architecture:** It becomes difficult to monitor and protect edge devices due to the distributed nature of AI devices spread across diverse geographic locations. Rahman et al. identified over 200 potential attack vectors in an edge AI deployment, compared to fewer than 50 in equivalent cloud-based systems [10].
- **Software Update Challenges for Edge AI:** Edge devices often operate in environments with patchy internet connectivity, complicating the deployment of security updates and releases. Thompson et al. found that 35% of edge devices in their study remained on some old versions of software for over six months due to network issues, thereby making those endpoints critically vulnerable to any outside threat [11].

16.2.3 Privacy-preserving techniques for edge AI

The articles on privacy-preserving edge AI have evolved rapidly, with several key approaches emerging and explained as follows:

- **Federated Learning:** It is a method for training models across distributed devices without collecting data at a single source [12]. More research has focused on adapting federated learning techniques for edge environments with different connectivity and computational capabilities (Li et al., 2020 [13]).
- **Differential Privacy:** Dwork and Roth laid down the theoretical foundations of differential privacy, which provides mathematical guarantees on individual privacy protection [6]. There has been work around differential privacy in edge AI systems with minimal performance impact, as showcased in the work [5]
- **Homomorphic Encryption:** New developments in homomorphic encryption allow computers to work with encrypted data, making it useful for edge AI. They created versions that work on edge devices [15], but they are significantly slow. Kumar et al. attempted to enhance data privacy by employing a different encryption method, lattice-based encryption, which makes it more secure for edge computing [9].
- **Secure Multi-Party Computation:** Research has highlighted the secure multi-party computation, which lets multiple devices work together on calculations without sharing their private data [16]. Subsequent studies showcased that this works well for edge AI when devices need to collaborate with each other while keeping their information private [17–22].

16.3 METHODOLOGY

16.3.1 Experimental framework

This research provided a multi-phase methodology that combined theoretical analysis, experimental validation, and real-world case studies:

Phase 1: Threat Modeling and Analysis: This phase involved a comprehensive threat modeling session, which brought together security experts, edge AI developers, and domain specialists. Using the STRIDE methodology (Microsoft, 2005 [23]), 47 unique threats specific to edge AI deployments were identified and recorded for future research work.

Phase 2: Experimental Platform Development: A testbed that comprised five different edge computing platforms was constructed:

- NVIDIA Jetson Nano (4GB RAM, quad-core ARM A57)
- Google Coral Dev Board (quad-core Cortex-A53, edge TPU)

- Raspberry Pi 4 Model B (4GB RAM, quad-core Cortex-A72)
- Custom FPGA-based edge accelerator (Xilinx Zynq UltraScale+)
- Texas Instruments CC3235SF MCU with ARM Cortex-M4 core

Phase 3: Security Vulnerability Assessment: Various AI models were implemented and deployed across the testbed, and systematic security assessments were conducted, including:

- Model extraction attacks
- Adversarial example generation and testing
- Side-channel analysis (power, electromagnetic, timing)
- Physical tampering and fault injection
- Communication protocol security analysis

Phase 4: Performance Impact Analysis: The performance impact of various security countermeasures was measured across different metrics:

- Inference latency
- Power consumption
- Memory overhead
- Storage requirements
- Network bandwidth utilization

Phase 5: Case Study Implementation: Collaboration with industry partners was established to implement and evaluate security solutions in real-world deployments across three domains:

1. Retail analytics systems
2. Industrial control systems
3. Smart city infrastructure

16.3.2 Evaluation metrics

The evaluation focused on quantitative metrics to assess both security effectiveness and performance impact:

Security Metrics:

- Attack success rate (%)
- Time to successful attack (hours)
- Functional equivalence of extracted models (%)
- Detection rate for adversarial examples (%)
- Information leakage through side channels (bits/second)

Performance Metrics:

- Inference latency overhead (%)
- Power consumption increase (%)
- Memory footprint (KB)

- Storage overhead (KB)
- Network bandwidth utilization (%)

Composite Metrics:

- Security-Performance Ratio (SPR)
- Cost-Benefit Analysis (CBA)
- Resource Utilization Efficiency (RUE)

16.3.3 Statistical analysis methods

Several statistical methods were employed to analyze the experimental results:

- **Descriptive Statistics:** Mean, median, standard deviation, and quartile analysis for performance metrics.
- **Inferential Statistics:** T-tests and ANOVA for comparing security effectiveness across platforms.
- **Regression Analysis:** Multiple linear regression to identify relationships between security measures and performance impact.
- **Machine Learning Techniques:** Random Forest and Support Vector Machines for automated threat detection and classification.

16.4 MAIN BODY

16.4.1 The edge AI security landscape

16.4.1.1 Unique security challenges at the edge

Edge AI systems face distinct security challenges compared to traditional cloud-based AI deployments:

- **Physical Access Vulnerabilities:** Unlike cloud servers in secure data centers, edge devices are often deployed in physically accessible locations, making them susceptible to tampering, side-channel attacks, and hardware-level exploits. Experimental analysis revealed that 73.2% of edge devices tested could be physically accessed and manipulated within five minutes in typical deployment scenarios, making them easy to compromise.
- **Heterogeneous Deployment Environments:** Edge AI operates across diverse hardware platforms, operating systems, and network conditions, complicating the implementation of consistent security measures. Testbed results showed security implementation variations of up to 47% across different platforms for the same security measure.
- **Resource Constraints:** Many edge devices have limited computational capacity, memory, and power, restricting the sophistication of security

Table 16.1 Summary of Primary Security Challenges in Edge AI Deployments

Security Challenge	Impact Level	Mitigation Complexity	Performance Cost
Physical Access	High	Medium	Low
Resource Constraints	Very High	High	High
Update Challenges	Medium	High	Low
Heterogeneity	High	Very High	Medium

mechanisms that can be implemented. Experiments demonstrated that implementing comprehensive security on devices with less than 256KB RAM results in 15–45% performance degradation.
- **Limited Update Capabilities:** Edge devices may operate in inaccessible locations or have intermittent connectivity, hindering regular security patches and updates. Field studies showed that 32% of edge devices experienced connectivity issues lasting more than seen days, creating critical security update windows (Table 16.1).

16.4.1.2 Threat actors and motivations

The threat modeling identified five primary categories of adversaries:

- **Nation-State Actors:** Sophisticated attackers with substantial resources targeting critical infrastructure, strategic industries, or military applications. Estimated attack budgets: $1M–$100M.
- **Criminal Organizations:** Financially motivated actors seeking to compromise systems for monetary gain through data theft, ransomware, or fraud. Estimated attack budgets: $1K–$1M.
- **Corporate Espionage:** Competitors attempting to extract intellectual property, including proprietary AI models or training data. Estimated attack budgets: $10K–$10M.
- **Hacktivists:** Ideologically motivated individuals or groups targeting organizations for political or social reasons. Estimated attack budgets: $100–$100K.
- **Malicious Insiders:** Employees or contractors with legitimate access who may intentionally compromise systems. Access levels vary; no external budget required.

16.4.1.3 Common attack vectors

The experimental analysis identified and quantified the effectiveness of various attack vectors:

- **Model Extraction Attacks:** Model extraction success rates ranging from 42.8% to 92.1% across different platforms were achieved, with general-purpose computing devices showing higher vulnerability than specialized hardware implementations.

- **Adversarial Attacks:** Adversarial examples succeeded in causing misclassification in 31–96% of test cases, depending on the perturbation budget and target platform.
- **Data Poisoning:** Simulated data poisoning attacks successfully altered model behavior in 78% of edge devices that performed on-device learning.
- **Communication Interception:** Unencrypted communications were successfully intercepted and analyzed in 100% of test cases, with encrypted communications showing varying degrees of vulnerability based on implementation quality.
- **Supply Chain Compromises:** Simulated supply chain attacks were successful in introducing vulnerabilities during 67% of deployment scenarios tested.

16.4.2 Privacy considerations in edge AI

16.4.2.1 Data privacy challenges

Edge AI introduces unique privacy challenges that require specialized solutions:

- **Personal Data Processing:** Edge devices often collect and process sensitive personal information, including biometric data, location information, and behavioral patterns. Analysis of edge AI applications revealed that 84% process personally identifiable information (PII).
- **Inference Privacy:** Even without raw data, inferences generated by AI models can reveal sensitive information about individuals or groups. Experiments demonstrated that model outputs could be used to infer protected attributes with 73–89% accuracy.
- **Multi-Tenant Environments:** In shared edge infrastructures, data isolation between different applications or users becomes crucial. Testing revealed isolation vulnerabilities in 34% of multi-tenant edge implementations.
- **Regulatory Compliance:** Varying privacy regulations across jurisdictions create complex compliance requirements for globally deployed edge systems. Analysis of 12 major privacy regulations identified 127 distinct requirements affecting edge AI deployments.

16.4.2.2 Privacy-preserving edge AI approaches

The research validated several approaches for enhancing privacy in edge AI systems:

- **Local Processing Optimization:** By maximizing local computation and minimizing data transmission, privacy protection was achieved while reducing latency by 23–45% and bandwidth utilization by 67–82%.

- **Federated Learning Implementation:** Federated learning implementations successfully maintained model accuracy within 2–5% of centralized training while eliminating the need for data centralization.
- **Differential Privacy Integration:** Differential privacy mechanisms were implemented with privacy budgets (ε) ranging from 0.1 to 10, achieving acceptable accuracy trade-offs for most applications (accuracy reduction of 1–12%).
- **Homomorphic Encryption Deployment:** While computationally intensive, optimized homomorphic encryption implementations enabled secure computation on encrypted data with 3–8x performance overhead on capable edge devices.
- **Secure Multi-Party Computation:** Distributed computation across edge devices using secure MPC protocols showed promise for collaborative AI applications while preserving individual privacy.

16.4.3 Securing the edge AI lifecycle

16.4.3.1 Secure model development and deployment

Secure Training Environments: Training environment security measures were implemented, including:

- Hardware security modules for key management
- Isolated training environments with strict access controls
- Comprehensive audit logging of all training activities
- Automated vulnerability scanning of training pipelines

Model Verification Protocols: The model verification framework includes:

- Statistical testing for backdoor detection
- Adversarial robustness evaluation
- Performance consistency validation
- Code review of model implementation

Secure Model Distribution: Secure distribution mechanisms were developed, featuring:

- Cryptographic signature verification (Ed25519 signatures)
- Encrypted model storage and transmission (AES-256-GCM)
- Version rollback prevention
- Integrity verification at deployment

Hardware-Based Attestation: Implementation of attestation mechanisms using:

- Trusted Platform Modules (TPM 2.0)
- ARM TrustZone technology

- Intel Software Guard Extensions (SGX) where available
- Custom hardware security elements

16.4.3.2 Runtime security for edge AI

Secure Execution Environments: The implementations utilized various secure execution technologies:

- Trusted Execution Environments (TEE) achieve 78–92% attack resistance
- Secure enclaves protecting model parameters and intermediate computations
- Hardware isolation features with a performance overhead of 8–15%

Continuous Monitoring Systems: Deployed monitoring solutions included:

- Anomaly detection for system behavior
- Network traffic analysis
- Performance deviation detection
- Real-time threat detection with response times under 100ms

Hardware Security Integration: Leveraging available hardware security features:

- Secure boot implementations achieved 100% detection of unauthorized modifications
- Hardware random number generators for cryptographic operations
- Physical unclonable functions (PUFs) for device authentication
- Side-channel resistant implementations

16.4.3.3 Secure communications

End-to-End Encryption: Implementation of communication security measures:

- TLS 1.3 for high-capability devices
- Lightweight protocols (CoAP with DTLS) for constrained devices
- Post-quantum resistant algorithms where feasible
- Perfect forward secrecy for all connections

Key Management Systems: Distributed key management solutions included:

- Hierarchical key derivation schemes
- Hardware-backed key storage
- Automatic key rotation policies
- Quantum-resistant key exchange protocols

16.4.4 Domain-specific security and privacy considerations

16.4.4.1 Healthcare and medical devices

Regulatory Compliance: HIPAA and similar healthcare privacy regulations require:

- Encrypted storage and transmission of health data
- Audit logging of all data access
- Patient consent management
- Data anonymization techniques

Safety-Critical Considerations: Security measures must not compromise medical device safety:

- Fail-safe security mechanisms
- Real-time performance guarantees
- Redundant security systems
- Emergency bypass protocols

16.4.4.2 Smart cities and public infrastructure

Public Privacy Considerations: Balancing surveillance capabilities with citizen privacy rights:

- Data minimization strategies
- Anonymization techniques
- Transparent data governance policies
- Citizen opt-out mechanisms

Infrastructure Protection: Securing critical urban systems:

- Isolated control networks
- Redundant security systems
- Physical security integration
- Cyber-physical security monitoring

16.4.4.3 Industrial IoT and manufacturing

Operational Technology Security: Integration of IT and OT security practices:

- Protocol-specific security measures
- Legacy system compatibility
- Real-time performance requirements
- Safety system integration

Intellectual Property Protection: Protecting manufacturing processes and know-how:

- Encrypted control algorithms
- Secure model deployment
- Access control for sensitive operations
- Anti-tamper hardware solutions

16.4.4.4 Autonomous vehicles

Safety-Security Integration: Ensuring security measures enhance vehicle safety:

- Authenticated control commands
- Secure sensor data processing
- Redundant decision-making systems
- Real-time threat detection

Over-the-Air Updates: Secure update mechanisms for vehicles:

- Signed firmware updates
- Delta update techniques
- Rollback capabilities
- Secure update channels

16.4.5 Emerging technologies and future directions

16.4.5.1 Hardware security innovations

AI-Specific Security Hardware: Development of specialized security hardware:

- Dedicated security processors for AI workloads
- Hardware implementations of privacy-preserving algorithms
- Integrated sensing and security processing
- Energy-efficient security accelerators

16.4.5.2 Post-quantum cryptography: preparing for quantum computing threats

- Implementation of quantum-resistant algorithms
- Hybrid cryptographic schemes
- Quantum key distribution for edge networks
- Resource-optimized post-quantum implementations

16.4.5.3 *Advanced security architectures*

Zero-Trust edge AI: Implementation of zero-trust principles:

- Continuous device verification
- Microsegmentation of edge networks
- Behavior-based access control
- Contextual authentication

Self-Defending edge AI Systems: Autonomous security capabilities:

- Adaptive security responses
- Self-healing security mechanisms
- Automated threat hunting
- Federated threat intelligence

16.5 DISCUSSION AND ANALYSIS

16.5.1 Security-performance trade-offs

The experimental results reveal several key insights regarding the balance between security and performance in edge AI systems:

Resource Allocation Strategies: The optimal allocation of computational resources between AI inference and security functions varies significantly across platforms and applications. The analysis identified three distinct performance-security profiles:

1. **Security-First Profile:** 25–40% of resources dedicated to security, suitable for critical infrastructure applications. Results in a 15–30% performance reduction but provides 78–94% protection against tested attack vectors.
2. **Balanced Profile:** 10–15% of resources for security, appropriate for enterprise applications. Achieved 5–15% performance impact while maintaining 62–78% attack resistance.
3. **Performance-First Profile:** 2–5% resource allocation to security, acceptable for non-critical consumer applications. Resulted in 1–5% performance overhead with 34–52% attack protection.

Scalability Considerations: Security implementation overhead scales differently across various edge device categories:

- **Ultra-constrained devices** (< 256KB RAM): Linear scaling with diminishing returns beyond 5% resource allocation
- **Moderately constrained devices** (256KB-4GB RAM): Optimal security-performance ratio at 10–15% resource allocation

Table 16.2 Security-Performance Profiles for Edge AI Deployments

Security Profile	*Resource Allocation*	*Performance Impact*	*Attack Resistance*	*Use Cases*
Security-First	25–40%	15–30% reduction	78–94%	Critical Infrastructure
Balanced	10–15%	5–15% reduction	62–78%	Enterprise Applications
Performance-First	2–5%	1–5% reduction	34–52%	Consumer Applications

- **High-capability edge devices** (> 4GB RAM): Sublinear scaling allowing sophisticated security with minimal impact

16.5.2 Implementation challenges and solutions

Heterogeneity Management: The diverse nature of edge computing platforms presents significant implementation challenges:

- **Hardware Variance:** Successfully addressed through hardware abstraction layers achieving 73–89% code reuse across platforms
- **Operating System Differences:** Resolved using containerization and standardized security APIs
- **Network Connectivity:** Managed through adaptive security protocols that adjust based on connection quality

Update and Maintenance: Field deployment challenges required innovative solutions:

- **Partial Updates:** Delta update mechanisms reduced bandwidth requirements by 67–82%
- **Offline Operation:** Security measures continued functioning for 7–30 days without connectivity
- **Remote Management:** Secure remote access enabled management of 98.7% of edge devices without physical access

16.5.3 Privacy-security interdependencies

The research identified critical relationships between privacy and security measures:

Complementary Effects: Privacy-preserving techniques often enhance security:

- Data minimization reduced the attack surface by 34–48%
- Local processing decreased communication vulnerabilities by 67%

- Differential privacy prevented model inversion attacks in 89% of test cases

Conflicting Requirements: Some privacy and security measures created tensions:

- Encryption overhead conflicted with real-time processing requirements
- Audit logging for security compliance challenged the privacy by design principles
- Security monitoring activities raised privacy concerns in consumer applications

16.5.4 Cost-benefit analysis

Economic analysis of security implementation revealed:

Implementation Costs

- **Development overhead:** 15–35% increase in project costs
- **Hardware costs:** $2–$50 per device for security enhancements
- **Operational costs:** 5–12% increase in maintenance expenses

Risk Mitigation Benefits

- Prevented intellectual property theft valued at $100K–$5M per incident
- Avoided regulatory penalties averaging $50K–$2M per violation
- Reduced incident response costs by 45–67%
- Maintained customer trust, preserving $500K–$10M in annual revenue

ROI Analysis: Security investments showed positive returns within:

- 6–12 months for enterprise applications
- 1–2 years for consumer applications
- 3–6 months for critical infrastructure

16.5.5 Future research directions

Based on the preceding analysis, varied research areas require further investigation:

Technical Research Priorities

1. Quantum resistant security for edge devices
2. Self-adaptive security mechanisms
3. Energy-efficient privacy-preserving algorithms
4. Cross-domain security standards

Applied Research Needs

1. Scalable security management solutions
2. Industry-specific security frameworks
3. Privacy-utility optimization techniques
4. Security testing automation for edge AI

16.6 CONCLUSION

16.6.1 Summary of key insights

In this part, the evolving world of edge AI security and privacy has been discussed. As computing power moves closer to the user, whether in the form of a sensor, a car, or a camera, the challenges shift as well. Thus, smart design and layered protection become the way forward for strong security.

Here are the main takeaways:

1. **Security Is Possible on Limited Devices:** Even resource-constrained platforms can run adequate protections, as long as they're optimized.
2. **Dedicated Hardware Helps:** Devices with built-in security features performed 40–60% better against attacks.
3. **Privacy Can Support Security:** Techniques like differential privacy and local processing don't just protect user data; they often improve system resilience.
4. **Performance Hits Are Manageable:** With proper tuning, most security setups reduce performance by only 10–15%.
5. **Tailored Approaches Work Best:** Domain-specific solutions consistently outperformed generic ones.

16.6.2 Actionable recommendations

For those building or managing edge AI systems, the following recommendations are provided:

For Developers:

- Build security into the design, not as an afterthought
- Use hardware security features wherever available
- Implement real-time monitoring and automated response tools

For Organizations:

- Choose edge hardware with built-in security capabilities
- Set up structured update and patch pipelines
- Customize security policies based on industry-specific needs

For Researchers:

- The focus should be on efficient, lightweight protection
- Need to develop tools to test and automate security in edge environments
- Explore privacy-preserving AI techniques for real-time applications

16.6.3 Limitations and future exploration

While being thorough, this research had some boundaries:

Technical Gaps:

- The experiments covered only a very small set of edge devices.
- Only known threats were discussed, and nothing about the unknown security issues.
- Privacy laws analyzed were primarily in English-speaking countries.

Research Directions:

- **Short-Term:** Test new hardware security features, automate configuration, and build sector-specific security toolkits
- **Medium-Term:** Deploy quantum-resistant encryption, build self-repairing systems, and optimize privacy trade-offs
- **Long-Term:** Create AI-powered security systems that adapt automatically, drawing on global threat intelligence

16.6.4 Future research prospects

Securing edge AI is not just a challenge but an opportunity as well. These edge systems are redefining the future of so many critical domains like healthcare, transportation, manufacturing, etc. Still, they can only thrive if security and privacy are implemented correctly and precisely, which will instill confidence and trust in the end user.

The central insight from this research is that organizations don't need to choose between protection and performance. Instead, both can be applied to the systems from the start through careful design and implementation. As use of edge AI continues to grow extensively, the incessant need to safeguard these systems must also grow in parallel. With regular innovation and collaboration between stakeholders, it's possible to build systems that are not only intelligent and efficient but also inherently secure.

In the future, edge AI security will likely involve closer integration of hardware and software security mechanisms, more automated security management systems, and sophisticated privacy-preserving computation techniques. It can only be successful if it requires continued collaboration between researchers, industry practitioners, and regulatory bodies to ensure that edge AI systems can deliver the performance while maintaining the security and privacy expectations of users and society as a whole.

REFERENCES

[1] Chen, M., & Ran, X. (2019). Artificial intelligence for internet of things. *IEEE Internet of Things Journal*, 6(3), 3887–3889.

[2] Al-Fuqaha, A., Guizani, M., Mohammadi, M., Aledhari, M., & Ayyash, M. (2018). Internet of Things: A survey on enabling technologies, protocols, and applications. *IEEE Communications Surveys & Tutorials*, 17(4), 2347–2376.

[3] Tange, K., De Donno, M., Fafoutis, X., & Dragoni, N. (2020). A systematic survey of industrial Internet of Things security: Requirements and fog computing opportunities. IEEE Communications Surveys & Tutorials, 22(4), 2489–2520. https://doi.org/10.1109/COMST.2020.3011208

[4] Carlini, N., & Wagner, D. (2017). Towards evaluating the robustness of neural networks. In *2017 IEEE Symposium on Security and Privacy (SP)* (pp. 39–57). IEEE. https://doi.org/10.1109/SP.2017.49

[5] Evans, D., Kolesnikov, V., & Rosulek, M. (2018). A pragmatic introduction to secure multi-party computation. *Foundations and Trends in Privacy and Security*, 2(2–3), 70–246.

[6] Dwork, C., & Roth, A. (2014). The algorithmic foundations of differential privacy. *Foundations and Trends in Theoretical Computer Science*, 9(3–4), 211–407.

[7] IIC. (2019). *Industrial Internet Security Framework*. Industrial Internet Consortium; ISO/IEC. (2020). *ISO/IEC 27001:2020—Information Security Management*. International Organization for Standardization.

[8] Johnson, R., & Patel, S. (2021). Resource-constrained cryptography for edge computing. *IEEE Transactions on Information Forensics and Security*, 16, 412–425.

[9] Kumar, S., Sharma, A., & Patel, R. (2020). Enhancing data privacy with lattice-based encryption technique. In *International Conference on Intelligent Human Computer Interaction* (pp. 435–443). Springer, Cham.

[10] Rahman, M., Ali, S., & Johnson, K. (2023). Identifying phishing attacks using URL-centric approaches with character-conscious transformer-based language model. In *International Conference on Cyber Security, Privacy in Communication Networks* (pp. 309–319). Springer Nature, Singapore.

[11] Thompson, R., Davis, M., & Wilson, J. (2019). Attack surface analysis in distributed edge computing. *IEEE Security & Privacy*, 17(3), 45–53.

[12] Meng, W., Li, Y., Kwok, L. F., & Choo, K. K. R. (2020). Towards power analysis based side-channel attacks for edge devices. *IEEE Transactions on Dependable and Secure Computing*, 18(6), 2718–2731.

[13] Li, T., Sahu, A. K., Talwalkar, A., & Smith, V. (2020). Federated learning: Challenges, methods, and future directions. *IEEE Signal Processing Magazine*, 37(3), 50–60.

[14] Xiao, Y., Jia, Y., Liu, C., Cheng, X., Yu, J., & Lv, W. (2019). Edge computing security: State of the art and challenges. Proceedings of the IEEE, 107(8), 1608–1631. https://doi.org/10.1109/JPROC.2019.2918437

[15] Papernot, N., McDaniel, P., Goodfellow, I., Jha, S., Celik, Z. B., & Swami, A. (2017). Practical black-box attacks against machine learning. In *Proceedings of the 2017 ACM on Asia Conference on Computer and Communications Security* (pp. 506–519). Association for Computing Machinery. https://doi.org/10.1145/3052973.3053009

[16] Alwarafy, A., Al-Thelaya, K. A., Abdallah, M., Schneider, J., & Hamdi, M. (2021). A survey on security and privacy issues in edge computing-assisted Internet of Things. IEEE Internet of Things Journal, 8(6), 4004–4022. https://doi.org/10.1109/JIOT.2020.3015432)

[17] Liu, Y., Kumar, S., Ma, J., & Liang, D. (2020). Edge AI security: A comprehensive survey. *ACM Computing Surveys*, 53(4), 1–36.

[18] Nigam, S., Sugandh, U., & Khari, M. (2022). The integration of blockchain and IoT edge devices for smart agriculture: Challenges and use cases. In *Advances in Computers* (Vol. 127, pp. 507–537). Elsevier.

[19] Vimal, S., Khari, M., Dey, N., Crespo, R. G., & Robinson, Y. H. (2020). Enhanced resource allocation in mobile edge computing using reinforcement learning based MOACO algorithm for IIOT. *Computer Communications*, 151, 355–364.

[20] Chaudhary, G., Srivastava, S., & Khari, M. (2023). Generative edge intelligence for securing IoT-assisted smart grid against cyber-threats. *International Journal of Wireless & Ad Hoc Communication*, 6(1).

[21] Gupta, J., Nayyar, A., & Gupta, P. (2015). Security and privacy issues in Internet of Things (IoT). *IJRCS—International Journal of Research in Computer Science*, 2(4), 18–22. ISSN: 2349-3828. Retrieved May 29, 2025, from https://researchscript.in/wp-content/uploads/2015/04/IJRCS020405.pdf

[22] Deepthi, B., Gupta, P., Rai, P., & Arora, H. (2022, May). Assessing the dynamics of AI driven technologies in Indian banking and financial sector. *Vision (ABDC Ranked Journal).* https://doi.org/10.1177/09722629221087371; ISSN: 0972-2629.

[23] Microsoft. (2005). *The STRIDE Threat Model.* Microsoft Security Development Lifecycle.

Index

For Product Safety Concerns and Information please contact our EU
representative GPSR@taylorandfrancis.com
Taylor & Francis Verlag GmbH, Kaufingerstraße 24, 80331 München, Germany

www.ingramcontent.com/pod-product-compliance
Ingram Content Group UK Ltd.
Pitfield, Milton Keynes, MK11 3LW, UK
UKHW022214190926
475702UK00003B/346

* 9 7 8 1 0 4 1 0 6 3 7 8 0 *